AF878796

# COSMOCHEMICAL EVOLUTION AND THE ORIGINS OF LIFE

VOLUME II

## CONTRIBUTED PAPERS

# COSMOCHEMICAL EVOLUTION AND THE ORIGINS OF LIFE

PROCEEDINGS OF THE FOURTH INTERNATIONAL CONFERENCE ON THE ORIGIN OF LIFE AND THE FIRST MEETING OF THE INTERNATIONAL SOCIETY FOR THE STUDY OF THE ORIGIN OF LIFE, BARCELONA, JUNE 25–28, 1973

## VOLUME II: CONTRIBUTED PAPERS

*Edited by*

J. ORÓ

*Departments of Biophysical Sciences and Chemistry, University of Houston, Houston, Texas, and Universidad Autónoma de Barcelona, Barcelona*

S. L. MILLER

*Department of Chemistry, University of California, San Diego, California*

C. PONNAMPERUMA

*Laboratory of Chemical Evolution, Department of Chemistry, University of Maryland, College Park, Maryland*

*and*

R. S. YOUNG

*Head of Planetary Biology, NASA, Washington, D.C.*

D. REIDEL PUBLISHING COMPANY

DORDRECHT-HOLLAND / BOSTON-U.S.A.

Library of Congress Catalog Card Number 74–77967

ISBN-13: 978-94-010-2284-2 e-ISBN-13: 978-94-010-2282-8
DOI: 10.1007/978-94-010-2282-8

Published by D. Reidel Publishing Company,
P.O. Box 17, Dordrecht, Holland

Sold and distributed in the U.S.A., Canada and Mexico
by D. Reidel Publishing Company, Inc.
306 Dartmouth Street, Boston,
Mass. 02116, U.S.A.

All Rights Reserved
Copyright © 1974 by D. Reidel Publishing Company, Dordrecht, Holland
and copyrightholders as specified on appropriate pages within
Softcover reprint of the hardcover 1st edition 1974
No part of this book may be reproduced in any form, by print, photoprint, microfilm,
or any other means, without written permission from the publisher

## TABLE OF CONTENTS

PART I / ORGANIC COSMOCHEMISTRY



PART II / ABIOTIC ORGANIC SYNTHESIS



PART III / ABIOTIC ORGANIC SYNTHESIS AND INTERACTIONS: POLYMERS

## PART IV / STRUCTURAL AND THERMODYNAMIC CONSIDERATIONS ON THE ORIGIN OF LIFE



## PART V / EARLY BIOCHEMICAL EVOLUTION

## PART VI / MISCELLANEOUS PAPERS

# PART I

# ORGANIC COSMOCHEMISTRY

# AMINO ACIDS IN CARBONACEOUS CHONDRITES

JAMES G. LAWLESS and ETTA PETERSON

*Ames Research Center, NASA, Planetary Biology Division, Moffett Field, Calif. 94035, U.S.A.*

**Abstract.** For almost 20 years laboratory experiments have advanced the concepts of chemical evolution, particularly with regard to formation of the amino acids. What has been generally lacking is concrete natural evidence for this chemical evolution hypothesis. The recent development of sophisticated analytical techniques and availability of carbonaceous chondrites with a minimum of terrestrial contamination has resulted in the identification of amino acids which provide strong evidence for a natural extraterrestrial chemical synthesis. Since the initial find in the Murchison meteorite (a type II carbonaceous chondrite) of both protein and nonprotein amino acids and amino acids with nearly equal abundances of D and L isomers, further studies have been carried out. These studies have revealed the presence of at least 35 amino acids; the population consists of a wide variety of linear, cyclic and polyfunctional amino acids which shows a trend of decreasing concentration with increasing carbon number. Investigations of the Murray meteorite (a type II carbonaceous chondrite) has produced similar results, but studies of the Orgueil meteorite (a type I carbonaceous chondrite) show only a limited suite of amino acids, some of which appear to be indigenous while others appear to be terrestrial contaminants. A sample of the Murchison meteorite was extracted with $D_2O$ and in addition to 'free' amino acids, showing no deuterium incorporation, some amino acids showed the presence of deuterium suggesting either a 'precursor(s)' or hydrogen-deuterium exchange which require(s) formation of carbon-hydrogen bonds.

## 1. Introduction

One of the most important classes of compounds in the scheme of chemical evolution is the amino acids. Since the initial identification of these compounds as products of an electrical discharge in an atmosphere of methane, ammonia, and water, many investigators have contributed to an extensive body of literature concerning their production in simulated prebiotic experiments. This abundance of data provides an impressive experimental base for the hypothesis of chemical evolution. Since the laws of chemistry are believed to be universal, there should be natural evidence for the chemical evolutionary process. Meteorites, particularly carbonaceous chondrites, appear to provide such evidence. Recently, Kvenvolden *et al.* (1970) identified amino acids – as well as other compounds – in the Murchison meteorite, a type II carbonaceous chondrite. The suite of amino acids found included protein and nonprotein amino acids, with many of these amino acids having nearly equal amounts of both the D and the L isomers. Subsequent investigations of the Murchison and other meteorites (Kvenvolden *et al.*, 1971; Cronin and Moore, 1971; Oro *et al.*, 1971a; Lawless *et al.*, 1971; Oro *et al.*, 1971b; Lawless *et al.*, 1972) have substantiated this finding and provided the strongest evidence to date for the indigenous nature of complex organic molecules in extraterrestrial samples. The recent development of sophisticated analytical techniques, particularly the combined gas chromatograph-mass spectrometer (GC-MS), has been instrumental in the success of these investigations. This paper reports on the use of GC-MS to characterize further the amino acids found in extracts of the Murchison meteorite.

## 2. Experimental

Two separate samples of the Murchison meteorite were analyzed. The first 5.1 g sample was refluxed with deuterated water ($D_2O$) for 20 hr. The $D_2O$ extract was charged on a Dowex 50 (H+) ion exchange column and eluted with $H_2O$ and 2*N* $NH_4OH$. A second 10 g sample of pulverized meteorite was refluxed with $H_2O$ for 20 hr. The $H_2O$ extract was evaporated to dryness and refluxed with 6*N* HCl for 20 hr. The hydrolyzate was evaporated to dryness, redissolved in $H_2O$, and charged on a Dowex 50 (H+) ion exchange column. The column was successively eluted with $H_2O$ and 2*N* $NH_4OH$. In each case the $NH_4OH$ eluate was evaporated to dryness, and the N-trifluoroacetyl-D-2-butyl ester derivatives of the amino acids were prepared (Pollock *et al.*, 1965; Pollock and Oyama, 1966). These compounds were analyzed by gas chromatography (Perkin-Elmer 900) using a flame ionization detector and a 150 ft × 0.02 in wall coated (UCON 75-H-90000) capillary column. The column was programmed from 100° to 150 °C at 1 °C min$^{-1}$. Procedural blanks showed no amino acids.

The same column and gas chromatographic conditions were used in the GC-MS analysis where the gas chromatograph was interfaced to a CEC 21-491 mass spectrometer through a membrane separator. As each compound eluted from the gas chromatograph and passed into the mass spectrometer, its mass spectrum was obtained using 70 eV electrons by taking successive scans at 4 s per decade. The covariant mass spectral peaks were then plotted and the identity of the compound determined. This GC-MS system is capable of identifying as little as $20 \times 10^{-9}$ g of the N-trifluoroacetyl-D-2-butyl ester derivative of an amino acid injected onto the gas chromatographic column.

## 3. Results and Discussion

Cronin and Moore (1971) observed a marked increase in the concentration of most of the amino acids as a result of acid hydrolysis of the Murchison meteorite aqueous extract. They pointed out that the state of the amino acids prior to extraction and the chemical reactions which they may undergo during extraction were not known. In order to understand further what might be happening during the extraction procedure, a sample of the Murchison meteorite was extracted with $D_2O$ and the resulting amino acids derivatized and analyzed by GC-MS. If carbon-hydrogen bond formation occurs during extraction and involves the extraction solvent, use of $D_2O$ would result in carbon-deuterium bond formation which could be detected in the mass spectra of the derivatized amino acids. The results of this experiment are summarized by the mass spectral data presented in Table I. First, it should be noted that with the exception of $\beta$-aminoisobutyric acid, all of the amino acids which were found in previous studies of the Murchison meteorite (Kvenvolden *et al.*, 1970; Kvenvolden *et al.*, 1971; Oro *et al.*, 1971a; Cronin and Moore, 1971) and the Murray meteorite (Cronin and Moore, 1971; Lawless *et al.*, 1971) have been identified in

## TABLE I

Major fragment ions of the amino acids found in the Murchison meteorite $D_2O$ extract†

| m/e | Isovaline | α-Aminoisobutyric Acid | Valine | N-Methylalanine | α-Amino-n-butyric Acid | Alanine | N-Methylglycine | N-Ethylglycine | Norvaline | β-Amino-n-butyric Acid | Glycine 1* | Glycine 2* | Glycine 3* | Glycine 4* | Pipecolic Acid | β-Alanine 1* | β-Alanine 2* | Proline | γ-Amino-n-butyric Acid | Aspartic Acid | Glutamic Acid | Deuterated Alanine Direct Probe** | Deuterated Alanine GC-MS** |
|---|---|---|---|---|---|---|---|---|---|---|---|---|---|---|---|---|---|---|---|---|---|---|---|
| 100 | 8 | | 6 | | | | | | | | | | | | | | | | | | | | |
| 102 | | | | | | | | | | | | 3 | 1 | 2 | | | | | | | | | |
| 103 | | | | | | | | | | | | 3 | 4 | 3 | | | | | | | | | |
| 104 | | | | | | | | | | | | | 3 | 6 | | | | | | | | | |
| 110 | | | | 13 | | | | | | | | | | | | | | | | | | | |
| 112 | | | | | | | | | | 10 | | | | | | | | | 17 | | | | |
| 113 | | | | | | | | | | 5 | | | | | | | | | | | | | |
| 114 | 11 | 9 | 15 | | | | | | | 12 | | | | | | | | | | | | | |
| 115 | | | | | | | | | | 14 | | | | | | | | | | | | | |
| 116 | | | | | | | | | | 8 | | | | | | 7 | | | | | | | |
| 120 | | | | | | | | | | | | | | | | 10 | 27 | | 12 | | | | |
| 126 | | | 6 | | 11 | | 7 | 48 | 40 | | 100 | 75 | 17 | 21 | 4 | 24 | 31 | | 19 | | | | |
| 127 | | | 2 | | 3 | | 3 | 34 | 16 | | 87 | 100 | 45 | 44 | 2 | 5 | | | | | | | |
| 128 | | | | | | | | 10 | 8 | | 14 | 52 | 100 | 100 | 4 | | | | | | | | |
| 129 | | | | | | | | | | | 5 | 13 | 46 | 49 | 2 | | | | | | | | |
| 130 | | | | | | | | | | | 5 | 2 | 3 | 3 | | | | | | | | | |
| 138 | | | | | | | | | | | | | | | | 10 | 10 | | | | | | |
| 139 | | | | | | 1 | | | | | | | | | | 18 | 16 | | 11 | 42 | 21 | | |
| 140 | | | 3 | 3 | | 100 | 100 | 64 | 38 | 96 | | | | | | 8 | 11 | | 16 | | 27 | | |
| 141 | | | 3 | | | 87 | 99 | 19 | 24 | 13 | | | | | | 4 | 8 | | | | | 1 | 1 |
| 142 | | | | | | 20 | 42 | 2 | | 4 | | | | | | 3 | 5 | | | | | 4 | 4 |
| 143 | | | | | | 2 | 11 | | | | | | | | | | | | | | | 25 | 25 |
| 144 | | | | | | | | | | | | | | | | | | | | | | 100 | 100 |
| 145 | | | | | | | | | | | | | | | | | | | | | | 41 | 47 |
| 146 | | | | | | | | | | | | | | | | | | | | | | 3 | 3 |
| 147 | | | | | | | | | | | | | | | | | | | | | | 1 | 1 |
| 152 | | | 2 | | 3 | | | 6 | | | 6 | 2 | 1 | | 2 | | | | | | 51 | | |
| 153 | | | 25 | 8 | 5 | | | 10 | 12 | 19 | 6 | 2 | 2 | | 1 | | | | 10 | | 33 | | |
| 154 | | 100 | 34 | 100 | 100 | | | 100 | 18 | 15 | 73 | 45 | 10 | 8 | | | | | 29 | | 17 | | |
| 155 | | 16 | 3 | 18 | 78 | | | 65 | 12 | 17 | 15 | 42 | 31 | 20 | | | | | 7 | | | | |
| 156 | | | | 2 | 19 | | | 18 | | 7 | 5 | 12 | 48 | 42 | | | | | | | | | |
| 157 | | | | | 3 | | | 2 | | 3 | 4 | 1 | 3 | 3 | | | | | | | | | |
| 164 | | | | | | | | | | | | | | | | | | 2 | | | | | |
| 165 | | | | | | | | | | | | | | | | | | 2 | | | | | |
| 166 | 9 | 7 | 3 | | | 2 | 4 | | 5 | | | | | | | 11 | 9 | 100 | | | | | |
| 167 | 4 | 3 | | | | 1 | 2 | | 5 | | | | | | | 7 | 6 | 22 | | | | | |
| 168 | 100 | | 100 | | | 21 | 29 | | 100 | | | | | | | 100 | 100 | 3 | | | | | |
| 169 | 12 | | 56 | | | 8 | 16 | | 95 | | | | | | | 21 | 77 | 1 | | | | | |
| 170 | | | 12 | | | 2 | 4 | | 24 | | | | | | | 4 | 25 | | | | | | |
| 171 | | | 10 | | | | | | 14 | | | | | | | | 3 | | | | | | |
| 172 | | | | | | | | | | | | | | | | | | | | | | | |
| 173 | | | | | | | | | | | | | | | | | | | | | | | |
| 180 | | | | 2 | 3 | | | 4 | | 5 | | | | | 100 | | | | 6 | | 73 | | |
| 181 | | | | | | | | 23 | | 5 | | | | | 14 | | | | 16 | | 52 | | |
| 182 | | 14 | | 14 | 15 | | | 15 | | 100 | | | | | 2 | | | | 100 | | 17 | | |
| 183 | | 2 | | 2 | 5 | | | 5 | | 33 | | | | | | | | | 14 | | | | |
| 184 | 12 | 5 | | | | | | | | 16 | | | | | | | | | 4 | 70 | | | |
| 185 | 1 | 2 | | | | | 13 | | | 6 | | | | | | 7 | 8 | | | 75 | | | |
| 186 | | | | | | | 14 | | | | | | | | | 12 | 14 | | | | | | |
| 187 | | | | | | | 5 | | | | | | | | | | 10 | | | | | | |
| 188 | | | | | | | | | | | | | | | | | 3 | | | | | | |
| 194 | 1 | | 3 | | | | | | | 8 | | | | | | | | 6 | | | | | |
| 195 | 2 | | 1 | | | | | | | | | | | | | | | 1 | | | | | |
| 196 | 5 | | 14 | | | | | | 18 | | | | | | | | | | | | | | |
| 197 | 1 | | 3 | | | | | | 9 | | | | | | | | | | | | | | |
| 198 | | | | | | | | | | | | | | | | | | | | | 100 | | |
| 199 | | | | | | | | | | 11 | | | | | | | | | 30 | | 62 | | |
| 200 | | | | | | | | | | 17 | | | | | | | | | 9 | | 22 | | |
| 201 | | | | | | | | | | 4 | | | | | | | | | | | | | |
| 207 | | | | | | | | | | | | | | | 2 | | | | | | | | |
| 208 | | | | | | | | | | | | | | | 6 | | | | | | | | |
| 209 | | | | | | | | | | | | | | | 1 | | | | | | | | |
| 211 | | | | | | | | | | | | | | | | | | 4 | | | | | |
| 212 | | | | | | | | | | | | | | | | | | 1 | | 100 | | | |
| 213 | | | | | | | | | | | | | | | | | | | | 24 | | | |
| 214 | | | | | | | | | | | | | | | | | | | | 20 | | | |
| 225 | | | | | | | | | | | | | | | | | | | | | 17 | | |
| 226 | | | | | | | | | | | | | | | | | | | | | 100 | | |
| 227 | | | | | | | | | | | | | | | | | | | | | 33 | | |
| 228 | | | | | | | | | | | | | | | | | | | | | 16 | | |

†Each mass spectrum is normalized to the most abundant ion above m/e 100.

*Data from successive scans of the GC peak showing partial chromatographic separation of nondeuterated and deuterated species.

**A synthesized sample of alanine with approximately 75% $D_4$ shows no mass spectral differences between direct probe and GC-MS methods of sample introduction.

TABLE II

Amino acids in the Murchison meteorite

| | | | |
|---|---|---|---|
| 1 | Isovaline | 27 | L-Pipecolic Acid |
| 2 | α-Aminoisobutyric Acid | 28 | Glycine |
| 3 | D-Valine | 29 | Neutral Cyclic |
| 4 | Linear Neutral ($C_6$) | 30 | Neutral Cyclic |
| 5 | L-Valine | 31 | β-Alanine |
| 6 | N-Methylalanine | 32 | Neutral Cyclic |
| 7 | D-α-Amino-n-butyric Acid | 33 | Polyfunctional Linear Aliphatic |
| 8 | D-Alanine | 34 | D-Proline |
| 9 | Linear Neutral ($C_5$) | 35 | L-Proline |
| 10 | L-α-Amino-n-butyric Acid | 36 | Linear Neutral ($C_5$) |
| 11 | L-Alanine | 37 | Unknown [b] |
| 12 | Linear Neutral ($C_5$) | 38 | Unknown [b] |
| 13 | $NH_3$ [a] | 39 | Linear Neutral ($C_5$) |
| 14 | Linear Neutral ($C_5$) | 40 | Unknown [b] |
| 15 | Linear Neutral ($C_6$) | 41 | γ-Aminobutyric Acid |
| 16 | N-Methylglycine | 42 | D-Aspartic Acid |
| 17 | N-Ethylglycine | 43 | L-Aspartic Acid |
| 18 | D-Norvaline | 44 | Polyfunctional Linear Aliphatic |
| 19 | L-Norvaline | 45 | Polyfunctional Linear Aliphatic |
| 20 | Linear Neutral ($C_5$) | 46 | Polyfunctional Linear Aliphatic |
| 21 | Linear Neutral ($C_6$) | 47 | Polyfunctional Linear Aliphatic |
| 22 | D-β-Aminoisobutyric Acid | 48 | Polyfunctional Linear Aliphatic |
| 23 | L-β-Aminoisobutyric Acid | 49 | Unknown [b] |
| 24 | β-Amino-n-butyric Acid | 50 | D-Glutamic Acid |
| 25 | Unknown [b] | 51 | L-Glutamic Acid |
| 26 | D-Pipecolic Acid | 52 | Unknown [b] |

[a] Present in blank

[b] Peaks labeled 'unknown' do not appear to be amino acids. Their identification awaits the use of high resolution mass spectrometry.

this Murchison meteorite specimen. This observation further demonstrates the ubiquity of this suite of amino acids in the original milieu. Comparison of the mass spectra of the 17 amino acids found in this sample (Table I) with those of known standards (Lawless, 1973a) show that every one of these amino acids have intact carbon-hydrogen bonds, indicating the existence of these compounds as 'free' amino acids in the meteorite. In addition to all of these compounds existing as indigenous 'free' amino acids, some amino acids show deuterium incorporation, indicating the formation of carbon-deuterium bonds during extraction in $D_2O$. Mass spectra of isovaline, α-aminoisobutyric acid, valine, N-methylalanine, pipecolic acid, proline, and γ-amino-n-butyric acid derivatives show no carbon-deuterium bonds, while the mass spectra of α-amino-n-butyric acid, α-alanine, N-methylglycine, N-ethylglycine, norvaline, β-amino-n-butyric acid, glycine, β-alanine, and glutamic acid show, in addition to carbon-hydrogen bonds, some deuterium incorporation. No conclusions can be reached concerning aspartic acid because of the weak mass spectra obtained. Two amino acids, glycine and β-alanine, showed mass spectral differences between successive scans of the respective GC peaks, indicating a gas chromatographic separation of the nondeuterated and deuterated species (see Table I). Other amino acids which show deuterium incorporation showed no indication of gas chromatographic fractionation. Subtle differences, however, would not have been detected due to the limited number (usually three) of scans obtained for each gas chromatographic peak. Catalytic activity of the meteorite matrix causing the replacement of hydrogen by deuterium in the 'free' amino acids cannot be eliminated. The existence in the meteorite of an amino acid precursor (or precursors) requiring carbon-hydrogen bond formation in addition to the 'free' amino acids is also consistent with the data.

A cursory glance at the gas chromatogram (Figure 1) of amino acids in the hydrolyzed Murchison meteorite extract reveals a complex suite of compounds; there appear to be many more amino acids than the 18 which have been positively identified. Indeed, at least 17 additional amino acids can be characterized on the basis of their mass spectral fragmentation patterns (Lawless, 1973b).

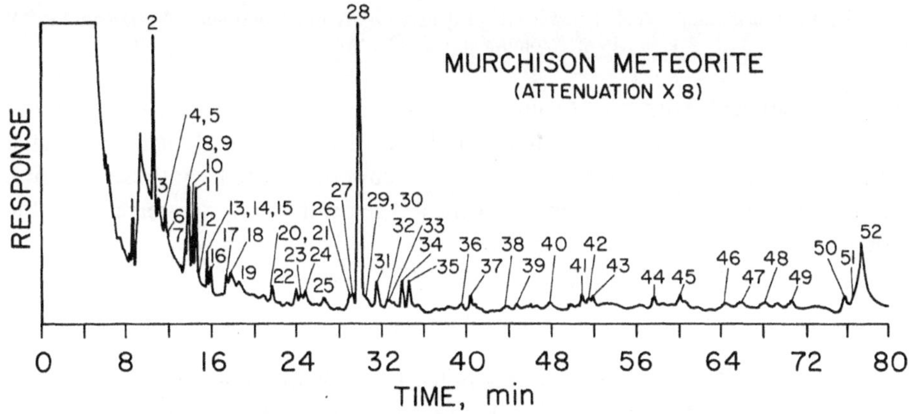


Fig. 1. The gas chromatogram of the N-trifluoroacetyl-D-2-butyl esters of amino acids in the acid hydrolyzed extract of the Murchison meteorite. The identifying numbers correspond to those in Table II.

Mass spectral characterizations show that six polyfunctional linear aliphatic amino acids (Figure 1, peaks 33, 44, 45, 46, 47, 48) are present, in addition to the two previously identified (aspartic acid and glutamic acid). They appear to be dicarboxylic amino acids which are isomeric with glutamic acid and higher homologs. Furthermore, three cyclic neutral amino acids (Figure 1, peaks 29, 30, 32) are present in addition to proline and pipecolic acid. Of the linear neutral amino acids, there appear to be 9* more (Figure 1, peaks 4, 9, 12, 14, 15, 20, 21, 36, 39) than the 14 reported previously.

Comparison of the linear neutral amino acids (Figure 2) present in the Murchison meteorite, in laboratory chemical evolution experiments, and in a terrestrial organism shows a marked similarity between the meteorite and laboratory experiments, and a significant difference between the meteorite and *E. coli.* Of the 103 theoretically

* This number could be 8 if peaks 9 and 12 in Figure 1 represent a D,L pair.

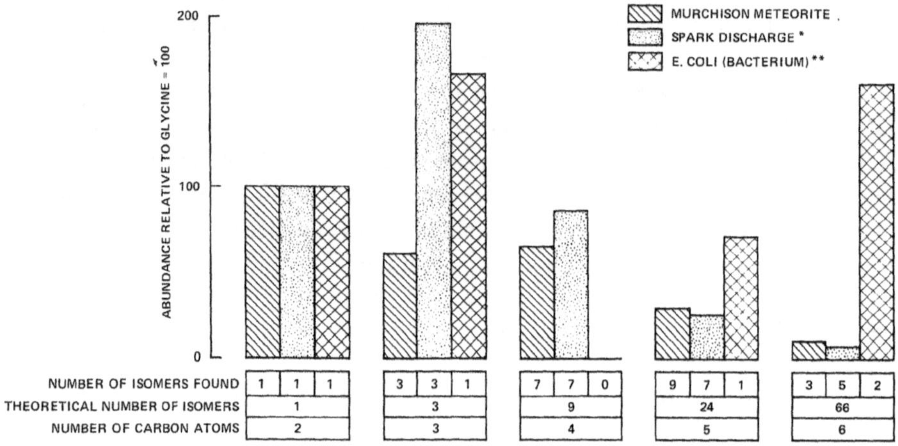


Fig. 2. Approximate amino acid (linear neutral) distribution from biotic and abiotic sources (*Ring *et al.*, 1972; Wolman *et al.*, 1972. **Luria, 1960.)

possible amino acids with two through six carbon atoms, only 5 are commonly found in *E. coli*, whereas the spark discharge and the Murchison meteorite each have at least 23 amino acids. These results lend further support to the hypothesis that amino acids are present in the Murchison meteorite as the result of an extraterrestrial abiotic synthesis.

## Acknowledgements

I thank Dr Keith A. Kvenvolden and Mrs Ruth Mack, of the Ames Research Center for their assistance. Thanks are also due to Dr Carleton Moore and the Center for Meteorite Studies, Arizona State University, for the sample used in this work.

## References

Cronin, J. R. and Moore, C. B.: 1971, *Science* **172**, 1327–1329.

Kvenvolden, K. A., Lawless, J. G., and Ponnamperuma, C.: 1971, *Proc. Nat. Acad. Sci.* **68**, 486–490.

Kvenvolden, K. A., Lawless, J. G., Pering, K., Peterson, E., Flores, J., Ponnamperuma, C., Kaplan, I. R., and Moore, C.: 1970, *Nature* **228**, 923–926.

Lawless, J. G.: 1973a, *Geochim. Cosmochim. Acta*, in press.

Lawless, J. G.: 1973b, *Geochim. Cosmochim. Acta* **37**, 2207–2212.

Lawless, J. G., Kvenvolden, K. A., Peterson, E., Ponnamperuma, C., and Moore, C.: 1971, *Science* **173**, 626–627.

Lawless, J. G., Kvenvolden, K. A., Peterson, E., Ponnamperuma, C., and Jarosewich, E.: 1972, *Nature* **236**, 66–67.

Luria, S. E.: 1960, in I. C. Gunsalus and R. Y. Stanier, (ed.), *The Bacteria-A Treatise on Structure and Function*, Vol. 1, Academic Press, New York, pp. 1–34.

Oró, J., Gibert, J., Lichtenstein, H., Wikstrom, S., and Flory, D. A.: 1971a, *Nature* **230**, 105–106.

Oró, J., Nakaparksin, S., Lichtenstein, H., and Gil-Av, E.: 1971b, *Nature* **230**, 107–108.

Pollock, G. E. and Oyama, V. I.: 1966, *J. Gas Chrom.* **4**, 126.

Pollock, G. E., Oyama, V. I., and Johnson, R. D.: 1965, *J. Gas Chrom.* **3**, 174.

Ring, D., Wolman, Y., Friedmann, N., and Miller, S.: 1972, *Proc. Nat. Acad. Sci.* **69**, 765–768.

Wolman, Y., Haverland, W. J., and Miller, S.: 1972, *Proc. Nat. Acad. Sci.* **69**, 809–11.

# REMARKS ON THE CHEMICAL CONDITIONS ON THE SURFACE OF THE PRIMITIVE EARTH AND THE PROBABILITY OF THE EVOLUTION OF LIFE

HANS E. SUESS

*Physikalisches Institut der Universität, Bern, Switzerland**

**Abstract.** The inner planets were formed from smaller objects that had no gases associated with them. These objects contained relatively small amounts of water and carbon in a form similar to that found in carbonaceous chondrites. The first forms of life must have originated during the time when the water reacted with the carbon (and also with nitrides, phosphides, etc.), while the hydrogen formed by this reaction was continuously lost from the gravitational field of the Earth. About $10^{44}$ atoms of carbon reacted with water during less than $10^{17}$ s. The crucial question is whether some form of life will always develop under these conditions, or whether the origin of life is an improbable, perhaps an immensely improbable event. At present it is still impossible to answer this question.

Over the past few decades two fundamental, long-disputed biological questions have been answered: First, it was found that all life goes back to one single line of ancestry, always using the same optical isomers, the same genetic code, and the same biochemical and biophysical principles, such as mechanisms of reduplication and of energy transfer. Second, scientists seem to agree now that an understanding of this singular origin of all terrestrial life does not require an extension or a modification of the laws of physics as we know them today.

However, a third question, as yet unanswerable and independent of the preceding ones, is whether such evolution was an inevitable process that will always occur within a realistic span of time, whenever appropriate environmental conditions prevail (Eigen, 1971), or whether the evolution of life was an improbable, perhaps immensely improbable occurrence (Elsasser, 1968).

Obviously, we, as individuals, necessarily had to be born in a universe that obeyed laws of nature of a kind that made our terrestrial life and its development possible, and, within this universe, we had to be born on the surface of a planet with appropriate past and present environmental conditions. These conditions were necessary, but were not necessarily sufficient for the evolution of life; some unlikely accident, or possibly a sequence of many such accidents, may also have been necessary to start an evolutionary process on Earth.**

Undoubtedly there exist in our universe a large number of earthlike planets on the surface of which life could have evolved. According to Brown (1964) more than $10^{10}$ such planets can be expected to exist in our galaxy alone. The probability that

* On leave from the University of California, San Diego, La Jolla, Calif. 92037, U.S.A.

** In the post-session discussion Professor Harold Urey expressed this in the following way: The evolution of life is not a miracle in the ordinary sense of the word, but it might be a 'Fermi-Miracle'. A first order Fermi-Miracle is defined as an event that occurs with a probability of 1/10. A Fermi-Miracle *n*-th order is an event that occurs with the probability of $(1/10)^n$.

one of these planets is, or ever was, inhabited by beings of a species like ourselves is immensely small, perhaps of the order of $10^{-100}$. The probability of finding some kind of life on some or all of these planets will obviously be much larger, but it is impossible to say how large it might be. The knowledge of this probability is certainly an important scientific question, perhaps the most important one of our time. There is no other quantity in science for which our ignorance spans as many orders of magnitude as this one, and especially no other of comparable philosophical and metaphysical significance.

There are two ways by which one could, in principle, learn about the probability of the evolution of life: (1) By observing and recognizing an independent transition of prebiotic macromolecules into living organisms either in the laboratory or on extraterrestrial objects. Such observations could immediately show that the origin of life was not an improbable process. (2) By quantitative considerations of the individual biochemical steps that led to the beginning of life and of the conditions that must have prevailed on the surface of the Earth when these first steps occurred several billion years ago. Our knowledge of the basic quantities necessary for these considerations is still rather limited, and at present this approach is far from leading to a final answer. It may yet be useful to summarize what might be needed to obtain such an answer.

Molecular biology should eventually make it possible to derive *lower* limits for the population of self-replicating units, the number of generations, the energy flux, and the time required for the evolution of a lasting biosphere. Cosmochemistry should be in a position to establish *upper* limits for the total amount of carbon and other biologically important elements, for the free energy fluxes, and for the length of geological time available on the Earth for the evolution of life. The crucial point of these considerations will then be whether or not a biological system which was capable of adjusting to a new form of free energy before the most primitive one was exhausted, could have evolved with reasonable probability during the available time. The purpose of this note is to present some examples of considerations that might allow estimates of these upper limits.

One of the few facts that can be considered firmly established is that the planet Earth formed from objects that were almost completely deprived of their gaseous constituents (including almost all water) at a time before they became a part of the planet. This can be seen from the fact that the Earth's atmosphere contains only a very small fraction of rare gases; of neon, for example, only $10^{-10}$ parts of what had been associated with the original primordial material remained. But even this neon, just as xenon, does not have the isotopic composition of the neon in the sun and in the primordial gas cloud from which the solar system formed. The Earth can never have had a dense atmosphere of solar matter, because, otherwise, rare gases undoubtedly would have been occluded and, as a consequence of degassing at elevated temperatures, would, just as the water of the oceans, have reached the surface of the Earth and would be present in the atmosphere.

Hence, it must be assumed that practically all the carbon on the Earth was originally in solid form, similar to that present in carbonaceous chondrites. The material on the surface of the early Earth was most heterogeneous and not in thermodynamic equilibrium, as the carbonaceous chondrites are not. It contained carbonaceous material, consisting of hydrocarbons and some 'organic' nitrogen- and sulphur-compounds, as well as nitrides, phosphides, sulfides, and, at the same time, hydrates that later provided water of the oceans.

There are approximately $10^{44}$ carbon atoms on the surface of the Earth. This is only a very small fraction (less than $10^{-5}$) of the carbon that was originally associated with the Earth's matter. Today, most of this carbon is oxidized and present in the form of carbonates. At some time prior to the formation of the Earth, all this carbon must have been present in a reduced state, because of the large excess of hydrogen in the matter from which the solar system formed. The oxidation of the carbon in early geological time was due to a reaction of the carbon with water and subsequent loss of hydrogen from the gravitational field. This reaction not only occurred in the planet Earth, but also on Mars and Venus, and, to a limited degree on parent bodies of carbonaceous meteorites, as can be seen from the presence of carbonates.

During the process of gradual oxidation, random chemical reactions led to the formation of high polymer molecules, some of them of the type of prebiotic macromolecules. In the presence of ammonia, amino-acids form (Miller, 1953, 1955, 1957; Miller and Urey, 1959). It would be important to know what fraction of carbon at any stage of this development might have combined to form molecules that could have played a role in the origin of life. One can expect that this fraction had reached a maximum at a time when comparable amounts of oxidized and unoxidized carbon were present (Suess, 1962). In the presence of liquid water, larger amounts of such molecules than those directly supplied by gas phase or surface reactions (such as Fischer-Tropsch reactions (Anders, 1973)) can be expected.

Prebiotic macromolecules that might have been present in the planetesimals from which the Earth formed would hardly have survived the process of degassing of the primordial material and the transport to the surface of the Earth.

Nitrides in the surface rocks of the primitive Earth must have reacted rapidly with water to give ammonia. It is quite possible that in the very early stages most of today's atmospheric nitrogen was in the form of ammonia, and that an appreciable partial pressure of ammonia was present in the primitive atmosphere (Bada and Miller, 1968; Sagan and Mullen, 1972). Soon, however, exhalations from the mantle must have supplied CO, $CO_2$, and some methane. Methane was probably present only in small amounts, such as corresponded to equilibrium concentrations in the reaction of C (solid) with $H_2O$. Ammonia combines with $CO_2$ in the presence of water to give $(NH_4)_2CO_3$ and a small amount of urea. Aldehydes and other simple compounds then contributed to the formation of prebiotic macromolecules in a way that has been extensively discussed in the literature (Miller and Orgel, 1973).

The concentrations of these prebiotic molecules, if evenly distributed on the surface and throughout the oceans, may have been quite small, but mechanisms by which these molecular species were concentrated, e.g., by adsorption on certain surfaces, were certainly possible. It is generally assumed that under such conditions the probability of accidental formation of self-replicating structures is high.

As the surface material of the primitive Earth was far from being in thermodynamic equilibrium, ample chemical energy was available for biological processes, even before any photochemical reactions became important. The most obvious energy source was the reaction of carbon, graphite or amorphous carbon, with water. Nearly all the carbon now on the surface of the Earth underwent this reaction, releasing some $10^{22}$ kcal, but only a very small fraction of this could have been available for organic metabolism. A considerable fraction of the carbon reached the surface of the Earth in the form of CO (Abelson, 1966), which can, under the influence of uv-light, react with water to form formaldehyde. Even after the source of CO was exhausted, formaldehyde could have formed in small amounts through a photochemical reaction of $CO_2$ with $H_2O$ (Groth and Suess, 1938). Formaldehyde in the presence of liquid water polymerizes under certain conditions to saccharides. Hence, many different reactions could have served as energy sources. It seems improbable, however, that life started in a form capable of making use of a variety of such processes. One particular reaction must have supplied the primary energy flux, a reaction such as perhaps the hydrolysis of a polymerization product of formaldehyde. In the mechanism of energy transfer, polyphosphates apparently played a role at a very early time, although it is strange that pyrophosphates do not seem to occur in meteorites.

We are well informed about the maximum length of time it took life to develop. It must have been less than some $10^{16}$ s. As long as we do not know whether such a length of time is adequate to lead to the evolution of living organisms in an environment as outlined above, and as long as we have not discovered an independent, extraterrestrial evolution of some kind of life, there is no way to tell whether or not one might anticipate life on another planet, no matter how large the number of such planets might be.

## Acknowledgements

Financial support through NASA Grant NGL-05-009-005 for the author's research on meteorites is gratefully acknowledged. Also, thanks are due to Johannes Geiss and Hans Oeschger, and especially to Stanley Miller for extended inspiring discussions. This note was written while the author was enjoying the hospitality of the University of Bern, Switzerland.

## References

Abelson, P. H.: 1966, *Proc. Nat. Acad. Sci.* **55**, 1365.
Anders, E.: 1968, *Accounts of Chem. Res.* **1**, 296.
Anders, E.: 1973, *Proc. of 4th Int. Conf. on Origin of Life*, Barcelona, Spain.
Bada, J. L. and Miller, S. L.: 1968, *Science* **159**, 423.
Brown, Harrison: 1964, *Science* **145**, 1177.

Eigen, Manfred: 1971, *Naturwiss.* **58**, 465.
Elsasser, W.: 1966, *Atom and Organism*, Princeton Univ. Press.
Groth, W. and Suess, H.: 1938, *Naturwiss.* **26**, 77.
Miller, S. L.: 1953, *Science* **117**, 528.
Miller, S. L.: 1955, *J. Am. Chem. Soc.* **77**, 2351.
Miller, S. L.: 1957, *Biochem. Biophys. Acta* **23**, 480.
Miller, S. L. and Urey, H. C.: 1959, *Science* **130**, 245.
Miller, S. and Orgel, L.: 1973, *The Origins of Life on the Earth*, Prentice Hall, Englewood Cliffs, N.J.
Sagan, C. and Mullen, G.: 1972, *Science* **177**, 52.
Suess, H. E.: 1962, *J. Geophys. Res.* **67**, 2029.
Weiss, A.: 1973, *Proc. of 4th Int. Conf. on Origin of Life*, Barcelona, Spain.

# MOLTEN EARTH AND THE ORIGIN OF PREBIOLOGICAL MOLECULES

MIKIO SHIMIZU

*Institute of Space and Aeronautical Science, University of Tokyo, Komaba, Meguro-ku, Tokyo, Japan*

**Abstract.** Evidence for the molten Earth at its accretion time has been accumulated through the geochemical investigations and the observations of the surfaces of planets by space probes such as Venera 8, Mariner 9, Surveyor, Luna, and Apollo. The primitive terrestrial atmosphere might have been derived from the volcanic gases, as suggested by Rubey, but of a higher temperature than so far assumed. A thermochemical calculation of the composition of the volcanic gas suggests the following possibilities:

(1) Large amounts of $H_2$ and CO were present in the primitive atmosphere. This gives a theoretical basis for the HCN-production experiment by Abelson.

(2) HCHO and $NH_3$ existed in the primitive oceans, of the amount comparable with the weight of the present biosphere.

(3) Plenty of $NO_3^-$, $SO_4^{--}$, and $PO_4^{---}$ were expected in the primitive oceans. The $NO_3^-$ ions might have been useful for the nitrate respiration advocated by Egami.

In an appendix, it is argued, on the basis of the observational evidence of the exospheric temperatures of planets by space probes, that a highly reducing atmosphere would (if it existed on the primitive Earth) have disappeared very quickly due to the thermal escape of hydrogen from its exosphere.

## 1. Introduction – Molten Earth

Recently, biologically important molecules such as HCN, HCHO, $NH_3$ etc. (abiotic precursors of nucleic acids, proteins, and sugars) have been found in the dense interstellar clouds in many compact H II regions where new stars are forming. The origin of these molecules might correlate with the formation of the 'solar systems' there (Sagan, 1972; Shimizu, 1973a). A direct relation of these interstellar molecules with life on the Earth is unlikely due to the catastrophic conditions at the time of planetary formation (Rank *et al.*, 1971). Since these molecules can form in such extreme conditions as those occurring in interstellar space, however, they might also survive in the severe environment of the primitive Earth.

Evidence has accumulated for the molten Earth at its accretion time through many geochemical investigations (Fanale, 1971; Oversby and Ringwood, 1971; Armstrong and Hein, 1973). Furthermore, exploration of other planets by space probes has provided support to the differentiation of planets, at least in the vicinity of the surfaces: Gamma ray analysis of Venera 8 suggested the presence of granite on the Venus surface; Mariner 9 TV experiments found many Martian craters of the caldera type; and data gathered by Surveyor, Luna, and Apollo showed the differentiation of the Moon through the findings of basalt in Maria and of anorthosite in Terra. It has been theoretized that the gravitational energy for the accretion of the Earth would have been larger by two orders of magnitude than the total radioactive energy in the interior of the Earth. Consequently, the Earth might easily have been molten if it accreted in a short time. The difficulty of the cooling of the molten Earth (McDonald, 1963) may be avoided by taking into account mantle convection.

In the following sections, I shall adopt the assumption of a molten Earth in its earliest period and point out the possibility that such a severe environment might be favorable for the formation of prebiological molecules and the evolution of the energy yielding metabolism of living systems.

## 2. Prebiological Molecules in the Primitive Atmosphere and Ocean

The main atmospheric constituents on the primitive Earth were suggested by Rubey (1951) to have been $H_2O$, $CO_2$, and $N_2$, which converted later to the present ocean, carbonated rocks, and atmosphere, respectively. The temperatures in the interior of the molten Earth were much higher than those of the present Earth, possibly 2000–3000 K (Ringwood, 1966). Consequently I shall attempt to determine the composition of minor constituents in the volcanic gases at the molten stage of the Earth by carrying out a thermochemical calculation on the population of various molecules that occur in gases at high temperatures (1500–6000 K) and pressures (10–1000 atm), including six elements in the ratios of

$$H:C:N:O:P:S = 1:1\times10^{-1}:3\times10^{-3}:7\times10^{-1}:1\times10^{-6}:1\times10^{-3},$$

which reflect a possible composition of Rubey's primitive atmosphere (Ingersoll and Leovy, 1971). In Figure 1, I have shown some examples of the variation with temperature of the molecular population of biologically important molecules in a gas with a pressure of 100 atm.

As can be seen from Figure 1, the main components of volcanic gases at high temperatures (2000–3000 K) are still $H_2O$, $CO_2$, and $N_2$. The primordial volcanic gas also contains

(1) $H_2$ and CO whose amounts are comparable or more than that of $N_2$;

(2) some soluble gases important for 'chemical evolution' such as HCHO and $NH_3$, whose amounts reach the total weight of the present biosphere; and

(3) other soluble gases correlated with the development of metabolism during biological evolution such as the oxides of nitrogen, sulphur, and phosphorus. (As to the oxide of phosphorus, I calculated the amount of PO alone, because of the lack of the molecular constants for $P_2O_5$ etc. However, the great stability of $PO_4^{---}$ suggests that this species might form in great amount in the primitive oceans.)*

After the solidification of the surface of the molten Earth, heavy rain may fall on the surface, accompanied with the eruption of volcanic gas. Soluble molecules might be dissolved in the rain drops, be accumulated in the primitive ocean, and be concentrated through evaporation due to heat from the interior. Gases from the marine volcanos at the bottom of shallow seas might also be dissolved in sea water. Under this circumstance, the primitive oceans might be very acidic at first; however, its neutralization might have occured in a short time, since strong convection in the

* A referee gave an interesting comment that the $PO_4^{---}$ concentration in the primordial oceans was probably limited by the very small solubility of apatite $Ca_5(PO_4)_3(OH, F)$. On the other hand, it might be also true that the soup of $PO_4^{---}$ was required on the primitive Earth to form ATP and other phosphate necessary for living systems.

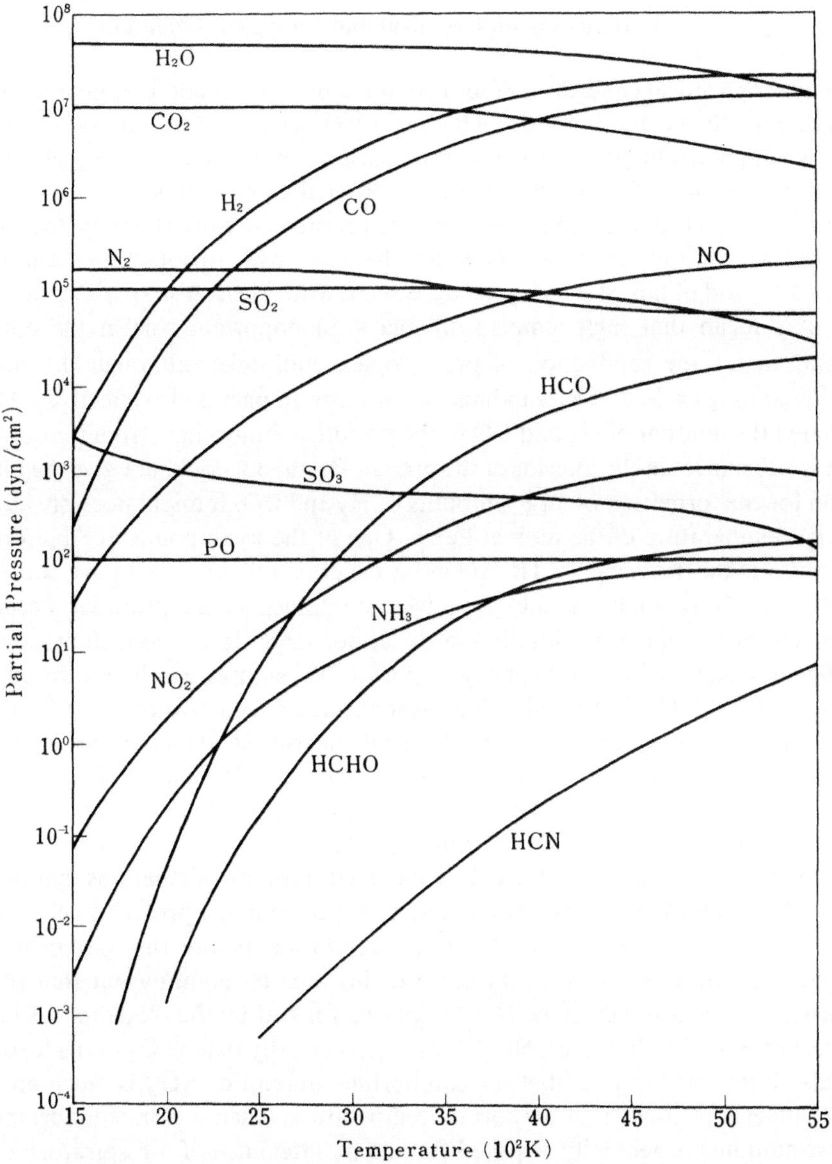


Fig. 1. Variation of the composition of Rubey's primordial atmosphere with temperature at the pressure of 100 atm.

mantle would have brought a great amount of fresh basalt from the interior to the bottom of oceans. On the other hand, $N_2$, $H_2$, and CO might have remained in the primitive atmosphere. The atmosphere might have been so turbulent due to the release of the latent heat of water that strong thunder storms may have lasted for a long time throughout the whole atmosphere during the continuous eruption of volcanic gases of high temperature.

## 3. A Discussion on Chemical and Biological Evolution

The physicochemical conditions described in the previous sections appear to be quite favorable for the evolution of life: Abelson (1966) performed a discharge experiment to show a high production of HCN in the mixed gas of $H_2$, CO, and $N_2$ which might be the composition of the early atmosphere after the precipitation of water and the absorption of $CO_2$ and soluble gases into the primitive oceans. (I conducted thermochemical calculations at 2000–5000 K for the gases with various ratios among $H_2$, CO, and $N_2$ and obtained a very good agreement with Abelson's experimental results. This may mean that high temperature plays an important role in the discharge mechanism for the production of prebiological molecules, although the chemical reaction among excited species induced by electron impact is also effective.) Abelson estimated the amount of $H_2$ and CO in the primitive atmosphere from that of buried organic substances in the interior of the present Earth, but could not give the ultimate reason for the formation of large amounts of $H_2$ and CO. It might possibly be due to the high temperature of the molten Earth. One of the weak points in Abelson's discussions was the deficiency of HCHO in his experiments. In cases I have considered, HCHO and $NH_3$ might already have been contained in the primitive oceans, independent of discharge (or thunder) in the atmosphere. It has been shown by many biochemists that various biologically important substances can be produced in the 'soup' of HCN, HCHO, and $NH_3$. The heat of numerous active volcanos in the early history might be very effective for forming proteins from solid amino acids which were formed by evaporation at the seashore, as pointed out by Harada and Fox (1965). The high concentration of salts in the primitive oceans might help to 'salt out' the coacervate droplets from such a polymer (Oparin, 1924).

As to biological evolution, it may be concluded from my calculations that plenty of $NO_3^-$, $SO_4^{--}$, and $PO_4^{---}$ etc. would also be expected in the primitive ocean. Egami (1974) advocated the existence of nitrate respiration before the establishment of oxygen respiration. Broda (1970) criticized this idea by pointing out that the time sequence was inadequate, since $NO_3^-$ might be formed by the oxidation of $NH_3$ by $O_2$. As I have shown, however, $NO_3^-$ does not necessarily require $O_2$ for its formation. It is also interesting to note that the equilibrium amount of $NO_3^-$ is much greater at high temperature than that at normal temperature. Such a non-equilibrium state (excess amount) is generally required for the maintenance of a respiratory system.

The oxidizing state of vocanic gases depends critically on the composition of the melt with which the gases are in equilibrium. Highly reducing volcanic gases might have been formed, if the gases were in contact with, say, metallic iron or lunar basalts. However, the theories by Ringwood (1966) and by Anders (1968) on the accretion processes of the Earth favored very quick formation of the Earth's core, thus excluding the possibility of uniform distribution of metallic iron in the mantle at the earliest time of the Earth's history. Furthermore, the Moon might have been captured or been formed in the outer region of the primordial Earth-Moon complex (Anderson, 1972). In both cases, the chemical composition of the Moon was different from the primordial material of the Earth. Of course, no one knows the true primordial substance of the Earth at present and, in this meaning, there is no *a priori* evidence for the gas composition chosen by Rubey and for its modification adopted in this paper. As I shall argue in the appendix, however, a highly reducing atmosphere might have disappeared very quickly from the primitive Earth, even if it existed once. The biochemical approach to the evolution of energy yielding metabolism by Egami (1974) suggested also the existence of $NO_3^-$ on the primitive Earth. In our opinion, therefore, it may be worthy of further investigation to seek the formation of life in a manner, as is described in the text.

## Appendix: The Exospheric Temperature of the Primitive Earth

Many scientists appear to believe that the primitive terrestrial atmosphere was composed of $CH_4$ and $NH_3$ and that the life was formed under this circumstance. The major constituent of the thermosphere in this period should have been hydrogen due to the photodissociation of atmospheric constituents. I shall point out here that the exospheric temperature which determines the escape rate of hydrogen, and eventually the evolution of the atmosphere, should be very high for a hydrogen atmosphere; such an extremely reducing atmosphere might therefore disappear in an instant (in a geological sense), leaving a 'coal tar" behind. Abelson has frequently pointed out that no one has found any geological evidence for the presence of enormous amounts of carbon compounds in the early history of the Earth.

Table I summarizes observational results of space probes such as Mariner 4, 5, 6, 7, and 9, Mars 3, sounding rockets etc. on the exospheric temperatures of the planets, arranging them in order of (1) magnitudes of the mass of the major atmospheric constituents and (2) distance of the planets from the Sun. It can easily be seen that, the larger the mean mass of the atmospheric constituents, the lower the exospheric temperature. This is due to the fact that, the larger the mass, the smaller the scale height $H = kT/\bar{m}g$ of the atmosphere (namely, the extension of the atmosphere to space) and that the distance between the heated atmospheric level (fire point) and the

TABLE I

The exospheric temperatures of Planets with various atmospheric constituents (K)

←Mean mass of major component

| Distance from the Sun→ | Venus | Earth | Mars | Jupiter |
|---|---|---|---|---|
| $\bar{m}=1$–$2$ ($H$, $H_2$) | (43000)[a] | (15000)[a] | (18000)[a] | 100–300 |
| $\bar{m}=28.8$ ($N_2$, $O_2$) | | 1000 | | |
| $\bar{m}=44$ ($CO_2$) | 650 | | 350 | |

[a] Estimated roughly by Gross (1972).

cooled mesopause (water pump) becomes shorter (Shimizu, 1969). The exospheric temperature is also very sensitive to the solar radiation flux which is inversely proportional to the square of the distance from the sun. Consequently Table I suggests an extremely high value of the exospheric temperature for the reducing atmosphere on the primitive Earth. Indeed, an estimate by Gross (1972) for an atomic hydrogen atmosphere at the orbit of the Earth resulted in the exospheric temperature of 15000 K.

Rasool and McGovern (1966) argued that (1) high emission rate in the infrared of methane dissociation products and (2) high conductivity of hydrogen due to its small mass should decrease the exospheric temperature of the primitive atmosphere ($CH_4$—$NH_3$ atmosphere) to 500–900 K. This argument was not satisfactory in the following points: The primitive atmosphere should in all probability be strongly turbulent due to the release of the latent heat of water; the photodissociation products should therefore be so efficiently diluted by eddy transport as in the present atmospheres of Earth, Mars, and Venus (Shimizu, 1972, 1973b). Furthermore, the mesopause temperature may be restricted in the range of 100–200 K in any circumstance. Then, however efficiently the infrared cooling mechanisms might work, the temperature gradient between the exosphere and the mesopause does not change appreciably. Secondly, as far as the mass effect is concerned, the scale height effect described above works more effectively than the effect of mass on the heat conductivity. As can readily be derived from the heat conduction equation, the exospheric temperature depends on the mass in the form of $H^2/k$, where $H$ is the scale height and $k$ the heat conduction coefficient, namely proportional to $1/m$. Consequently, contrary to the conclusion of the previous investigators, the smaller the mass of the major atmospheric constituent, the higher the exospheric temperature of the planets. The reader may frequently find in McGovern's paper (1969) that he obtained high exospheric temperatures, if he did not artificially restrict the amount of hydrogen in the thermosphere.

**Note added in proof.** A surprizingly low intensity of Lyman alpha airglow observed recently by Pioneer 10 and OAO 3 ultraviolet photometer experiments indicated a strong atmospheric mixing effect in the upper atmosphere of Jupiter as well as of terrestrial planets (M. Shimizu, 17th COSPAR meeting report, II–VII 1–7, 1974). This evidence endorses our arguments on the instability of $CH_4$ atmosphere on the primitive Earth (and, then, the necessity of a *non-stationary* atmosphere of $H_2$ and CO for the generation of living system).

## References


Abelson, P. H.: 1966, *Proc. N.A.S.* **55**, 1365.
Anders, E.: 1968, *Acc. Chem. Rev.* **1**, 289.
Anderson, D.: 1972, *Nature* **239**, 263.
Armstrong, R. L. and Hein, S. M.: 1973, *Geochim. Cosmochim. Acta* **37**, 1.
Broda, F.: 1970, *Prog. Biophys. Mol. Biol.* **21**, 143.
Egami, F.: 1974, *Origins of Life* **5**, 405.
Fanale, F. P.: 1971, *Chem. Geol.* **8**, 79.

Gross, S. H.: 1972, *J. Atmospheric Sci.* **29**, 214.
Harada, K. and Fox, S. W.: 1965, in S. W. Fox (ed.), *The Origin of Prebiological Systems*, Academic Press, New York, p. 289.
Ingersoll, A. P. and Leovy, C. B.: 1971, *Ann. Rev. An. Astrophys.* **9**, 147.
McDonald, G. J. F.: 1963, *Space Sci. Rev.* **2**, 473.
McGovern, W. E.: 1969, *J. Atmospheric Sci.* **26**, 623.
Oparin, A. I.: 1924, *Origin of Life*, Izd. Moslowski Rabokii, Moscow.
Oversby, V. M. and Ringwood, A. E.: 1971, *Nature* **234**, 463.
Rank, D. M., Townes, C. H., and Welch, W. J.: 1971, *Science* **174**, 1083.
Rasool, S. I. and McGovern, W. E.: 1969, *Nature* **212**, 1225.
Ringwood, A. E.: 1966, *Geochim. Cosmochim. Acta* **30**, 41.
Rubey, W. W.: 1951, *Bull. Geol. Soc. Amer.* **62**, 1111.
Sagan, C.: 1972, *Nature* **238**, 77.
Shimizu, M.: 1969, *Proc. 24th meeting Phys. Soc. Japan, Kanazawa* (in Japanese).
Shimizu, M.: 1972, *Space Res.* **12**, 293.
Shimizu, M.: 1973a, *Prog. Theor. Phys.* **49**, 153.
Shimizu, M.: 1973b, *J. Geophys. Res.* **78**, 6780.

# SOIL AND WATER AND ITS RELATIONSHIP TO THE ORIGIN OF LIFE

DUWAYNE M. ANDERSON and AMOS BANIN*

*U.S. Army Cold Regions Research and Engineering Laboratory, Hanover, N.H., U.S.A.*

**Abstract.** Soils of the terrestrial planets form at the boundaries between lithosphere, atmosphere and hydrosphere. Biogenesis occurred in these zones; thus, it is axiomatic that some, perhaps many, stages of biogenesis occurred in intimate association with the mineral constituents of soils. Because of a high surface to mass ratio and, consequently, a high surface reactivity, the layer lattice clay minerals are the most important of these. According to the geological record, clay minerals appeared very early on the primordial Earth. Recent investigations have confirmed their presence in carbonaceous meteorites and have indicated their occurrence on Mars.

In this paper we collect pertinent physico-chemical data and summarize the organic reactions and interactions that are induced or catalyzed by clays. Many clay-organic reactions that do not occur readily at high water contents proceed rapidly at adsorbed water contents corresponding to surface coverages of one or two molecular layers. One or two monolayers of adsorbed water correspond to extremely dry or cold planetary environments. Some consequences of these facts *vis à vis* biogenesis on Mars are considered.

## 1. Introduction

It now seems certain that the multiplicity of organic compounds required in primitive life forms could have arisen through the 'normal' sequences of abiotic synthesis (Oró, 1972, Ponnamperuma and Gabel, 1971). Consideration of the reaction-kinetics that must have been involved has produced general agreement on the necessity of a liquid phase. Beginning with Bernal, it has often been observed that zones where lithosphere and hydrosphere come in contact (or, on the molecular level, where liquid and solid phases join) were critical locations (Bernal, 1967; Fox, 1965; Rutten, 1972). Others, however, have argued from time to time that the entire sequence might have occurred in an essentially open aqueous system. The basic biochemical unity of terrestrial life has fostered the view that surviving life-forms, despite great morphological diversity, evolved from a single primitive precursor. Even though a number of concurrent abiotic sequences originally may have developed, the common view is that eventually one prevailed. In principle, it should be possible to deduce the environmental locale in which the prevailing sequence originated; in actuality, this information probably will never be apprehended.

It is axiomatic in evolutionary thought that complexity has, over time, arisen step by step out of simplicity; thus, research in chemical evolution began with efforts to establish how the building blocks of molecular biology could have arisen through abiotic synthesis from such simple molecules as $CH_4$, $CO$, $NH_3$, $H_2$, $N_2$ and $H_2O$. This has been followed by investigations of abiotic syntheses of organic molecules of increasing complexity. In order to guard against ambiguity in the interpretation of experimental data, most investigators have endeavored to keep the number of the

* On leave from the Department of Soil and Water Science, The Hebrew University of Jerusalem, P.O. Box 12, Rehovot, Israel.

reactants and the complexity of the experimental design to a minimum. For this reason, until relatively recently, heterogeneous and imperfectly characterized solid substrates such as clays and other soil constituents were avoided although the possible critical roles of mineral surfaces often implicitly were recognized. Now, sufficient knowledge pertaining to the general aspects of biogenesis is in hand so that an increasing number of systematic experiments involving mineral substrates are being made. In this paper, we collect and examine recent experimental evidence bearing on the involvement of soil particulates in biogenesis.

## 2. Clays, the Active Constituent of Soil

Soils form where the lithosphere contacts atmosphere and hydrosphere. In the most primitive condition soil consists entirely of mineral matter derived by simple comminution of parent rock together with the products of primary chemical alteration and resynthesized secondary minerals. In the present context, the important processes occur at solid-liquid and solid-gas interfaces. Hence, the extent of these interfaces is of primary importance. Because of an exceedingly large surface area to mass ratio, the smallest particles constitute the most active fraction of planetary soils. By definition, this is the clay fraction. Although seldom, if ever, in true equilibrium with surrounding environments, the clay minerals can be very stable, as indicated by their presence in the oldest known sedimentary rocks and a high abundance throughout geologic history (Ronov *et al.*, 1973). According to one possible sequence (Eitel, 1966) the zeolites appear first, following the late magmatic-hydratogenic state. Subsequent hydrothermal stages and low temperature diagenesis lead to the appearance of the several clay mineral families. Following synthesis and continuing diagenesis, representative species rapidly became distributed throughout the atmosphere, hydrosphere and, much of the lithosphere.

For present purposes, it is permissible to employ the term clay in its most general sense, i.e. materials that exhibit the familiar properties of clays. This includes naturally occurring, fine-grained, heterogeneous mineral matter exhibiting a strong affinity for water, plasticity and base exchange. This includes all the crystalline clay minerals of the kaolinite, halloysite, montmorillonite, vermiculite, illite, chlorite, attapulgite, sepiolite and palygorskite families and, in addition, others present in a state of subdivision sufficiently small to exhibit colloidal properties such as amorphous, allophane. The crystalline clay minerals for the most part are platy alumino-silicates composed of alternating layers of silicon in tetrahedral coordination, and aluminum in octahedral coordination, with oxygen. Stacked in 2:1 or 1:1 sequences these form elementary crystalline sheets; several elementary sheets then stack to form packets (tactoids) or quasi-crystals. Without going into the details, because of isomorphous substitution of metallic ions within the lattice, such as aluminum for silicon and iron for aluminum, a residual net negative charge usually is imparted to each elementary sheet. This charge must be balanced by adsorbed cations, usually sodium, calcium, etc. Taking montmorillonite as an example, the fundamental particle is an alumino-silicate sheet 10 Å in thickness and up to several thousand Å in length and width. Faces of the sheets consist of planes of oxygen atoms in planar hexagonal array. Tactoids normally contain two to fifteen elementary sheets depending on the exchangeable cation(s) present. The principal effect of variations in tactoid size is to bring about a change in the ratio of external to internal surface area (Banin and Lahav, 1968). Other clay minerals differ in structural detail, but in general possess similar characteristics. A summary of some of the best characterized clay minerals is given in Table I. Interestingly, it is a general observation that regardless of mineralogy, clay-size particles, produced by comminution tend toward platy shapes, even quartz (Krinsley and Smalley, 1973). Clays exist in interstellar space in the form of small ($<0.1\ \mu$) 'dust' particles and have been identified as constituents of comets and carbonaceous meteorites (Bass, 1971; Oró, 1972; Rutten, 1972). Clays thus are

TABLE I

Surface characteristics of the major alumino-silicate clay minerals

| Characteristics | Montmorillonite | Vermiculite | Illite | Kaolinite |
|---|---|---|---|---|
| Lattice stacking units (Si tetrahedral layers: Al octahedral layers) | 2:1 | 2:1 | 2:1 | 1:1 |
| Total specific surface area ($m^2\ g^{-1}$) | 700–800 | 700–800 | 100–200 | 25–50 |
| Percentage of internal surface | 10% to 90%[a] | 80% to 90% | 0 | 0 |
| Cation exchange capacity (meq $g^{-1}$) | 0.8–1.2 | 1.2–2.0 | 0.15–0.40 | 0.02–0.10 |
| Average surface charge density (esu $cm^{-2}$) | $3.3–4.3 \times 10^4$ | $5.0–7.2 \times 10^4$ | $4.3–5.8 \times 10^4$ | $2.3–5.9 \times 10^4$ |
| Area per elementary charge ($Å^2$) | 110–150 | 60–100 | 80–110 | 180–200 |
| Free swelling in water ($cm^3\ g^{-1}$) | 0.5–20[a] | 0.5–15[a] | 0.1–1.2 | 0–0.3 |
| Range of interlammelar expansion | 6–240 Å[a] | 6–200 Å[a] | – | – |

[a] Depending upon exchangeable cation.

ubiquitous in time and space and have been present at virtually all stages of planetary accretion and subsequent evolution. Interactions between clay particles and the reactants and products of prebiotic sequences of organic synthesis therefore must have been of significance in primordial chemical evolution as well as subsequent stages of speciation and population of the Earth.

## 3. Clay-Organic Reactions

Not all the interactions possible would have facilitated biogenesis: many would have tended to hinder or interrupt the process. On the other hand, many might have aided and some could have been vital. Concentration of reactants is required at certain steps. Adsorption and preferential orientation of significant precursors by mineral substrates is a process by which this function may have been accomplished. Another crucial process that could have been operative is the juxtaposition of reactants and the catalysis of surface reactions, the stabilization of intermediates and catalysis of subsequent polymerizations. Clay substrate-organic reactions that can lead to the synthesis of larger and more complex molecules from simple precursors may be categorized as:

(1) Simple bonding by interaction at the surface (physical and chemisorption)

(2) Initiation of (or rate enhancement of) reactions, including polymerizations (Catalysis)

(3) Suppression of randomness in recombination, polymerization, and condensation reactions (Epitaxy)

In many cases, the first may occur without leading to the second and the first and second occur without the third.

Clay-organic reactions investigated to date are summarized in Table II. Organic

TABLE II

Representative known clay-organic interactions

| | Clay type | Mechanism | References |
|---|---|---|---|
| I. ADSORPTION | | | |
| n-Alkyl-ammonium ions | M | Cationic adsorption | Jordan (1949a, b) |
| | V | | Walker (1950) |
| | V | | Garret and Walker (1962) |
| | M | | Raman and Mortland (1969) |
| n-Alkyl-pyridinum ions | M | Cationic adsorption | Greenland and Quirk (1962) |
| Pyridines, purines, nucleosides | M | Cationic adsorption | Lailach *et al.* (1968a, b) |
| | M | and coordination | Lailach and Brindly (1969) |
| | M | complex | Farmer and Mortland (1966) |
| Diquat and paraquat ions | Soils | Cationic adsorption | Knight and Tomalinson (1967) |
| (herbicides) | M, K, I | and coordination complex | Weed and Weber (1968, 1969) |
| Quinoline | M, K, I | Surface interaction with polar groups | Doehler and Young (1962) |
| Polyalcohols (various) | M | Surface interaction | Tettenhorst *et al.* (1962) |
| | M, N, V | with polar groups | Brindly (1966) |
| | M | | Greenland (1963) |
| | M | | Dowdy and Mortland (1968) |
| Hystidine | M | Polar groups; coordination | Heller-Kallai *et al.* (1972) |
| Proteins (various) | M | Polar groups and | Ensminger and Giesking (1941) |
| | K | hydrogen bonding | McLaren (1954) |
| | M, K, I, V | | Harter and Stotzky (1971) |
| | M, K, I, V | | Albert and Harter (1973) |
| $NH_3$ adsorption | M, V | Coordination complex formation and cationic adsorption following protonation | Mortland *et al.* (1963) |
| Urea and urea derivatives | M | Coordination complex | Mortland (1966) |
| | M | formation and cationic | Tahoun and Mortland (1966 a,b) |
| | M | adsorption following protonation | Farmer and Ahlrichs (1969) |
| Pyridine, methylamine, | M, S | Cationic adsorption | Farmer and Mortland (1966) |
| 3-aminotriazole | M | following protonation | Raman and Mortland (1969) |
| Ethylenediamine | M | Protonation and hydrogen bonding | Cloos and Laura (1972) |

*Table II (Continued)*

| | Clay type | Mechanism | References |
|---|---|---|---|
| II. ACID CATALYSIS REACTIONS | | | |
| Ethyl-acetate hydrolysis; sucrose inversion | M | Bronsted acidity (Adsorbed H+) | Coleman and McAuliffe (1955) |
| Trans-alkylation of alkylammonium cations | M | Bronsted acidity (partially dissociated adsorbed water) | Calvet *et al.* (1964) |
| | M | | Durand *et al.* (1972) |
| Decomposition of amines | M | Bronsted acidity (partially dissociated adsorbed water) | Chaussidon and Calvet (1965) |
| Decomposition of Co (III) hexamine cations | M | Bronsted acidity (partially dissociated adsorbed water) | Chaussidon *et al.* (1962) |
| Coloration reactions (benzidine-benzidinium) | M | Lewis acidity (aluminum ions exposed at edges) | Hauser and Leggett (1940) |
| Decomposition of S. Triazines | M | Protonation and hydrolysis | Russell *et al.* (1968) |
| | M | | Cruz *et al.* (1968) |
| III. POLYMERIZATION REACTIONS | | | |
| Methacrylates | M | | Blumstein (1965) |
| | M | | Blumstein *et al.* (1969) |
| | M | | Solomon and Loft (1968) |
| | M | | Malhotra *et al.* (1972) |
| Vinyls | M | | Dekking (1967) |
| Styrene | M, K, A | | Solomon and Rosser (1965) |
| IV. CATALYSIS AND POLYMERIZATION REACTIONS OF BIOLOGICAL SIGNIFICANCE | | | |
| Amino acid formation from $NH_2$, | M | | Yoshino *et al.* (1971) |
| $H_2$, $CO_2$ and $ND_2$, $D_2$, $CO_2$ | M | | Hayatusu *et al.* (1971) |
| Amino acids and purines formation from CO and $NH_3$ | Z | | Fripiat *et al.* (1972) |
| Formation of trioses, tetroses, pentoses and hexoses from formaldehyde | K | | Gabel and Ponnamperuma (1967) |
| Polymerization of amino acid phosphate anhydrides | M | | Paecht-Horowitz (1971) |
| Polymerization of amino acids | M | | Cloos *et al.* (1966) |
| | | | Fripiat *et al.* (1966) |
| Formation of polypeptides from amino acid adenylates | M | | Paecht-Horowitz *et al.* (1970) |
| Oligonucleotide formation from thymidylic acid | M | Pentanucleotides formed in the presence of montmorillonite | Ibanez *et al.* (1971) |
| ADP synthesis | K | | Steinman *et al.* (1965) |
| | | | Steinman (1966) |

M = Montmorillonite A = Attapulgite
S = Saponite K = Kaolinite
N = Nontronite Z = Zeolites
I = Illite

adsorption on clays involves direct, electrostatic adsorption and ion-dipole and coordination bonding to adsorbed inorganic ions on the clay. Also of importance are van der-Waals attraction, hydrogen bonding, covalent bonding and entropy effects (Mortland, 1970). Possible combinations of these bonding mechanism explain the observation that for large organic ions of the same valence, organic cations are more strongly adsorbed than inorganic cations.

The catalytic activity of clays and zeolites derives in part from the moderation of molecular displacements by adsorption forces and the ability of clays to exchange ionic species and expose adsorbed molecules to strong electrostatic forces at a variety of localized charge sites. Naturally occurring surfaces are heterogeneous with regard to sorption sites, but the density of sites of surface charge and sites of potential catalytic activity is surprisingly uniform, the total number per unit weight depending principally on the specific surface area. Thus it can be suggested in passing that in many in-vitro syntheses, surface catalysis by the apparatus may have contributed significantly to the results, especially in reactions of low yield, requiring relatively high temperatures. In media such as beach sand and basalt, moreover, clays usually are present to some extent unless specifically removed. In such experiments, the question of reaction rates and yield should be addressed keeping the total surface areas exposed by both apparatus and media in mind.

Catalysis of reactions by clay surfaces in most cases is related to a 'surface acidity' resulting from one or more of the following mechanisms:

(a) Protons donated from the adsorbed, exchangeable hydrogen ions (Coleman and McAullife, 1955); Bronsted acidity.

(b) Protons transferred from adsorbed water molecules polarized by the electrostatic fields of exchangeable, metallic cations; e.g. protonation of pyridine to pyridinium cation on partially hydrated montmorillonite (Farmer and Mortland, 1966; Touillaux *et al.*, 1968): Bronsted acidity.

(c) Protons transferred from other adsorbed ions or molecules; e.g. proton transfer from adsorbed $NH_4^+$ to 3-Aminotrizole forming 3-Aminotriazolium cation (Russel *et al.*, 1968); Bronsted acidity.

(d) An adsorbed ion or molecule acting as an electron acceptor; e.g. electron transfer from benzene, xylene, etc. to Cu (II) on montmorillonite; (Donner and Mortland, 1970); Lewis acidity.

(e) The clay lattice acting as an electron acceptor; e.g. electron transfer to lattice aluminum exposed at an edge, transforming colorless benzidene to blue, monovalent benzidinium cation; (Hauser and Leggett, 1940); Lewis acidity.

Mechanisms (b) and (c), commonly referred to as 'protonation' reactions, are now well characterized. Regarding mechanism (e), Lewis basicity may operate in some circumstances. It has been suggested that a lattice component in a relatively low oxidation state, such as $Fe^{++}$, may act as an electron donor (Solomon, 1968); experimental confirmation of this mechanism, however, presently is lacking.

Polymerization on clay substrates has been demonstrated in a number of cases. According to Solomon (1968) and Solomon *et al.* (1971) polymerizations involving free radicals generally are catalyzed by clay surfaces only if a suitable 'initiator' is present; otherwise they are inhibited. Reactions involving cationic radicals generally are catalyzed, but reactions involving anionic radicals are catalyzed only if certain conditions involving polymer type, expandability of the clay, and the presence of a transition metal of low oxidation state in the lattice (e.g. $Fe^{++}$), are fulfilled. Consequently, most anionic-radical polymerizations appear to be inhibited by clay mineral surfaces. The polymerization of hydroxy methacrylate on montmorillonite surfaces, however, is an exception (Solomon, 1968). It is of particular interest to note that 'insertion polymers' produced from methyl methacrylate monomers sorbed by montmorillonite (Blumstein, 1965; Blumstein *et al.*, 1969), formed aggregates, "... the architecture (of which) is reminiscent of the elementary crystallite of montmorillonite." (Blumstein *et al.*, 1969). An example of the ability of mineral substrates to confer orientational order on condensed surface phases is the orientation of the nematic phase of p-azoxyanisol on glass. This 'liquid crystal' displays sensitivity both to crystallographic orientation of the substrate and to subtle variations in surface topography (Bikerman, 1970). The full range of such epitaxial phenomena is immense but undefined; it merits systematic evaluation.

## 4. Clay-Water-Organic Interactions

Clay surfaces are commonly hydrophyllic; isosteric heats of water vapor adsorption of the order of 75 kcal mole$^{-1}$ characterize the first increments. Values for subsequent increments decline with increasing surface coverage to the normal latent heat of liquefaction. The presence of water in a definite quantity and position is essential for some of the processes discussed above, particularly those in the category of acid catalysis. Optimal water contents for acid catalysis are in the range of 5–10% by weight for montmorillonite (Durand *et al.*, 1972). This corresponds approximately to a monolayer of water separating adjacent lamella. The enhanced catalytic activity can be attributed to a perturbation of water molecules by the electrostatic fields surrounding the exchangeable ions and covering the silicate surfaces that results in localized increases in proton activity. Higher water contents interfere by decreasing the rate of interaction between the reacting species and the finite number of perturbed water molecules, reducing conversion rates. On the other hand, desication of the surface below the optimum value reduces conversion rates because of reductions in the number of perturbed water molecules.

At high water contents, clays can form gels that, in many significant respects, mimic the gels that constitute portions of intracellular media. Depending on clay and electrolyte concentration, these gels exhibit thixotropy, and syneresis, and have water activities and rheological properties comparable to those required for biochemical processes and characteristic of present day intracellular fluids. Although mere rheological mimicry cannot be advanced as proof that inorganic gels played essential roles in the origin of life, neither is it without significance. It is highly probable that many adsorbed organic molecules were conveyed by this means into aqueous environments, complete with electrolytes, like those characteristic of living cells.

## 5. Expanding Clay Lattices

The possible role played by clay substrates in orienting organic molecules and bringing about juxtapositions favorable for charge or proton transfer reactions or effective in reducing randomness in addition reactions and polymerizations is a particularly striking possibility in the case of the expanding-lattice clay minerals montmorillonite and vermiculite. In these minerals, individual lamella separate, expand and contract according to the degree of hydration. A full treatment is precluded but a brief discussion of an expanding vermiculite is given for purposes of illustration.

Vermiculites exist as powders, small flakes and large single crystals. As shown in Table I, they have high ion exchange capacities. When saturated with lithium or butylammonium ions and subsequently exposed to water, they expand by imbibition in the (001) dimension. The rate and extent of expansion of lithium saturated Kenya vermiculite in water is illustrated by the time-lapse photographs presented in Figure 1. It is seen that the crystal expands in the (001) direction to about 20 times its original dimension. Since expansion is uniform and the initial thickness of the swelling units is about 10 Å, it follows that the thickness of the imbided interlamellar water layers averages about 200 Å. Gradual dehydration brings about collapse of the expanded crystal and complete dehydration is only partially reversed. Semiclosure of the interlamellar space can also be accomplished, without dehydration, by freezing. In the case of the various ionic forms of montmorillonite (a comparable mineral) the interlamellar space contracts on freezing to about 9 Å at −5 °C and to about 6 Å at −10 °C (Anderson and Hoekstra, 1965). On subsequent thawing of this mineral, re-expansion occurs. The lamellar ice crystals are preferentially oriented, presumably due to interaction with the mineral surface; the same effects have been observed with lithium-vermiculite. It is significant that from −10 °C to −180 °C the thickness of the interface separating adjacent lattice layers remains nearly constant, at 6 Å. Calorimetric investigations of both pure and mixed clay substrates has shown that this 6 Å thickness of interlamellar water remains liquid-like down to at least −30 °C (Anderson and Morgenstern, 1974). The interface thickness as a function of temperature is shown in Figure 2. The ordinate, given in $g\ H_2O\ m^{-2} \times 10^2$, is equivalent when multiplied by 100, to interface thickness in Å (thus $0.10\ g\ H_2O\ m^{-2} \times 10^2 \approx 10$ Å thickness). Specific data for lithium-vermiculite are not yet available, but inasmuch as its other properties parallel those of montmorillonite it is safely concluded that its interface thickness vs. temperature relationship is encompassed by the curves shown in Figure 2. It is seen then that liquid-like interfaces separating lamellar ice sheets from the silicate surfaces range in thickness from 5 or 6 Å to more than 80 Å. Due to solute exclusion during ice crystallization, the concentration of solutes can change over wide ranges during freezing and thawing. At about −40 °C the interface undergoes a secondary exothermic phase change, perhaps due to solute precipitation or a transition to the glassy state (Anderson and Tice, 1971) with no perceptible change in clay lattice spacing. Because of these characteristics, expanding lattice clay minerals exhibit properties and behavior well adapted to perform the functions of concentration, orientation and positioning of reactant molecules. The process may be regulated by such ordinary environmental phenomena as floculation, sedimentation, coagulation, wetting and drying, or freezing and thawing.

With respect to systematic experimental investigations designed to demonstrate or clarify the possible roles of substrates at various stages of biogenesis, the system just discussed has much to recommend it. Expanded lithium or butylammonium vermiculite has a very large specific surface; its interlamellar space is easily opened to more than 100 Å; potential reactants may be introduced during imbibition of water or by diffusion after the mineral has been expanded by exposure to aqueous media; and crystal sizes are completely adjustable from macroscopic to submicron

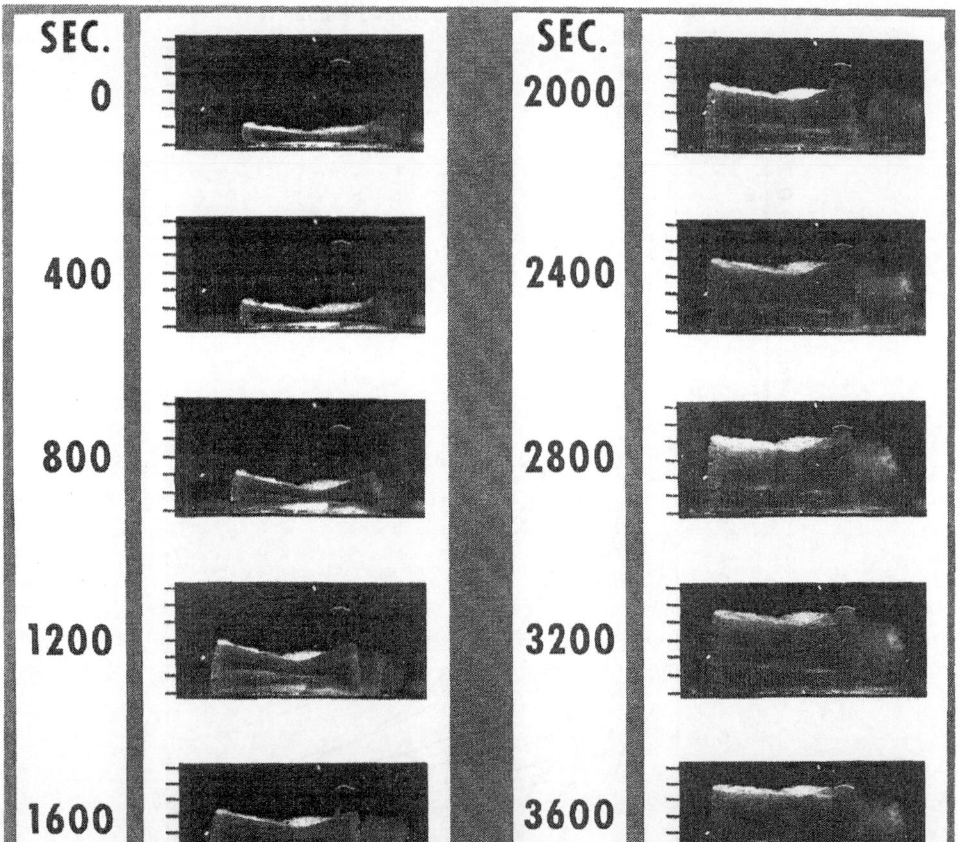


Fig. 1. Selected time-lapse photos following the process of swelling in Li-Vermiculite freshly immersed in water. Scale in mm on left-hand side of each photo gives relative sizes. Initial thickness of the dry vermiculite particle was about 1 mm.

dimensions. The possibility of molecular manipulations by means of controlling changes in interlamellar lattice spacing by wetting and drying or by freezing and thawing is useful and by transforming portions of the interfacial liquid phase to ice, concentration to nearly any desired degree can be accomplished. Thus it should be

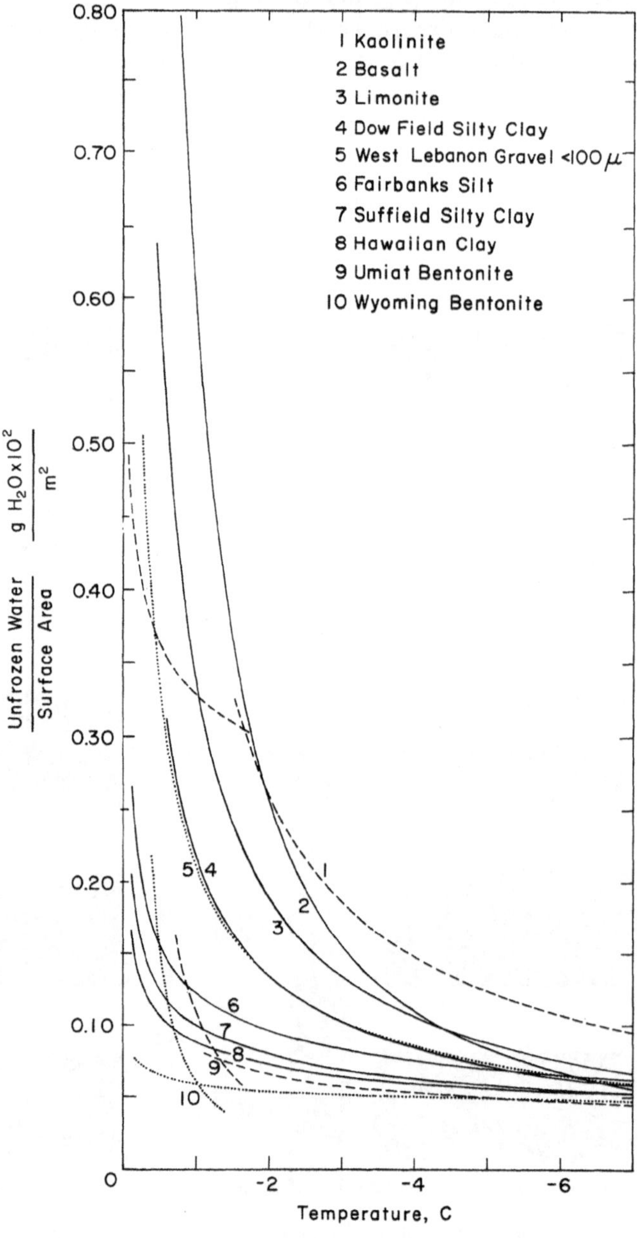


Fig. 2. The amount of water remaining unfrozen in partially frozen clay-water suspensions as a function of temperature for ten representative substrates. (Generalized from published data, Anderson and Tice, 1972.)

possible to bring reactants sufficiently near sites of surface and ionic exchange and active protons to accomplish protonation and acid catalysis and preferred ordering due to epitaxy becomes a possibility not only on the silicate surfaces but on the sandwiched ice surface as well (Anderson and Morgenstern, 1974). Evidence of a characteristic preferred orientation of the proton-proton vector of the water molecules in the interfacial regions seems to be conclusive (Woessner, in press). Thus claywater systems offer favorable environments for a multiplicity of prebiotic reactions.

## 6. Biogenesis on Mars

We turn briefly now to some remarks *vis à vis*, the possibility of biogenesis on Mars. Assuming that the planet was formed by low temperature accretion from cosmic dust, each submicron particle would have had, let us say, a monomolecular layer of adsorbed molecules representative of those present in the primordial gaseous envelope. Among these are water, ammonia, formaldehyde, hydrogen cyanide, methyl alcohol, formic acid, formamid, acetonitrite, methyl-acetylene, acetaldehyde, thioformaldehyde and larger organic molecules yet to be positively identified. Assuming, for simplicity, spherical particles of radius 1 $\mu$, a density of 3, and taking the average density of the adsorbed monomolecular complex as unity, a rough estimate of the total quantity of organic and inorganic volatiles that could have been collected by this means of about $2 \times 10^{24}$ g for a planet such as Earth (planetary mass $\approx 6.0 \times 10^{27}$ g) is obtained. For Mars, the estimate is about $2 \times 10^{23}$ g (planetary mass $\approx 6.4 \times 10^{26}$ g). Rubey (1955) estimates the total excess volatiles in the present terrestrial atmosphere, hydrosphere and sedimentary rocks to be about $1.7 \times 10^{24}$ g. This agreement is intriguing.

These estimates could be as much as an order of magnitude smaller if the adsorbed phase consisted originally of less than a monomolecular layer due to degassing prior to or during accretion; on the other hand they could be as much as a factor of three larger if chemisorption was active and as much as an order of magnitude larger if the original particles were as small as 0.1 $\mu$ in radius. In this connection, recent evidence has shown that the hydroxylation and subsequent production of water from oxygen in the silicate lattice occurs in the solar wind (Cadenhead and Buergel, 1973). Carbonaceous meteorites contain relatively large amounts of hydrated platy silicates, mainly magnesium and iron silicates, similar to terrestrial clay minerals. Bass (1971) claims, for example, to have identified montmorillonite and serpentine as constituents. Carbonaceous meteorites thus are of interest not only because of the relatively high amount of carbon contained in them but their relatively high water contents as well. According to Wiik's classification, type I carbonaceous meteorites contain 3–7% carbon, up to 22% water bound up in laminar magnesium and iron silicates. Isotope ratio studies of the meteorites (mainly of the Orgueil meteorite) have shown beyond reasonable doubt that the carbon compounds in them are of extraterrestrial origin. It is still debated whether or not the organic compounds found in them are of biotic or abiotic origin; however, the evidence more and more favors the latter view

(Kvenvolden *et al.*, 1971; Oró *et al.*, 1971). Without debating this long argument (see for example the review of Baker, 1971), we point out that accompanying these compounds and making up the most prominent group of mineral constituents (constituting up to 50% of it) are flakey, silicate minerals closely resembling terrestrial clay. This combination graphically portrays the intimate relationships possible between the platy silicate minerals and the abiotic precursors involved in biogenesis.

Allowing for considerable latitude in estimates, the quantities of volatiles that could have been 'gettered' as adsorbates are large. Whereas on Earth, it is possible that many, if not all, the significant steps in abiotic synthesis and biogenesis may have occurred in the hydrosphere with only sporadic and possibly insignificant contact with particulates, the situation is greatly different on Mars where the present hydrosphere is composed principally of ice and subsurface water. Although a number of geomorphic features revealed by Mariner 9 have been interpreted as indicative of the possibility of a liquid hydrosphere in the past, its existence surely was very limited in aerial extent and brief. Aeolian processes predominate and the atmosphere for aeons has been combed by circulating and sedimenting dust. Thus, extensive deposits of fine grained particulate matter whose surfaces have no doubt equilibrated with successive atmospheric environments, have been laid down. If biogenesis is proceeding at the present time, the focus of activity is in the martian soil at soil-water interfaces. The dimensions of the adsorbed water interface in which the vital reactions must occur if they are to be accomplished were computed in an earlier paper (Anderson *et al.*, 1967). Adsorption isobars for montmorillonite were given for possible partial pressures of water vapor in the martian atmosphere. For a time, these were regarded with some scepticism because the possibility of montmorillonite on Mars was not generally accepted. This has changed since Hunt *et al.* (1973) showed that infrared spectra returned by Mariner-71 provide evidence for the presence of montmorillonite on Mars. The water vapor adsorption isobars show that under present conditions of temperature and pressure, the quantity of water adsorbed normally does not exceed a monolayer. In circumstances where frost or ice is in contact with mineral surfaces, an unfrozen liquid-like interface may approach 5–6 Å, as shown in Figure 2. Thus, the situation is not favorable. However, the presence of salts and other soluble substances at this interface may increase the thickness, *d*, at a given temperature by (Banin and Anderson, 1974):

$$d = \frac{ae}{10^{-4}\, S\theta} \tag{1}$$

where $d$ is in Å, $e$ is the salt content, meq $g^{-1}$ soil, $S$ is the specific surface area of the soil ($m^2\ g^{-1}$), $\theta$ is the temperature in °C below zero and $a$ is a coefficient that depends on the salt type. The latter is 3.72, 2.79 and 1.86 deg meq$^{-1}$ ml, for the 1-1, 1-2 and 2-2 salt types, respectively. Because of this effect it is conceivable that circumstances might exist on Mars in which the activity, $P/P_0$, of the interfacial water and the thickness of the unfrozen, liquid-like interface might exceed 0.5 and 6 Å, respectively. In exotic, or emphemeral environments where the water vapor pressure might rise to, or even exceed 10 millibars, more favorable situations might exist and the reactions vital to biogenesis become correspondingly more favored. On a planet-wide basis it must be concluded that if biogenesis has proceeded to the stage of bringing forth organisms, it is virtually certain that most are now sustained in aqueous interfaces of the types discussed above. If biogenesis is now occurring, the probability that its vital processes are proceeding in these locations is high. If biogenesis has been initiated, but temporarily interrupted by hostile environmental conditions, its remains and dormant fragments must be present there.

## References


Albert, J. T. and Harter, R. D.: 1973, *Soil Sci.* **115**, 130.
Anderson, D. M. and Hoekstra, P.: 1965, *Soil Sci. Soc. Amer. Proc.* **29**, 498.
Anderson, D. M. and Morgenstern, N. R.: 1973, *Permafrost: The North American Contribution to the Wash, D.C. 2nd Int. Conf. on Permafrost, NAS-NRC Pub.*
Anderson, D. M. and Tice, A. R.: 1971, *Soil Sci. Soc. Amer. Proc.* **35**, 47.
Anderson, D. M. and Tice, A. R.: 1972, *Highway Research Record* **393**, 12.
Anderson, D. M., Gaffney, E. S., and Low, P. F.: 1967, *Science* **155**, 319.
Baker, B. L.: 1971, *Space Life Sci.* **2**, 472.
Banin, A. and Anderson, D. M.: 1974, *Water Resour. Res.* **10**, 124.
Banin, A. and Lahav, N.: 1968, *Nature* **217**, 1146.
Bass, E. N.: 1971, *Geochim. Cosmochim. Acta* **35**, 139.
Bernal, J. D.: 1967, *The Origin of Life*, World Publishing Company, Cleveland, 345 p.
Bikerman, J. J.: 1970, *Physical Surfaces*, Academic Press, New York, 476 p.
Blumstein, A.: 1965, *J. Polymer Sci.* **A3**, 2653.
Blumstein, A., Blumstein, R., and Vanderspurt, T. H.: 1969, *J. Colloid Interface Sci.* **31**, 236.
Brindley, G. W.: 1966, *Clay Minerals* **6**, 237.
Cadenhead, D. E. and Buergel, W. G.: 1973, *Science* **180**, 1166.
Calvet *et al.*: 1964, *Bull. Grpe. Fr. Argiles XIV*, 59.
Chaussidon, J. and Calvet, R.: 1965, *J. Phys. Chem.* **69**, 2265.
Chaussidon, J., Calvet, R., Helsen, J., and Fripiat, J. J.: 1962, *Nature* **196**, 161.
Cloos, P., Calicis, B., Fripiat, J. J., and Makay, K.: 1966, *Proc. Intern. Clay Conf. Jerusalem* **I**, 233.
Cloos, P. and Laura, R. D.: 1972, *Clays Clay Minerals* **20**, 259.
Cruz, M., White, J. L., and Russel, J. D.: 1968, *Israel J. Chem.* **6**, 315.
Coleman, N. T. and McAuliffe, C.: 1955, *Clays Clay Minerals* **3**, 282.
Dekking, H. G.: 1967, *J. Appl. Polymer Sci.* **11**, 23.
Doehler, R. W. and Young, W. A.: 1962, *Clays Clay Minerals* **9**, 468.
Doner, H. E. and Mortland, M. M.: 1970, *Science* **166**, 1406.
Dowdy, R. H. and Mortland, M. M.: 1968, *Soil Sci.* **105**, 36.
Durand, B., Pelet, R., and Fripiat, J. J.: 1972, *Clays Clay Minerals* **20**, 21.
Eitel, W.: 1966, *Silicate Science IV*, Academic Press, New York, p. 461.
Ensminger, L. E. and Gieseking, T. E.: 1941, *Soil Sci.* **51**, 125.
Farmer, W. J. and Ahlrichs, J. L.: 1969, *Soil Sci. Soc. Am. Proc.* **33**, 254.
Farmer, V. C. and Mortland, M. M.: 1966, *J. Chem. Soc. (London) A*, 344.
Fox, S. W. (ed.): 1965, *The Origins of Prebiological Systems*, Academic Press Inc., New York, p. 167.
Fripiat, J. J., Cloos, P., Calicis, B., and Makay, K.: 1966, *Proc. Intern. Clay Conf. Jerusalem* **I**, 223.
Fripiat, J. J., Poncelet, G., Van-assche, A. T., and Mayaudon, J.: 1972, *Clays Clay Minerals* **20**, 331.
Gabel, N. W. and Ponnamperuma, C.: 1967, *Nature* **216**, 453.
Garrett, W. G. and Walker, G. F.: 1962, *Clays Clay Minerals* **9**, 557.
Greenland, D. J.: 1963, *J. Colloid Sci.* **18**, 647.
Greenland, D. J. and Quirk, J. P.: 1962, *Clays Clay Minerals* **9**, 484.
Harter, D. and Stotzky, G.: 1971, *Soil Sci. Soc. Am. Proc.* **35**, 383.
Hauser, E. A. and Leggett, M. B.: 1940, *J. Am. Chem. Soc.* **62**, 1811.
Hayatsu, R., Studier, M. H., and Anders, E.: 1971, *Geochim. Cosmochim. Acta* **35**, 939.

Heller-Kallai, L., Yariv, S., and Riemer, M.: 1972, (preprints) *Proc. Intern. Clay Conf. Madrid* **II**, 381.
Hunt, G. R., Logan, L. M., and Salisbury, J. W.: 1973, *Icarus* **18**, 459.
Ibanez, J. D., Kimball, A. P., and Oró, J.: 1971, *Science* **173**, 444.
Jordan, J. W.: 1949a, *J. Phys. Chem.* **53**, 294.
Jordan, J. W.: 1949b, *Min. Mag.* **28**, 598.
Knight, B. A. G. and Tomlinson, T. E.: 1967, *J. Soil Sci.* **18**, 233.
Krinsley, D. H. and Smalley, I. J.: 1973, *Science* **180**, 1277.
Kvenvolden, K. A., Lawless, J. G., and Ponnamperuma, C.: 1971, *Proc Nat. Acad Sci.* **68**, 486.
Lailach, G. E. and Brindley, G. W.: 1969, *Clays Clay Minerals* **17**, 95.
Lailach, G. E., Thompson, T. D., and Brindley, G. W.: 1968a, *Clays Clay Minerals* **16**, 285.
Lailach, G. E., Thompson, T. D., and Brindley, G. W.: 1968b, *Clays Clay Minerals* **16**, 295.
Malhotra, S. L., Parikh, K. K., and Blumstein, A.: 1972, *J. Colloid Inter face* **41**, 318.
McLaren, A. D.: 1954, *J. Phys. Chem.* **58**, 129.
Mortland, M. M.: 1966, *Clay Minerals* **6**, 143.
Mortland, M. M.: 1968, *9th Int. Congr. Soil Sci.* **1**, 691.
Mortland, M. M.: 1970, *Adv. Agron.* **22**, 75.
Mortland, M. M., Fripiat, J. J., Chaussidon, J., and Uytterhoeven, J. B.: 1963, *J. Phys. Chem.* **67**, 248.
Oró, J., Gibert, J., Lichtenstein, H., Wikstrom, S., and Flory, D. A.: 1971, *Nature* **230**, 105.
Oró, J.: 1972, *Space Life Sci.* **3**, 507.
Paecht-Horowitz, M.: 1971, in R. Buvet and C. Ponnamperuma (eds.) *Chemical Evolution and the Origin of Life*, North-Holland Pub. Co., Amsterdam, p. 245.
Paecht-Horowitz, M., Berger, J., and Katchalsky, A.: 1970, *Nature* **228**, 636.
Ponnamperuma, C. and Gabel, N.: 1971, in C. Ponnamperuma (ed.), *Exobiology*, North-Holland Pub. Co., Amsterdam, p. 81.
Raman, K. V. and Mortland, M. M.: 1969, *Soil Sci. Soc. Am. Proc.* **33**, 313.
Ronov, A. B., Migdisov, A. A., and Yaroshevsky, A. A.: 1973, *Proc. Symp. Hydrogeochemistry and Biogeochemistry*, Tokyo, 1970, Vol. I, p. 40.
Ruby, W. W.: 1955, *Geol. Soc. Amer. Spec. Paper*, No. 62.
Russel, J. D., Cruz, M., and White, J. L.: 1968, *J. Agr. Food Chem.* **16**, 21.
Rutten, M. G.: 1971, *The Origin of Life*, Elsevier Pub. Co., Amsterdam, London, New York.
Solomon, D. H.: 1968, *Clays Clay Minerals* **16**, 31.
Solomon, D. H. and Loft, B. C.: 1968, *J. Appl. Polymer Sci.* **12**, 1253.
Solomon, D. H. and Rosser, M. J.: 1965, *J. Appl. Polymer Sci.* **9**, 1261.
Solomon, D. H., Swift, J. D., and Murphy, A. J.: 1971, *J. Macromol. Sci. Chem.* **A5**, 587.
Steinman, G.: 1966, *Science* **154**, 1344.
Steinman, G., Kenyon, D. H., and Calvin, M.: 1965, *Nature* **206**, 707.
Tahoun, S. and Mortland, M. M.: 1966a, *Soil Sci.* **102**, 248.
Tahoun, S. and Mortland, M. M.: 1966b, *Soil Sci.* **102**, 314.
Tettenhorst, R., Beck, C. W., and Brunton, G.: 1962, *Clays Clay Minerals* **9**, 500.
Touillaux, R., Salvador, P., Van Der Meersche, C., and Fripiat, J. J.: 1968, *Israel J. Chem.* **6**, 337.
Walker, G. F.: 1950, *Nature* **166**, 695.
Walker, G. F.: 1960, *Nature* **187**, 312.
Weed, S. B. and Weber, J. B.: 1968, *Am. Mineral* **53**, 478.
Weed, S. B. and Weber, J. B.: 1969, *Soil Sci. Soc. Am. Proc.* **33**, 379.
Woessner, D. E.: 1974, *Symp. on NMR and Mass Spectroscopy in Pesticide Chemistry*, Plenum Press, New York, in press.
Yoshino, D., Hayatsu, R., and Anders, E.: 1971, *Geochim. Cosmochim. Acta* **35**, 927.

# MICROBIAL CONTRIBUTIONS TO THE EVOLUTION OF THE 'STEADY STATE' CARBON DIOXIDE SYSTEM

RICHARD Y. MORITA

*Dept. of Microbiology and School of Oceanography, Oregon State University, Corvallis, Ore. 97331, U.S.A.*

**Abstract.** Various processes for the production of carbon dioxide by microorganisms are presented. It is postulated that a 'microniche' developed in a reducing environment; a symbiotic relationship between alga-like organisms and bacterium-like organisms in the 'microniche' governed the production of carbon dioxide resulting in the establishment of the steady state carbon dioxide system in the sea.

## 1. Introduction

The carbon dioxide system in the sea can be depicted as follows:

$$
\begin{array}{l}
CO_2\ \text{(atmosphere)} \\
\updownarrow \\
CO_2 + H_2O \leftrightarrow H_2CO_3 \leftrightarrow HCO_3^- \leftrightarrow CO_3^-\ \text{(aqueous)} \\
\qquad\qquad\qquad\qquad\qquad\qquad\quad \updownarrow \\
\qquad\qquad\qquad\qquad\qquad\qquad\quad CaCO_3\ \text{(sediment)}
\end{array}
$$

This system can be described as a steady state because the influx and efflux of carbon are equal. This system is described in detail by Pytokowicz (1971). The steady state apparently arose at least as early as one billion years ago, and is the milieu in which the higher biological forms evolved.

The following are controversies I would like to avoid since I am in no position to defend any of them. Furthermore, I believe they have no bearing on my presentation.

(1) Presence or absence of $CO_2$ in the primitive atmosphere and hydrosphere.

(2) 'Thick soup' concept *versus* the concentration of organic matter by one means or another in order to initiate life processes.

(3) Primitive alkaline ocean *versus* primitive acid ocean.

(4) Large initial influx of $CO_2$ *versus* slow influx of $CO_2$ into the sea.

This paper will discuss the microbiological aspects of the formation and utilization of carbon dioxide. There are probably many loopholes in this presentation; however, I invite critical comments.

## 2. Inventory of Carbon

Rubey (1961) lists the following amounts of carbon (as $CO_2$) in the atmosphere, hydrosphere, and biosphere:

| | |
|---|---|
| Living organisms and undecayed organic matter | $14.5 \times 10^{18}$ g |
| Organic carbon | $25000 \times 10^{18}$ g |
| Coal, oil, etc. | $27 \times 10^{18}$ g |

The amount of organic matter totals $2.5 \times 10^{22}$ g while the total amount of carbon (organic and inorganic) is $9.2 \times 10^{22}$ g. The amount of organic carbon represents more than one-fourth of the total carbon.

Lasaga *et al.* (1971) presented calculations and experimental data which suggest that the methane atmosphere of the primordial Earth could have been polymerized by ultraviolet radiation in geologically short periods of time to produce an oil slick 1 to 10 m thick on Earth. This represents a tremendous amount of organic carbon, most of which probably was oxidized subsequently to carbon dioxide.

Many investigators have shown that constituents of the primordial atmosphere could have been combined abiologically to synthesize various organic compounds such as amino acids, polypeptides, proteins, carbohydrates, hydroxy acids, purines, pyrimidines, porphyrins, etc. by various means. Since there is no good method by which oxidation could occur in a reducing atmosphere to produce carbon dioxide, tremendous amounts of organic matter probably accumulated. If one accepts the 'thick soup' theory, then a mechanism is needed to produce the carbon dioxide – especially if one accepts the concept that initially there was no carbon dioxide present. On the other hand, if one accepts the concept that carbon dioxide *was* present in the primitive atmosphere, then some of the carbon dioxide itself would have undergone reaction to produce organic matter.

In the carbon cycle today, it is recognized that 90% of the carbon dioxide produced is formed by bacterial action (Stanier *et al.*, 1970). Rabinowitch (1951) has calculated that all carbon dioxide passes through the photosynthetic process in approximately 350 yr. If microbial functions ceased for a period of three months, all life on Earth would expire (Russell-Hunter, 1970).

We must therefore take into consideration the carbon dioxide produced by various methods from organic materials formed abiologically (from methane and, if present, carbon dioxide) on the primitive Earth.

## 3. Fossil Precambrian Microbial Forms

A excellent review of Precambrian microorganisms and early evolutionary events is presented by Schopf (1970); Margulis (1970) details the possible types of microorganisms that could have evolved from the first primitive cell 4 billion years ago.

Microfossils have been found in many Precambrian sedimentary rocks. The oldest materials examined contain microfossils. In the Fig Tree Series (3.1 billion years old) Barghoorn and Schopf (1966) found bacterium-like fossils (*Eobacterium isolatum*); they also described an alga-like organism (*Archaeosphaeroides barbertonenis*) (Schopf and Barghoorn, 1967).

In the Onverwacht Series (3.2 billion years old) Engel *et al.* (1968) reported spheroidal and cup-shaped, carbonaceous, alga-like bodies. Because the Onverwacht Series includes the oldest, little-altered sedimentary rocks on Earth, the authors postulate that evidence relating to the origin of unicellular life is buried in older rocks now obliterated by igneous and metamorphic events.

In the Onverwacht material, carbon isotopic analyses ($C^{13}/C^{12}$) indicate that biological fractionation had occurred. After reviewing data obtained from hydrocarbon analyses of various ancient sedimentary materials, Calvin (1969) stated,

the degree of specificity is the heart of the matter and has really persuaded most investigators that these patterns do indeed represent the residues of biological activity in the ancient sediment and could not have been arrived at in any other way.

In microbiological terms, it simply means that microbial metabolism has been active at least since the oldest fossil record (3.2 billions years ago).

## 4. Microbial Formation of Carbon Dioxide in the Primordial Reducing Environment

The environment during the early Precambrian is thought to have been reducing; therefore, the microbiological processes utilizing organic matter formed by abiological processes would have been anaerobic. Hence, we must think in terms of anaerobic respiration and fermentation processes. The former can be ruled out probably since the hydrogen acceptor in the metabolic process must be an oxidized inorganic molecule such as sulfate or nitrate. Trimethylamine oxide can also act as a hydrogen acceptor (H. Larsen, personal communication). If trimethylamine oxide was abiotically synthesized in the primordial system, then anaerobic respiration could be considered. However, a review of the literature does not indicate its existence. In fermentation processes, the hydrogen donor and acceptor are organic molecules.

Amino acids have been identified as one of the major constitutents formed by abiotic means. The fermentation of amino acids (Stickland reaction) can produce carbon dioxide (Stickland, 1934). The Stickland reaction between alanine and glycine illustrates the formation of carbon dioxide:

$$1 \text{ alanine} + 2 \text{ glycine} + \text{water} = 3 \text{ acetic acid} + CO_2 + 3\ NH_3.$$

Stickland reactions also occur between other amino acids.

Certain clostridia (anaerobic organisms) can also metabolize single amino acids with the liberation of ammonia, carbon dioxide and hydrogen; the single amino acids that can be metabolized in this fashion are serine, aspartic acid, glutamic acid, histidine, tyrosine, methionine and cysteine. Other clostridia are known to metabolize glutamic acid to yield acetic acid, butyric acid, carbon dioxide, hydrogen and ammonia; or to form propionic acid from fermentation of alanine, or to ferment serine to produce acetic acid, carbon dioxide, hydrogen and ammonia. Even certain propionic bacteria that can ferment single amino acids to yield propionic acid as the main end product, yield carbon dioxide also.

The decarboxylation of amino acids is also known to occur within certain anaerobes as well as facultative anaerobes. The amino acids known to be decarboxylated are tryptophane, leucine, lysine, serine, arginine, ornithine, histidine, tyrosine, phenylalanine, aspartic acid and glutamic acid. Other types of decarboxylation are known to occur in the fermentation industry by other organisms, many of which are anaerobes (Wallen *et al.*, 1959).

In many carbohydrate fermentation reactions, carbon dioxide and organic acids are the end product. These organic acids could help leach carbon dioxide from shale. Bacterial alcoholic fermentation results in production of carbon dioxide and ethyl alcohol. Many types of fermentation reactions yield carbon dioxide. The more well known reactions that yield carbon dioxide are (i) decarboxylation of pyruvate to carbon dioxide and acetaldehyde; (ii) dismutation of pyruvate to acetate, lactate and carbon dioxide; and (iii) the phosphoroclastic reaction of pyruvate to acetate and formate, the latter forming hydrogen and carbon dioxide. Fermentation reactions by the enteric group of organisms yield copious amounts of carbon dioxide, depending upon the substrate as well as the organism in question. In many fermentation reactions, hydrogen is evolved. This hydrogen could then escape into the atmosphere, thereby leaving the organic matter, as a whole, more oxidized than it was before the fermentation reactions occurred.

## 5. Other Possible Mechanisms of Microbial Carbon Dioxide Production in the Precambrian and Microbial Diversity

Figure 1 gives the time sequence of events here postulated to have occurred during the Precambrian Eon. Engel *et al.* (1968) suggest that the first unicellular prokaryotic cell originated before 3.2 billion years ago. If one accepts the 'thick soup' concept, then it would have been possible to form many different physiological types of unicellular organisms simultaneously. However, if organic matter were concentrated at different loci, diverse types of microorganisms might similarly arise. In this situation, the 'microniche' would contain various types of organic matter and ultraviolet light would help bring about genetic changes. Although many geochemists do not like the idea of ultraviolet light among life forms, it should be recognized that it is a good mutagenic agent. Furthermore, in an organically rich 'microniche' at the surface of the Earth, oxygen could be produced by ultraviolet photolysis of water. However, if ultraviolet light did not penetrate through the primitive atmosphere, then suitable 'microniches' could have occurred in droplets of water suspended in the atmosphere after storms.

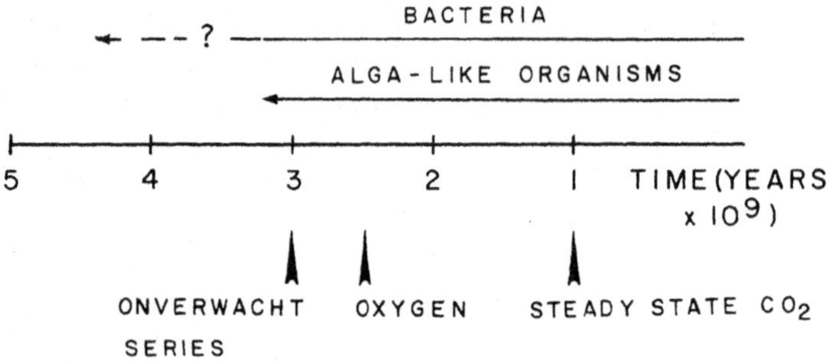


Fig. 1. Approximate time scale of events postulated in the Precambrian Eon.

Since the amount of oxygen produced by this method would not be great, it would be sufficient for a few cells to evolve and generate an aerobic type of respiration. Although the rest of the environment would have been reducing in nature, the oxygen produced in the 'microniche' would have been metabolized immediately. As the genetic material mutated, the fittest organisms would have survived and might have produced pigments protecting them from further ultraviolet radiation. This 'microniche' concept could also work in a 'thick soup', floating on the surface. If the metabolic mechanism were either fermentative or were based on aerobic respiration, carbon dioxide would result. Thus, alga-like organisms could have evolved at this time due to the occurrence of genetic changes coupled with the existence of abiotically formed porphyrins.

In the Onverwacht Series, Engel *et al.* (1968) described alga-like organisms. Figure 2 depicts this situation prior to the development of a 'macroniche' where most of the photosynthetic organisms are floating on the surface. This initial 'microniche' would have been influenced by its immediate environment where carbon dioxide

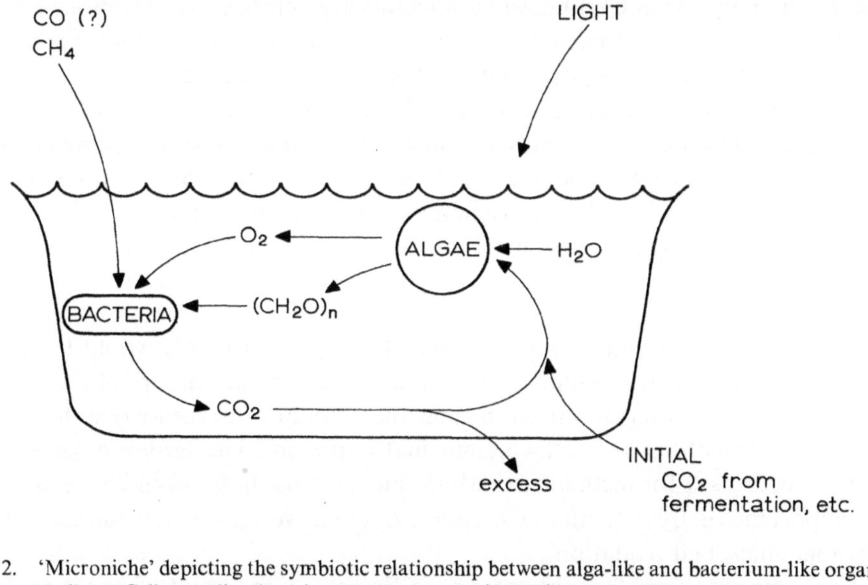


Fig. 2. 'Microniche' depicting the symbiotic relationship between alga-like and bacterium-like organisms during day-light. Cells are dissolved in organic matter; $(CH_2O)_n$ represents synthesized organic matter present as polysaccharide or dead alga-like cells.

diffused into the system from the metabolism of anaerobic organisms or from leaching of shale. The alga-like organisms would have carried on photosynthetic activity employing the energy from sunlight to produce more organic material in the form of new cells or polysaccharide materials. If some of the alga-like cells expired, then the carbon molecules of the dead cells could have been utilized by bacterium-like organisms to form more carbon dioxide. Polysaccharide materials would have undergone the same process. The limiting factor initially might have been availability of carbon dioxide. The oxygen produced could have been used in aerobic respiration which would have produced large amounts of carbon dioxide. If the bacterium-like cells were facultative anaerobes, they might have been able to switch metabolic processes from fermentation to aerobic respiration depending upon availability of oxygen (e.g., during the night). The probable reasons for the development of alga-like organisms within a 'microniche' include the lack of a utilizable energy source by the bacterium-like cell, thereby giving rise to an autotrophic alga-like cell; or that within the 'microniche', the amount of carbon dioxide became too great for the organisms present to metabolize. In the latter situation, the alga-like organisms evolved to fit the environment. The genesis of algae could also have resulted from mutation of a photosynthetic bacterium when the availability of hydrogen sulfide became depleted in a microenvironment. In this situation, the organisms switches from the use of hydrogen sulfide to water in its metabolism (the photosynthetic structure, the thylakoid, in photosynthetic bacteria resembles the chloroplast in of green plants). The other bacterial cells within the 'microniche', having been exposed to oxygen formed by photosynthesis, might then have mutated to an aerobic form of metabolism. These preliminary forms could have been facultative aerobes. As conditions became more favorable, the 'microniche' became a 'macroniche' – a condition apparently established by the time of deposition of the Onverwacht material.

As a result of these 'microniches', carbon dioxide would have been produced and if the amount were excessive, would have then diffused into the surrounding environment – especially at night. The excess carbon dioxide would have diffused into the hydrosphere and then into the atmosphere or eventually to carbonates. The amount of oxygen produced by photosynthetic processes would have been used immediately by the surrounding facultative anaerobes and aerobes to produce more carbon dioxide.

On the basis of the 'thick soup' concept, the oxygen available would have been used for oxidation of the accumulated organic matter. In the concept of the concentration of organic matter by various means, the amount of oxygen would have been more than sufficient to oxidize the organic matter present. The surplus oxygen could then have been used for methane oxidation; methane oxidizers would have assumed great importance in this situation. If carbon monoxide were present, it could also have undergone microbial oxidation.

It is recognized today that the oxygen on Earth was the result of photosynthetic processes (Cloud, 1972). Such processes could not have come about without the availability of carbon dioxide. If the carbon dioxide of the primitive atmosphere was close to the amount currently in the atmosphere, then the question as to the source of the carbon dioxide for photosynthesis (oxygen production) must be answered. This would require production of additional carbon dioxide by microbial means and geochemical processes. Is the leaching rate of carbon dioxide from shale sufficient?

The 'microniche' idea can give rise to an influx of carbon dioxide to the ocean until the ocean system is saturated sufficiently to allow photosynthetic organisms to produce more oxygen than is needed to degrade various types of organic matter present (including methane) – eventually producing the oxygen atmosphere as we know it today. As 'microniches' become the major environment, there would have been a slow process of changing the reducing primitive atmosphere to an oxidizing atmosphere. Over geological time it is possible, through trial and error, that the influx of carbon dioxide into the sea would equal the efflux of carbon dioxide – a steady state system.

If the microfossils in the Onverwacht Series are not truly alga-like, then the 'microniche' concept I have presented may not have started until a later time (i.e., when certain oxygen-producing algae made their appearance). One of the main reasons for the survival of algal-like material in the various shales may be due to reducing conditions at the time of sedimentation. True algal-like organisms would produce oxygen resulting in an oxidizing environment which would make it very difficult to find remains of such organisms. If pockets of oxidizing 'microniches' did occur, they probably would have been decomposed before any geological process could be initiated to preserve them.

The concept that I have presented is not consistent with that of Margulis (1970). She states that carbon dioxide was synthesized into more complex molecules at the expense of other types of organic molecules. Energetically, this appears unsound. In addition, there is no gradual change from a reducing environment to an oxidizing environment.

The above ideas may help to explain the microbial contribution to the evolution of the steady state carbon dioxide system of the sea. Other geochemical processes for producing carbon dioxide must be taken into consideration, and both processes go hand in hand.

## Acknowledgements

I thank R. M. Pytkowicz, Y.-H. Li, J. Baross, and C. Ponnamperuma for discussions concerning the subject presented. Thanks are also due to the Committee on the Origin of Life, Oregon State University, and Oregon State University Foundation. This paper was supported by NSF grant GA 38583X.

## References

Barghoorn, E. S. and Schopf, J. W.: 1966, *Science* **152**, 758–763.
Calvin, M.: 1969, *Chemical Evolution*, Oxford Univ. Press, N.Y., 278 pp.
Cloud, P.: 1972, *Amer. J. Sci.* **272**, 537–548.
Engel, A. E. J., Nagy, B., Nagy, L. A., Engel, C. G., Kremp, G. O. W., and Drew, C. M.: 1968, *Science* **161**, 1005–1008.
Lasaga, A. C., Holland, H. D., and Dwyer, M. J.: 1971, *Science* **174**, 53–55.
Margulis, L.: 1970, *Origin of Eukaryotic Cells*, Yale Univ. Press, New Haven and London, 349 pp.
Pytkowicz, R. M.: 1971, in D. Dyrssen and D. Jagner (eds.), 'The Changing Chemistry of the Oceans', *Nobel Symp.* **20**, Wiley Interscience, New York, London, Sydney.
Rabinowitch, E. I.: 1951, *Photosynthesis and Related Processes*, Interscience, N.Y.
Rubey, W. S.: 1961, *Bull. Geol. Soc. Amer.* **62**, 1111–1147.
Russell-Hunter, W. E.: 1970, *Aquatic Productivity*, The Macmillan Co., 306 pp.
Schopf, J. W.: 1970, *Biol. Rev.* **45**, 319–352.
Schopf, J. W. and Barghoorn, E. S.: 1967, *Science* **156**, 508–512.

Stanier, R. D., Doudoroff, M., and Adelberg, E. A.: 1970, *The Microbial World*, Prentice-Hall, Inc., Englewood Cliffs, N.J.

Stickland, L. H.: 1934, *Biochem. J.* **28**, 1746–1759.

Wallen, L. L., Stodola, F. H., and Jackson, R. W.: 1959, *Type Reactions in Fermentation Chemistry*, Northern Utilization Research and Development Divisions, U.S.D.A. ARS-71-13, Peoria, Ill., 496 pp.

# STRUCTURES OF BIOGENIC ORIGIN FROM EARLY PRECAMBRIAN ROCKS OF EURO-ASIA

A. S. LOPUCHIN

*Institute of Geology of the Order of the Red Banner of Labor of the Academy of Sciences of Kirgiz S.S.R., Frunze, U.S.S.R.*

**Abstract.** Spheroidal microfossils mainly 20 to 100 μm in diameter and exhibiting granular surface textures have been recovered from Early Precambrian rocks by applying a new method of water separation in combination with thin chemical preparation. In contrast to the Acritarcha, these microfossils are characterized by a relatively low specific weight (close to one) and considerable fragility due to impregnation by mineral matter. They occur in Archean sediments of Hindustan, in rocks of the Baltic and Aldan Shields with ages of 3.0 to 3.5 billion ($10^9$) years, and in Proterozoic deposits in many regions of Euro-Asia. They commonly occur in great number in Precambrian sediments of West Africa, Australia and North America. These forms are here regarded as *Menneria* Lopuchin and are considered to be blue-green algae. *Menneria* resembles alga-like forms reported by Engel, Nagy and their co-workers from the Onverwacht Series and microfossils reported by Schopf and Barghoorn from the Fig Tree Series, both of the Swaziland System of southern Africa. In addition to spheroidal microfossils, ribbon-like and filiform microstructures are here reported from Archean deposits.

The biogenic structures here described from the Early Precambrian of Euro-Asia are considered to have been photosynthetic and planktonic. Their progressive evolution, intensive production of organic matter, and biogeochemical role in concentration of rare elements is discussed.

Much prominence has been given by the world scientific literature of the last decade to results of the thorough investigations of fossil organic matter and morphologically preserved residues of microorganisms detected in the most ancient rocks of the Earth's crust. The heightened interest in the study of the bio-organic component of Precambrian sediments has been primarily associated with realization of the importance of living systems in processes of hypergene sedimentation and the concentration of minerals. These investigations have also an exceptionally important theoretical significance as they lead to understanding of the origin and evolution of the organic world on the Earth and to the establishment of general laws of life in the universe. For, as suggested by Oparin, the origin of life is an integral component of the general natural process of the evolution of matter. Without knowledge of this, as Bernal wrote, we shall not be able to understand the life of today, we shall not be able to understand its essence, possibilities and limitations, and we shall not be able to control it.

The Precambrian Eon is known to encompass seven-eighths of the entire geologic history of the Earth and is characterized by many interesting and significant events. Perhaps the most important stage in the evolution of the Earth is that which occurred at the very beginning of Precambrian time, more than 4.0 billion ($10^9$) years ago, when the lithosphere and hydrosphere were formed and the first primitive organisms originated, leading to the formation of an oxygen-bearing atmosphere. It was during this stage that a specific environment, which A. Süss named 'biosphere', first appeared, an environment defined scientifically for the first time by Vernadsky.

"Biosphere consists of the complex of living organisms, a living matter, scattering in myriads of individuals, continuously dying and originating, possessing a colossal effective (biogeochemical) energy and representing a mighty geologic strength.... Concentration of definite chemical elements by the living matter in the biosphere is evidently its prevailing biogenic-geological process" (Vernadsky, 1965).

Vernadsky's ideas, having much in common with the theory of the origin of life of Oparin and Choldein, have become the basis for scientific thought regarding the first stages of the evolution of the Earth; they have found many followers among the scientific community. The hypothetical 'organic pellicle' of the primary ocean has subsequently found scientific support both in works of organic geochemists and in paleobiological investigations. Today we have data proving the existence of not only organic matter of biogenic nature but also microfossils, with preserved morphology and organogenic structure, in primary sedimentary rocks deposited in Early Precambrian basins.

In the U.S.S.R., study of microscopic plant fossils from sediments of Precambrian age was undertaken for the first time by Professor S. H. Naumova and her associates. During recent years, many investigators have begun to work in the field of 'micropaleophytology' of the Precambrian; as a result, considerable progress has been made. However, certain of this work requires redescription and systematization. In addition, the indigenous, *in situ*, nature of these reported Precambrian fossils needs to be confirmed, as the possibility of contamination is present at practically all stages of study, including the introduction of laboratory artifacts and even the possibility of contamination by microfossils of younger geologic age in the natural environment. Nevertheless, the validity of many reported microfossils, especially those known from rocks of the upper Precambrian, is beyond any doubt; these forms require only additional, more careful study and description (with observance of all rules of the International Code of Botanical Nomenclature) which will undoubtedly permit their use for biostratigraphic purposes. Menner, who for many years has promoted the development of a biostratigraphy for the Precambrian, especially dwelled upon these questions at the Third International Palynological Conference (Menner, 1973). I believe that the primary importance for the solution of this task would be the resulting unification of terminology used in the description of such fossils and the introduction of a perfocard index (Lopuchin, 1971a).

Another area in need of investigation is the paleobiological study of moderately and more often severely metamorphosed sedimentary rocks of the early Proterozoic and Archean. Up to the present, the possibility of finding organic residues in such ancient sediments has been considered absurd. Considerable accumulations of organic residues which may contain clues indicative of the evolution of life processes in the Early Precambrian have remained uninvestigated. Such pessimism to a great extent has been the result of the widespread conception that magmatic rocks predominate in the Precambrian and that extreme differences exist between the evolution of geologic processes in the Precambrian and the Phanerozoic. At present, however, there are many data indicative of the erroneousness of such opinions (Sidorenko and Tenjakov, 1973). It has been established, for example, that the most ancient sedimentary sequences contain practically all types of sedimentary rocks, including carbon-bearing varieties that sometimes make up entire formations. It is natural that ancient Precambrian epochs differed from more recent epochs by having a lesser variety of organisms; but, in my opinion, in quantitative relation, the evidence of life expresses itself almost in the same way as in the Late Precambrian. In favor of such an assumption, A. V. Sidorenko and Sv. A. Sidorenko (1963–1973) convincingly prove that the 'para-rocks' of the Archean and Proterozoic testify to the existence of a considerable content of solely biogenic carbon. Data indicating the intensive fractionation of sulphur isotopes 3.0 billion years ago comprise circumstantial evidence of a vast development of photosynthetic vegetation. It is known that this fractionation process takes place during an oxidizing-reducing reaction which can occur only in the presence of a sufficiently high content of free oxygen in the atmosphere (Vinogradov *et al.*, 1969).

Thus, if the concept of a photosynthetic source for the oxygen of the Earth's atmosphere is true, the comparatively widespread occurrence of biogenic carbon in primary sedimentary rocks of the Early Precambrian seems required. In connection with this, paleobiological findings of remnants of microorganisms (preserving their structure and chemical composition), which have brought to us the most important information about the character of the earliest forms of life on our planet, acquire special meaning.

The first findings of alga-like microfossils from rocks of Early Precambrian age were made by Schopf and Barghoorn (1967) and by Engel *et al.* (1968). These investigators made a study of relatively little metamorphosed rocks of the Swaziland system of South Africa, the age of which, determined by the radiogenic method, is between 3.1 and 3.3 billion years. Bacterium-like rods and spheroidal alga-like organisms were detected in petrographic thin sections of black cherts of Fig Tree series, the carbonaceous fraction of which has been shown to include the hydrocarbon alkanes pristane and phytane. American paleobiologists have also succeeded in detecting the most ancient microfossils now known, in rocks of the Onverwacht Series underlying the Fig Tree sediments. These alga-like fossils, described in Engel *et al.*, are much like *Menneria* Lopuchin, a genus known from Archean calciphyres of the Aldan Shield (discussed in more detail below). Like results obtained on rocks of Fig Tree Series, chromatograms of hydrocarbons extracted from Onverwacht sediments, according to Calvin (1969) are indicative of biologically active material.

A much more diverse microfossil assemblage has been described by Barghoorn and Tyler (1960) from sediments of the Gunflint Iron Formation of North America. The absolute age of this unit is about 1.9 billion years. Twelve species of microorganisms, belonging to eight genera, were described. Paleobiological study of these organisms in thin sections has been followed by paleobiochemical investigations, which made it possible to refer these microfossils to photosynthetic ancestors of modern blue-green algae.

Thus, American scientists have succeeded in showing quite convincingly the presence of primitive organisms in rocks of the Precambrian, with an age of 2.0 to 3.3 billion years, and to confirm their observations by paleobiochemical investigations. Investigating thin sections of the rock as they do to a considerable extent gives a guarantee against contamination; at the same time, however, it limits the investigator to examining only an extremely small quantity of rocks which, to a degree, makes the result of the investigation dependent on chance, and limits analysis to rocks of relatively high organic content. The search for microfossils in thin sections of usual 'para-rocks' of the Archean, however, is commonly not as rewarding. In such rocks, morphological remnants may not be detectable; here, paleobiochemical data refer not to specific organisms, but to the organic matter in general which is contained in the rock as a dispersed mixture. Even taking into consideration the most recent achievements of contemporary biochemistry on the synthesis of organic matter, it is hardly possible to determine with assurance the exact origin of fossil organic matter (chemical fossils) apart from paleobiological investigations. Data of paleobiochemistry and various physical-chemical methods provided by study of isolated, morphologically preserved microfossils and, probably better, by study of separate individuals (*e.g.*, by electron microprobing) will be the most trustworthy in this case. Such complex investigations and their validity depend first of all on the possibility of finding a sufficient quantity of microfossils in the rocks examined, and second, the ability to repeat such analyses many times under the same conditions.

During the past few years, due to application of a new technique based on water separation in combination with thin chemical preparation, I have discovered a new group of fossil phytoplankton (of the family Menneriaceae, and the genus *Menneria*); this group is characterized by general and sometimes by mass spreading, beginning from the rocks of the Early Precambrian and extending up to Carboniferous (Lopuchin, 1971b). The morphological peculiarities of these microfossils (as observed under the light microscope) underwent comparatively few changes in the course of time. In my opinion, this morphological conservatism, a characteristic of primitive microorganisms in general, was the result of the stability of their environment – a relatively unchanging world ocean, an environment that determined the possibility of the origin and the development of life on the Earth itself.

The specific weight of *Menneria*-type microfossils is approximately one; they can therefore be separated in distilled water and even in ethanol combining the separation process with thin chemical preparation. Under the microscope, these fossils are seen to be approximately spheroidal or discoid microforms and usually having folds due to crumpling or having scarred surficial thickenings. The size of the microfossils varies from 5 to 400 $\mu$m, principally between 20 and 100 $\mu$m. At magnifications greater than 1000 times, their surface appears markedly and evenly granular. The size of individual 'granules' (the individual structural elements) is of the order of 1 $\mu$m. Without exception, these 'granules' are interpreted as representing separate ultramicroscopic cells. Thus, the described microforms may be considered to be colonial organisms, a supposition well confirmed by the wide range of size of the fossils. They are also characterized by considerable fragility and can be destroyed rather easily during a rigid regime of palynological maceration of a rock sample. While making preparations, the scattered 'granules', becoming attached to air bubbles, sometimes form pseudospheres, which at low magnifications may lead an inexperienced observer to error. In spite of their fragility, with sufficient care unbroken forms can be extracted from the rock and carried into a separate preparation and, if it is necessary, turned and fixed on a microscope slide in a required position. During thin chemical preparation, specimens only partly released from the mineral matrix are often observed, once more confirming their indigenous, *in situ* occurrence in the rock. The presence of these fossils in primary sedimentary rocks of the Early Precambrian has been proved by more than one hundred analyses of rock samples from the Archean of the Aldan, Baltic and Hindustan Shields. It is necessary to stress that these findings have not been merely fortuitous; they are the result of careful searches involving the gradual tracing of these morphologically well-preserved microfossils backward through time, beginning with rocks of early Paleozoic age. In sediments of the Proterozoic, such forms have been detected in a number of rock suites from the Baltic shield, in the North Tien Shan and in the Gunflint cherts of North America. Large numbers occur in the Riphean of the Siberian Platform (Chekurovka), in the Riphean and Vendian of North Mongolia (Hubsugul), Kazachstan, Kirgizia, Usbekistan, the Russian Platform, Prednestrovje, and in the Taudeni Syncline of Northwest Africa. They have also been detected in the rocks of Bitter Springs Formation of central Australia. Similar microfossils have been detected in Brioverian phtanites of Normandy by Roblot, who discussed their plant nature; for this purpose a special microprobe and the method of carbon isotopic analysis were used. For these remnants, Roblot suggested the name of *Hymenophacoides* (Roblot, 1967). It is a pity that this work was received by me after the description and effective publication of *Menneria*; for this reason, the name proposed by Roblot is not here used.

Thus, for the fossils here studied, practically global distribution has been established during all the Precambrian. However, their distribution throughout the geologic column is far from being even, and primarily depends on paleogeographical conditions of their environment and on patterns of sedimentation. Large assemblages of these fossils occur as a rule only in dark-colored, bituminiferous, carbonaceous rocks and in coaly-siliceous and coaly-argillaceous shales; they are usually accompanied by increased concentrations of certain rare elements. I have also observed considerable quantities of such fossils in the cementing minerals of conglomerates, in aleurites and in quartzitic sandstones. Rocks of the shungite type are supersaturated with them. In this respect, calciphyres and graphitic rocks of Archean age were most rewarding. Paraamphibolites contain rare forms exhibiting poor preservation, and are thus similar to green-colored phyllitic shales of the Late Precambrian.

It is logical that the development of microscopic water vegetation in the Precambrian was subject to the same laws, and was controlled by the same factors, which define the life of contemporary phytoplankton. Information given on this question in the works of Bogorov (1971) and Bordovsky (1971) is in good agreement with this concept as based on the fragmentary data in our possession about the fossil ancestors of contemporary algae. Making similar analogy allows us, to a considerable extent, to direct our considerations about Precambrian vegetation from the field of guesswork into the realm of well-founded ideas, and makes it easier to understand the laws governing the development of the most ancient flora on the Earth and its influence on geologic processes (*e.g.*, rock and ore formation).

The transition from a non-oxygenic atmosphere to an oxygen-bearing atmosphere was perhaps the most significant stage in the evolution of biosphere, having had a definite influence on hypergene processes. The time of this transition, a major global episode in the history of the Earth, coincided with a mass evolution of photosynthetic phytoplankton. The establishment of the date of this transitional period has been of great scientific and practical interest, and may be determined only by complex paleobiogeochemical and paleobiological investigations. The detection of widespread microfossils in the Precambrian has justified logically only an *a priori* connection between the development of photosynthetic life, the formation of the oxidizing atmosphere, and the accumulation of organic matter in early sedimentary rocks.

My morphologic studies of microfossils have shown that at least three forms of primitive life existed in the Early Precambrian: spheroidal phytoplankton, ribbon-like forms, and filamentous, microscopic planktonic organisms. All of these forms are characterized by a low specific weight (close to one) and a granular (here interpreted as colonial) structure. In the Proterozoic, the diversity of phytoplanktonic organisms became greatly increased; by the Late Precambrian, diverse groups had evolved (forms referred now to the combined group 'Acritarcha' by the majority of investigators). These younger fossils are characterized by a more complex structure and by an increased specific weight, a factor which makes it necessary to apply a different method for their isolation (Lopuchin *et al.*, 1971). At the same time, more simply organized organisms, namely representatives of the genus *Menneria*, evidently prevailed in large numbers, and remained extant during the rise of the Acritarcha (this has been proved by many analyses of materials from the Riphean of the Siberian Platform, the Russian Platform and the North Tien Shan which have demonstrated that a complete set of vegetative remnants can be obtained from sediments of the upper Precambrian and lower Paleozoic only if various methods of isolation of fossils are applied sequentially to the rocks).

In conclusion, as a working hypothesis, I suggest that microfossils of the family Menneriaceae were the first photosynthetic organisms of the Earth; due to their simple organization, these forms were distinguished by high reproductive rates and high productivity and were characterized by global spreading. During the entire Precambrian, these organisms were the main producers of organic matter. As a result of their living processes, they created an oxygen-bearing atmosphere, thus providing the possibility for the subsequent appearance of more complex organisms. If this is so, then these organisms were the living matter that determined all geochemical conditions on the Earth (according to the concepts of V. I. Vernadsky and his followers, A. P. Vinogradov, A. V. Sidorenko, N. M. Strachov and other scientists) and that had a great influence on processes of sedimentation and ore formation during Precambrian time.

## References

Barghoorn, E. S. and Tyler, S. A.: 1960, *Science* **147**, (3658), 563–577.
Bernal, J.: 1967, *The Origin of Life*, London (1967); Moscow, publ. house 'Mir' (1969), pp. 13–249, 336–350.
Bogorov, V. G.: 1971, in *Organic Matter of Recent and Fossil Sediments*, Moscow, 'Nauka', pp. 12–15.
Bordovsky, O. K.: 1971, in *Organic Matter of Recent and Fossil Sediments*, Moscow, 'Nauka', pp. 104–119.
Calvin, M.: 1969, *Chemical Evolution*, Oxford (1969); Moscow, 'Mir' (1971), pp. 7–238.
Engel, A., Nagy, B., Nagy, L. A., Engel, C. G., Kremp, G. O. W., and Drew, C. M.: 1968, *Science* **161**, 1005–1008.
Gerling, E. K., Iskanderova, A. D., Levchenkov, O. A., and Michailov, D. A.: 1970, *Dokl. Akad. Nauk SSSR* **194**, N 6.
Lopuchin, A. S.: 1971a, in *Dispersional Remnants of Fossil Plants of Kirghizia*, Frunze, 'Ilim', pp. 11–34.
Lopuchin, A. S.: 1971b, in *Dispersional Remnants of Fossil Plants of Kirghizia*, Frunze, 'Ilim', pp. 80–89.
Lopuchin, A. S.: 1973, *Dokl. Akad. Nauk. SSSR*, *Izvestia*, N 6, pp. 42–45.
Lopuchin, A. S.: 1973, *Geologiska Föreningens i Stokholm Förhadlingar* **95**, 329–338. (Stockholm, September 30).
Lopuchin, A. S. and Moralev, V. M.: 1973, *Geology and Exploration* N 7, pp. 185–187.
Lopuchin, A. S. and Volkov, R. I.: 1973, *Geology and Exploration*, N 8, p. 153.
Lopuchin, A. S., Mareeva, V. I., and Malov, V. P.: 1971, in *Dispersional Remnants of Fossil Plants of Kirghizia*, Frunze, 'Ilim', pp. 36–44.
Lopuchin, A. S., Trophimov, D. M., and Borovsky, V. V.: 1972, *Geology and Exploration*, N 11, pp. 141–143.
Menner, V. V.: 1973, *Palynology in Stratigraphy: Problems of Palynology*, Moscow, 'Nauka', pp. 31–35.
Naumova, S. N.: 1968, in *Stratigraphy of the Lower Paleozoic of Central Europe*, Rep. of Soviet geologists, Moscow, 'Nauka'.
Oparin, A. I.: 1968, *Life: Its Nature, Origin and Development*, Moscow, 'Nauka', pp. 3–172.
Kenyon, D. H. and Steinman, G.: 1969, *Biochemical Predestination*, New York, London (1969); Moscow, 'Mir' (1972), pp. 9–333.
Roblot, M. M.: 1967, *Compt. Rend. Acad. Sci. Paris* **264**, 1–3.
Rudnik, V. A., Sabotovich, E. V., and Terentjev, V. M.: 1969, *Dokl. Akad. Nauk SSSR* **182**, N 4.
Schopf, J. W. and Barghoorn, E. S.: 1967, *Science* **156** (3774), 508–511.
Sidorenko, A. V.: 1972, *Lithology and Minerals*, N 6.
Sidorenko, A. V. and Sidorenko, Sv. A.: 1968, *Dokl. Akad. Nauk SSSR* **183**, N 1, 181–188.
Sidorenko, A. V. and Sidorenko, Sv. A.: 1970, *Dokl. Akad. Nauk SSSR* **183**, N 1, 184–187.
Sidorenko, A. V. and Sidorenko, Sv. A.: 1971, *Soviet Geology*, N 5, pp. 3–20.
Sidorenko, A. V. and Tenjakov, V. A.: 1973, *Z. angew. Geol.* **19**, N 4, 161–164.
Vassoevich, N. B.: 1971, in *Organic Matter of Contemporary and Fossil Sediments*, Moscow, 'Nauka', pp. 5–11.
Vernadsky, V. I.: 1960, *Selected Papers*, Moscow, 'Nauka', **5**, 100–302.
Vernadsky, V. I.: 1965, *The Chemical Structure of Biosphere of the Earth and its Environment*, Moscow, 'Nauka', pp. 5–374.
Vinogradov, A. P.: 1959, in *Origin of Life on the Earth*, Tr. Mezhdunar. Symp. Moscow.
Vinogradov, A. P.: 1963, in *Through the Eyes of a Scientist*, Moscow, 'Nauka'.
Vinogradov, A. P.: 1964, *Geochemistry of Organic Matter*, Moscow, 'Nauka', pp. 4–11.
Vinogradov, V. I., Ivanov, I. B., Litsarev, M. A., Pertsev, N. N., and Shonnin, L. L.: 1969, *Dokl. Akad. Nauk SSSR* **188**, N 5, 1144–1147.

## Explanation to Figures

Figures 1 and 2 illustrate biogenic residues from primary sedimentary rocks of the Archean of the Aldan Shield; the forms are found in calciphyres of the Yengrian Series (rock sample from G. B. Himmelfarb). In certain samples, microfossils occur in comparatively large quantities, (i.e., they were found in almost every field of microscope view under a magnification of 200 times). However, well-preserved forms, especially large specimens, were detected rarely and with great difficulty.

Figure 3 illustrates fossils from the Archean of the Baltic Shield (a), North Tien Shan (b), and South India (c–f) (rock samples were collected by G. B. Himmelfarb, O. Bakirov and V. M. Moralev). (g) and (h) convincingly show the surprising preservation of fossil microorganisms in rocks having a high level of metamorphism. The organic wall (g) is isolated from a graphite crystal (h). In the outline of the organic wall, an absolutely exact imitation of the form of a graphite crystal is seen that could only have resulted from processes occurring during graphitization of the dispersed organic matter in which more stable biogenic cell walls of disappeared microorganisms were found to be buried. During isolation from graphite, a part of the wall was found to be torn and turned back toward the outside. Under the microscope, the interior of the wall is seen clearly. Similar biogenic residues were also observed in graphite gneisses of the Predcharvarian Complex of South India (f).

Figure 4 shows representatives of 'light' (*i.e.*, low specific weight) fossils extracted by water separation from rocks of middle Proterozoic and upper Precambrian (rock samples were provided by J. W. Schopf, Sv. A. Sidorenko, D. M. Trofimov; samples were also collected by the author and his scientific co-worker, V. P. Malov).

As is seen in the photomicrographs, a granular surface structure is characteristic of all fossils shown, including the ribbon-like and filamentous forms. Certain fossils, however, especially those occurring in graphitic rocks, do not exhibit this texture.

FIGURE 1

(a) *Menneria* ex gr. *granosa* Lopuchin (diameter, 140 μm). This form is similar to alga-like fossils from the Onverwacht Series of the Swaziland System of South Africa described by A. Engel, B. Nagy and others.

(b) Photocopy of a microorganism from the Onverwacht Series (diameter, 106 μm). A. Engel, B. Nagy and others represent this individual as typical of the Onverwacht fossils detected by them. The difference in the quality of preservation of the objects shown in (a) and (b) is the result of the differing separation techniques used for the two specimens.

(c) *Microtaenia* sp. This form, found in considerable quantities, occurs as ribbon-like scraps (width of ribbon ranges from 10 to 100 μm).

(d) *Microtaenia* sp. A clumped ribbon similar to (c), above (width of ribbon ranges from 10 to 70 μm).

(e) A filamentous alga showing branching (width of filament is on the order of 1 to 2 μm).

FIGURE 2

(a) *Menneria primaeva* Lopuchin sp. nov. (diameter 400 μm). Paratype. A very large form, seldom found. Aldanium, Yengrian series, the Aldan Shield, the Aldan river. The absolute age of the Yengrian series rocks is 3500 million years (Gerling *et al.*, 1970; Rudnik *et al.*, 1969).

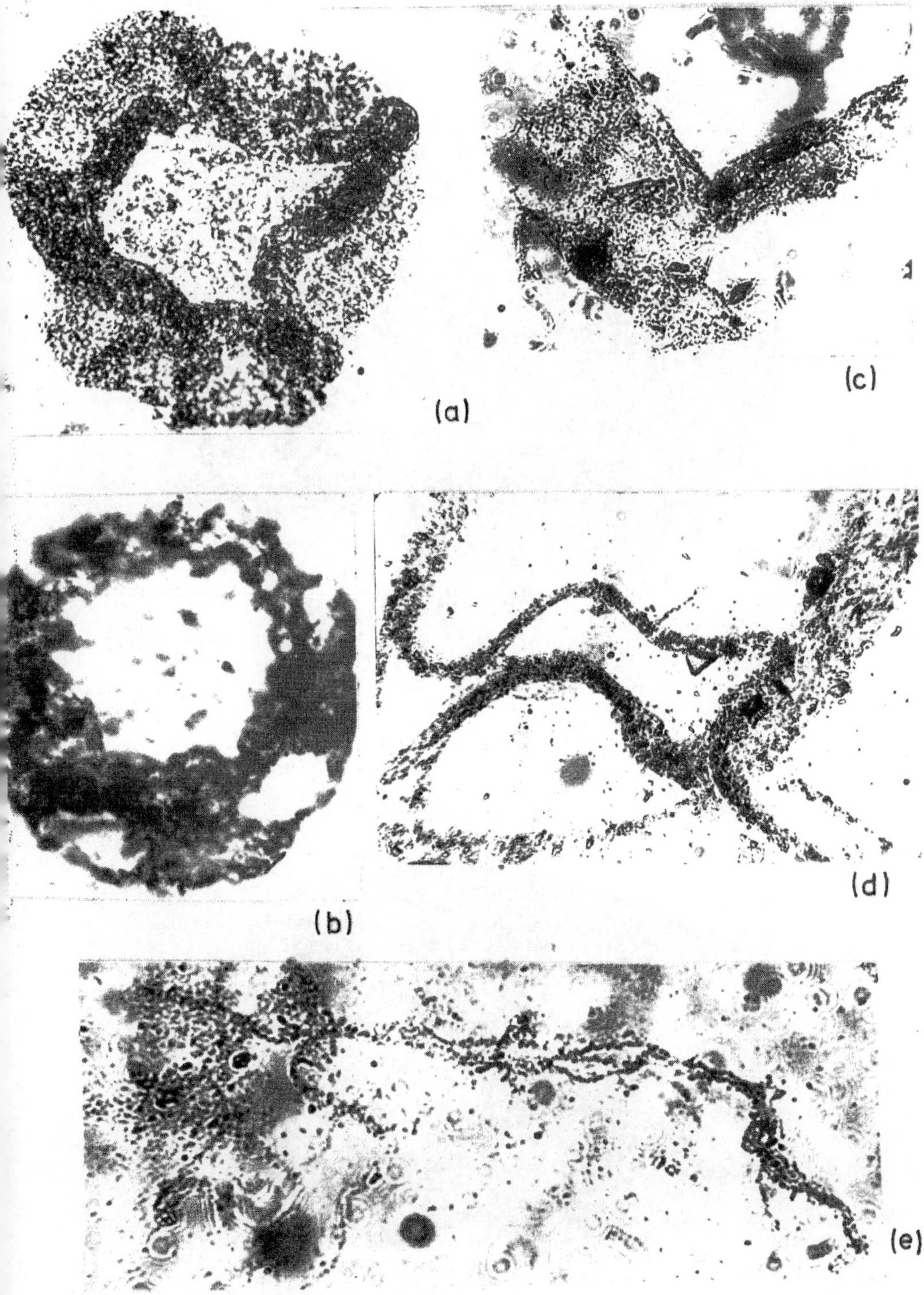


Fig. 1.

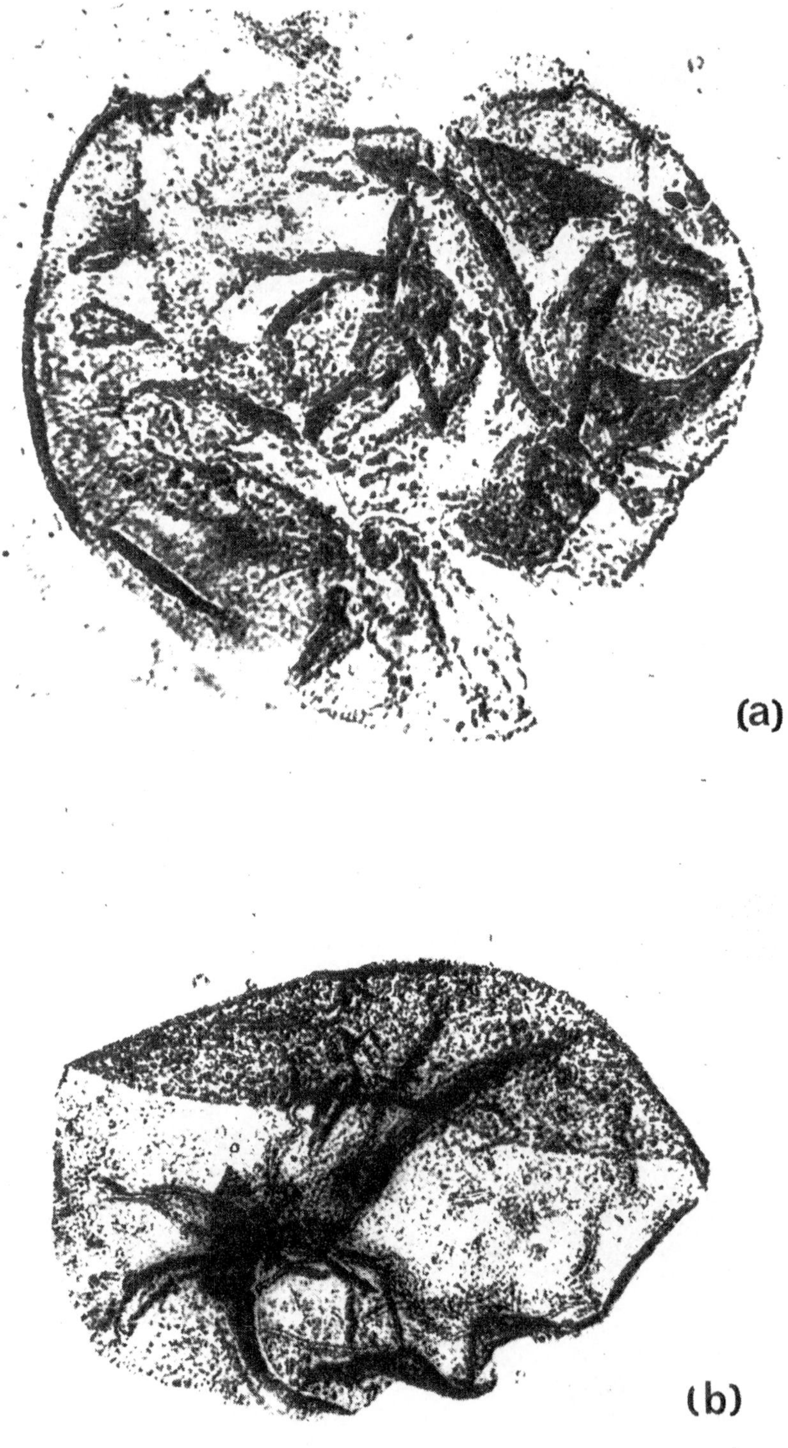


Fig. 2.

(b) *Menneria primaeva* Lopuchin sp. nov. (diameter 350 μm). Holotype. Organic spheroidal microformations with granular structure. The granule size is near 1 μm. As the result of deformation they can form scar-like or scalloped folds. Microformations are very easily subjected to destruction and there are formed a great number of fragments having the shape of scraps with a typical granular surface and also many individual small granules. The specific weight is close to 1. It is without exception that these biofossils were colonial organisms similar to the living blue-green algae of the Microcystidaceae Elenkin family. Originate from the stratatypical area of the Siberian Platform. The Aldan Shield, the Aldan river, calciphires of the Yengrian series, aldanium. Note the similarity with algal microfossils from the Onverwacht series of the Swaziland System, South Africa (Engel *et al.*, 1968).

FIGURE 3

(a) *Menneria* ex gr. *roblotae* Lopuchin (diameter, 100 μm). From the Archean of the Baltic Shield.

(b) *Menneria* ex gr. *roblotae* Lopuchin (diameter, 140 μm). From the suppositional Archean of the North Tien Shan.

(c), (d) *Menneria* sp. (diameter, 50 to 200 μm). From the Archean, Predcharvarian Complex of South India.

(e) *Microtaenia* sp. Ribbon-like alga with a width of 15 μm from the Predcharvarian Complex of South India.

(f) Portion of a graphite crystal with a cover (organic wall) contained inside.

(g) *Leiopsophosphaera mugodjarica* Lopuchin (diameter, 18 μm). From the Riphean of the South Urals. Note the turned back and torn portion of the form.

(h) A graphite crystal from which a cover (organic wall) was separated. The photomicrograph was made in one field of view of the microscope.

FIGURE 4

(a) *Menneria levis* Lopuchin (diameter, 56 μm). From chert of the Gunflint Iron Formation, Ontario, Canada.

(b) *Menneria levis* Lopuchin (diameter, 70 μm). From the Jatulian of Karelia, occurring in shungitic rock.

(c) *Menneria levis* Lopuchin (diameter, 115 μm). From the upper Precambrian Toudheni Syncline of Northwest Africa.

(d) *Menneria levis* Lopuchin (diameter, 200 μm). From the Riphean of northeast Prianabarje near the lower reaches of the Lena River in Chekurovka.

(e) *Menneria foraminis* Lopuchin (diameter, 100 to 150 μm). Forms are shown at a stage of cell division. From the upper Precambrian of North Tien Shan.

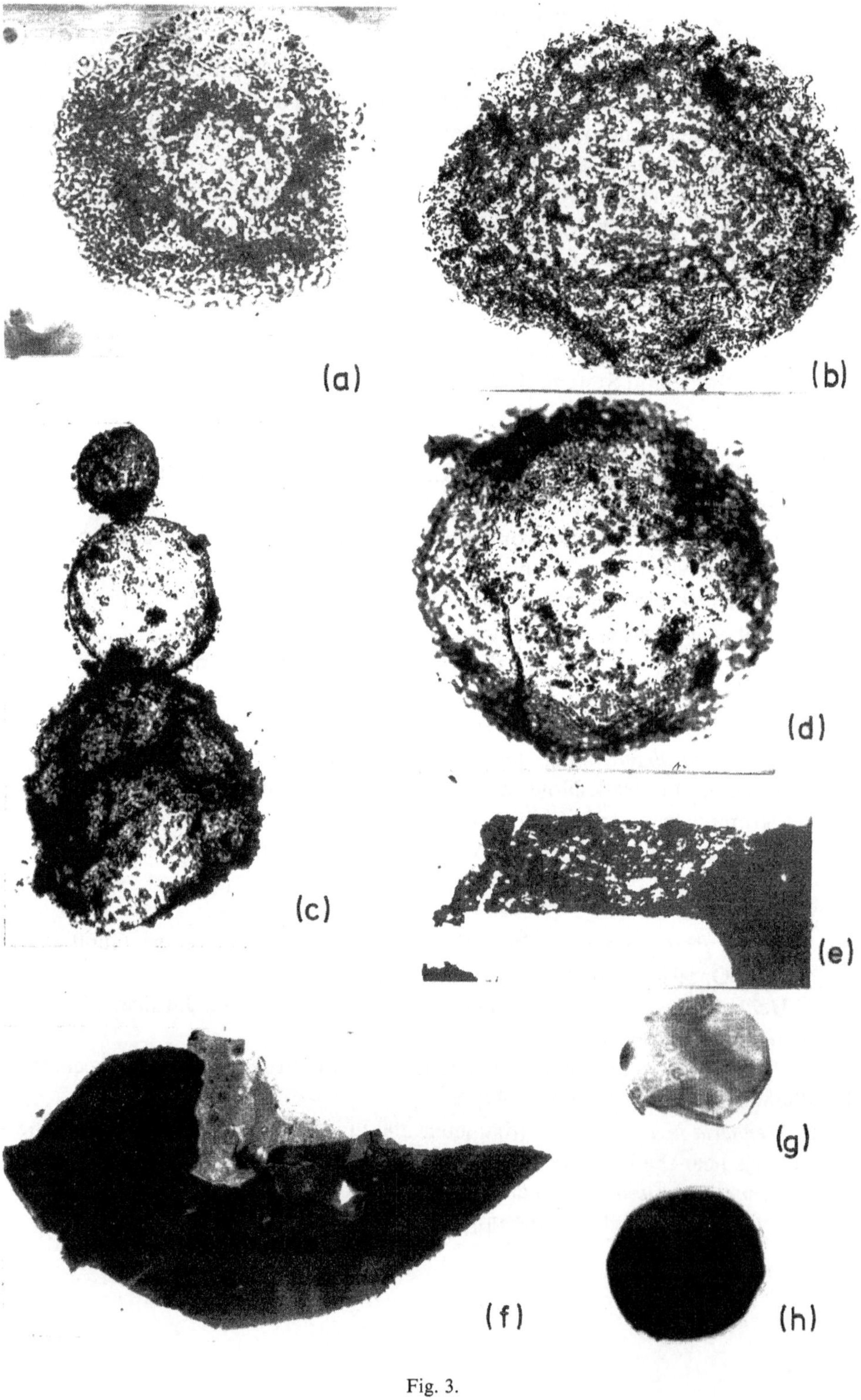


Fig. 3.

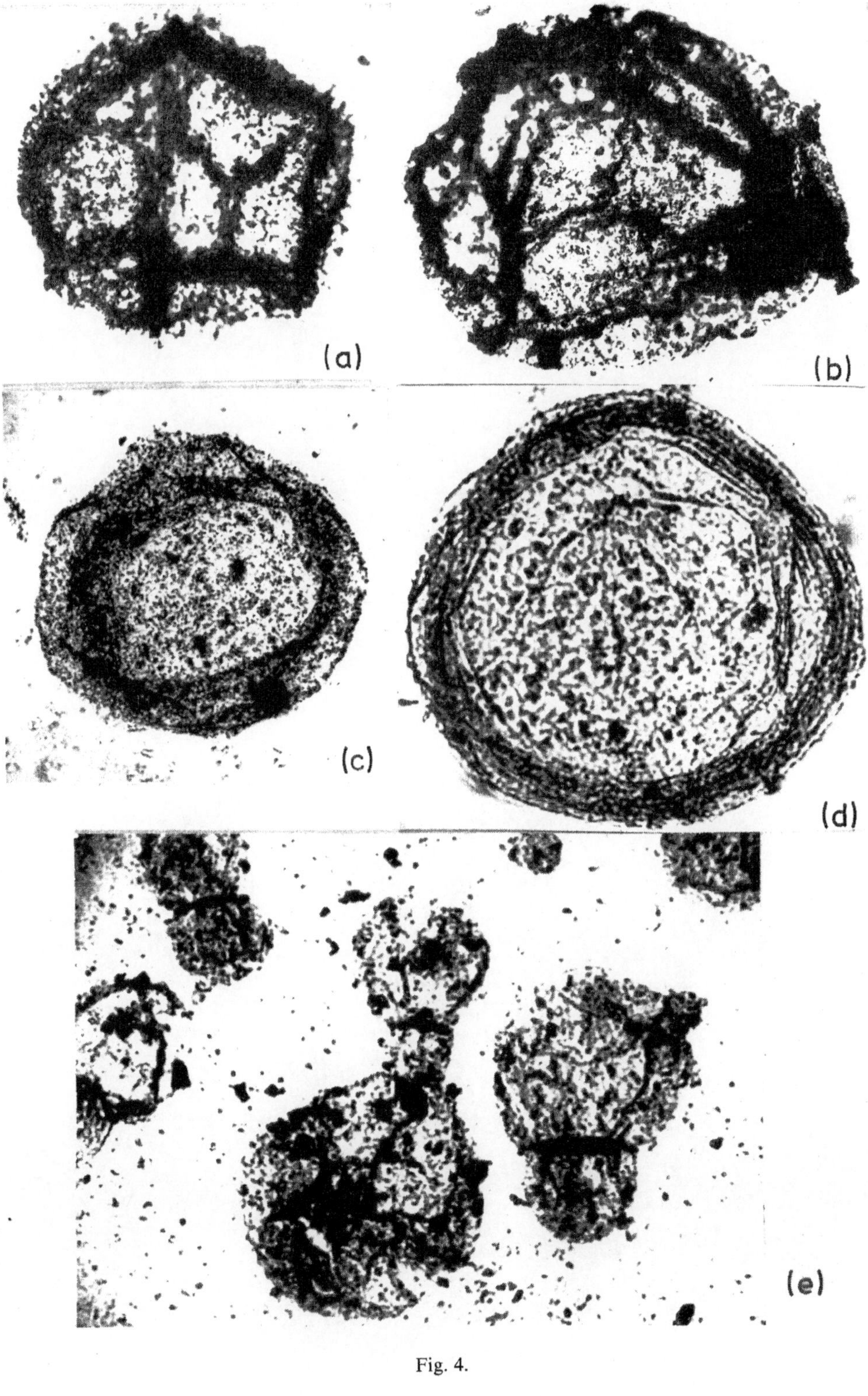


Fig. 4.

# PART II

# ABIOTIC ORGANIC SYNTHESIS

# PHOTOCHEMICAL SYNTHESIS OF SIMPLE ORGANIC FREE RADICALS ON SIMULATED PLANETARY SURFACES – AN ESR STUDY

SHIN-SHYONG TSENG and SHERWOOD CHANG

*Planetary Biology Division, Ames Research Center, NASA, Moffett Field, Calif. 94035, U.S.A.*

**Abstract.** Electron spin resonance spectroscopy provided evidence for formation of hydroxyl radicals during ultraviolet photolysis (254 nm) at −170 °C of $H_2O$ adsorbed on silica gel or of silica gel alone. The carboxyl radical was observed when CO or $CO_2$ or a mixture of CO and $CO_2$ adsorbed on silica gel at −170 °C was irradiated. The ESR signals of these radicals slowly disappeared when the irradiated samples were warmed to room temperature. However, re-irradiation of CO or $CO_2$, or the mixture CO and $CO_2$ on silica gel at room temperature then produced a new species, the carbon dioxide anion radical, which slowly decayed and which was identical with that produced by direct photolysis of formic acid adsorbed on silica gel. The primary photochemical process may involve formation of hydrogen and hydroxyl radicals by means of (1) photodissociation of $H_2O$ physically adsorbed on the silica gel, or (2) absorption of the excitation energy by the silica gel surface with subsequent cleavage of the silanol bonds, or (3) dissociation of $H_2O$ molecules through photosensitization by the surfaces or a combination of (1) to (3). Subsequent reactions of these radicals with adsorbed CO or $CO_2$ or both yield carboxyl radicals, $CO_2H$, the precursors of formic acid. Our results confirm the formation of formic acid under simulated Martian conditions and provide a mechanistic basis for gauging the potential importance of gas-solid photochemistry for chemical evolution on other extraterrestrial bodies, on the primitive earth and on dust grains in the interstellar medium.

## 1. Introduction

The future Viking Mission to Mars will seek to provide data on the possible existence of life and organic compounds (of abiotic or biotic origin, or both) on the surface of the planet. Laboratory data from abiotic organic synthesis experiments conducted under simulated Martian conditions can provide important guidelines regarding the nature of possible abiotic indigenous Martian organic compounds, the kinds of artifacts that may be encountered, and the interpretation of data sent back from the planet. Although a generally accepted hypothesis is that the synthesis of biologically important compounds in primitive planetary atmospheres required chemically reducing conditions, Hubbard *et al.* (1971, 1973) have recently reported the synthesis of organic matter in a gas mixture compositionally similar to the oxidized atmosphere of Mars. They showed that $^{14}CO_2$ and $^{14}C$-organic compounds were formed when a mixture of $^{14}CO$ and water vapor diluted with $^{12}CO_2$ or $N_2$ in the presence of soil or pulverized vycor substrata was irradiated with ultra-violet light of wave lengths (> 250 nm) considerably longer than the wavelengths absorbed by these gases. Since the question as to whether organic compounds could have been formed abiotically on the surface of Mars is of fundamental importance to the concept of chemical evolution and to the interpretation of results from the Viking molecular analysis experiment, we sought to confirm the results of the previous investigation by Hubbard *et al.* while focusing our attention on identifying the primary processes and the reactive intermediates involved in the simulated

Martian organic synthesis. Thus, in addition to gaining insight into potential Martian organic chemistry, conclusions might be drawn regarding the significance of heterogenous gas-solid organic photochemistry for the prebiotic Earth, the solar nebula, and other primordial environments.

Photochemical processes frequently involve homolytic cleavage of molecular bonds to form reactive radical intermediates. The use of electron spin resonance (ESR) to study the irradiated mixtures of CO, $CO_2$ and $H_2O$ adsorbed on silica gel can provide important data on such intermediates and on the processes involved in their production. This report constitutes a preliminary account of such an investigation.

## 2. Materials and Methods

The silica gel (60–80 mesh, Davison 08, gas chromatograph grade) used in these experiments was acid washed and kindly supplied by Drs. Hubbard and Voecks of Jet Propulsion Laboratory, California Institute of Technology. Carbon monoxide (99.99%, $O_2$ content less than 20 ppm), and carbon dioxide (99.995%, $O_2$ content less than 10 ppm) were Matheson research grade. The vacuum and gas-handling system was of conventional design and newly constructed with no prior exposure to mercury.

The ESR spectrometer was a Varian V-4500 X-band reflection type operated at 9.22 GHz and with 100 kHz magnetic field modulation. The rectangular cavity was operated in the $TE_{103}$ mode. The end walls allowed light transmission through the cavity. Low temperatures were achieved by passing cold nitrogen gas around the sample and were controlled by a Varian Variable Temperature Assembly. The sample was situated in a dewar insert that passed vertically through the cavity. The ESR sample tube ($20 \times 0.4$ cm, OD) were made of Suprasil quartz (which could be irradiated with no resulting radical production) with a pyrex section containing a high vacuum stopcock for sample introduction.

The source of ultraviolet light was a low pressure mecury lamp (Pen-ray Lamp Model 11SC-1C, Ultra-Violet Products, Inc., San Gabriel, California) with Vycor shield to eliminate the 184.9 nm line. The light intensity for the 253.7 nm line was 42 mW $cm^{-2}$ at a distance of 12 in. according to the manufacturer.

Unless otherwise indicated, the silica gel (200 mg) was placed in the ESR tube and was heated at 180° for 6 hr in a vacuum of $10^{-5}$ torr, after which time the sample was continuously pumped under vacuum for 16 hr at room temperature. CO (pressure, 15 torr) or $CO_2$ (pressure 15 torr) gases were introduced into the tube through the manifold, and were allowed to equilibrate with the silica gel for 30 min. The tube was then sealed. For experiments with gaseous mixture of CO and $CO_2$ a total pressure of 150 torr was used, of which 10% (i.e., 15 torr) was CO. For experiments with water (deionized, distilled), formic acid (90%, Baker analyzed reagent), or hydrogen peroxide (30%, Mallinckrodt Analytical Reagent), the silica gel in the tube was exposed to vapors of the above pre-degassed liquids for 30 min, and the tube was then sealed. Low temperature irradiation of the samples was conducted in the ESR cavity with the lamp placed at about 3 in. from the samples. ESR spectra were recorded before, during and after the irradiation. The $g$-factors were obtained by comparison with the value for diphenyl-picryl hydrazyl (DPPH). The $g$-factor of DPPH was taken as 2.0037 (Swartz *et al.*, 1972).

## 3. Results

Irradiation of the silica gel alone at −170 °C slowly produced a radical species. Its spectrum is shown in Figure 1a. The spectrum consists of a major peak with $g=2.0120$ and a minor peak at a field 27 G lower, i.e., at $g=2.0292$. The same spectrum (Figure 1b) was obtained when silica gel was first exposed at room temperature to the vapors (0.3 torr) of a dilute aqueous solution of hydrogen peroxide, and then irradiated at −170 °C. The ESR signal appeared much faster than when silica gel was irradiated alone. The $g$-factor of 2.0120 is consistent with that reported (2.01) for the hydroxyl radical produced in a frozen water medium at −196 °C (Kroh *et al.*, 1962). The growth of the radical signal with time is presented in Figure 2. These radicals were

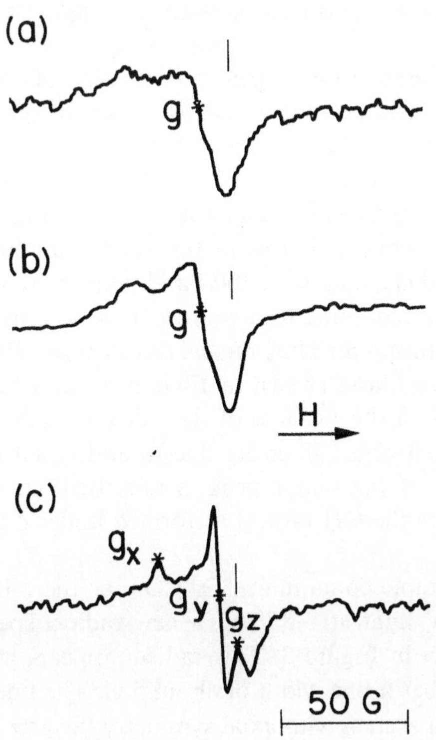


Fig. 1. ESR spectra of radicals produced at −170 °C (a) after 50 min irradication of silica gel; (b) after 5 min irradiation of $H_2O_2$ adsorbed on silica gel; (c) after 50 min irradiation of $O_2$ adsorbed on silica gel. The vertical line indicates the position of DPPH signal.

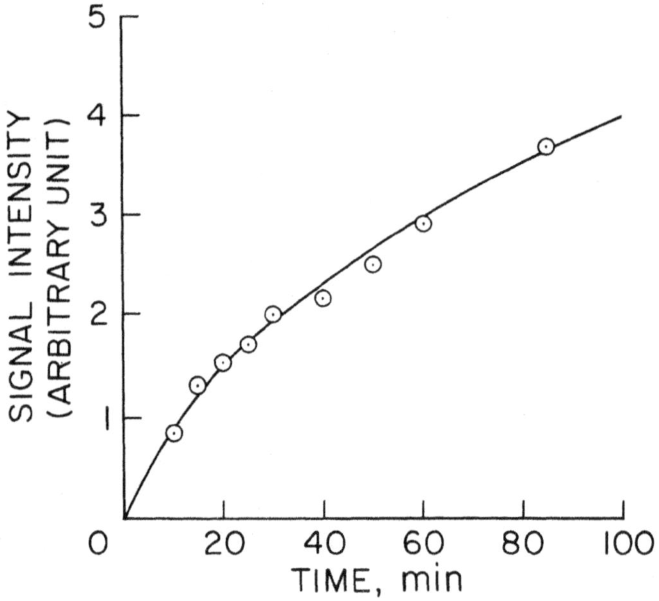


Fig. 2. Formation of OH radicals by UV irradiation of silica gel at −170 °C.

stable at −170 °C but began to disappear on warming of the silica gel, the minor peak decaying slower than the major one. At room temperature the ESR signal disappeared completely. The thermal decay behavior indicated that in addition to the radical responsible for the major peak, small amounts of another radical were present. When the silica gel was irradiated at −170 °C after prior exposure to $O_2$ (1 torr) a new spectrum appeared (Figure 1c) which exhibited $g$-factor anisotropy with $g_x = 2.0255$, $g_y = 2.0088$, and $g_z = 2.0037$ and doublet hyperfine splitting of 10 G for the peak at $g_z$ due to coupling to a proton. This new signal was identical with that reported in the literature for $HO_2$ radical (Kroh *et al.*, 1961; Smith and Wyard, 1961), but different from those shown in Figures 1a and b. Adsorption of water (vapor pressure, 5 torr) on the surfaces of the silica gel followed by irradiation did did not lead to enhanced signal intensity due to additional radical formation. Although the assignment of the major peak is unsettled (cf. Kroh *et al.*, 1961) we suggest that it represents the OH radical as formed under our adsorption and irradiation conditions.

When a silica gel sample containing CO, $CO_2$, or the mixture of CO and $CO_2$ was irradiated with UV light at −170 °C, a new radical species was formed. The ESR spectrum is shown in Figure 3. This radical appears at higher field than the hydroxyl radicals, and has a line width of about 5 G. The line shape shows $g$-anisotropy characteristic of a system with axial symmetry (Swartz *et al.*, 1972; Adrian *et al.*, 1962; Kneubuhl, 1960). The $g$ factors for this new radical are $g_x = g_y = 2.0024$ and $g_z = 1.9971$, and the isotropic or average $g$ factor is 2.0006. This latter value is in excellent agreement with the average $g$ factor of 2.0005 attributed by Chantry *et*

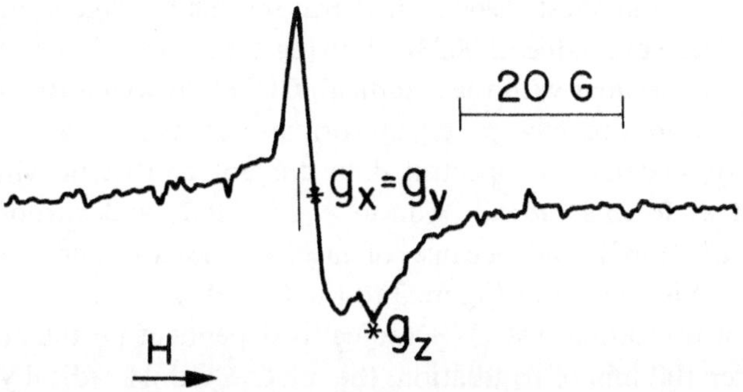


Fig. 3. ESR spectra of $CO_2H$ radicals produced by UV irradiation of a mixture of CO and $CO_2$ adsorbed on silica gel at $-170$°C.

*al.* (1962) to the $CO_2H$ radical formed by $\gamma$-irradiation of potassium bicarbonate crystals. It is also close to the liquid $g$ factor (2.0002) reported for the $CO_2H$ radical generated from oxidation of formic acid in the titanium (III)-hydrogen peroxide

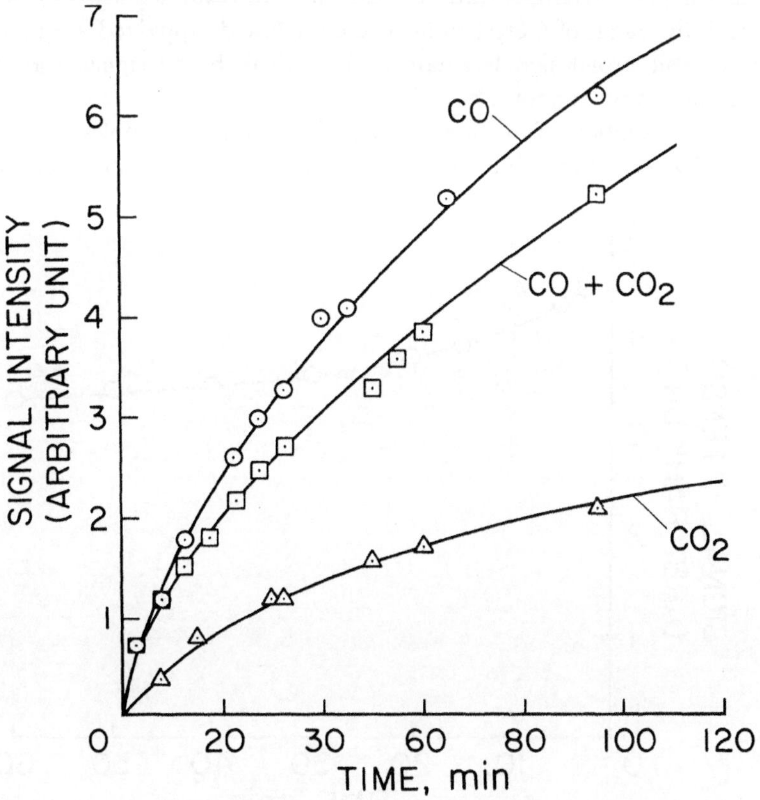


Fig. 4. Formation of $CO_2H$ radicals by UV irradiation of CO, $CO_2$, or mixture of CO and $CO_2$ adsorbed on silica gel at $-170$°C.

system (Norman and West, 1969). Furthermore, the average $g$-factor of this radical is below the free spin value (2.0023), showing that it is unlikely to be the $\pi$-radical, $HCO_2$, but is consistent with the $\sigma$-radical, $CO_2H$, in which the unpaired spin is on an orbital of approximately $sp^2$ type (Norman and West, 1969; Wertz and Bolton, 1972). Finally, because the spectral data are not consistent with other candidate species such as formyl radical (Adrian *et al.*, 1962) and carbonate radical anion (Chantry *et al.*, 1962), and because of mechanistic considerations (see below), we attribute the ESR spectra in Figure 3 to the $CO_2H$ radical.

The rate of formation of $CO_2H$ radical is dependent on the adsorbed gases. For example, after 100 min of irradiation, the relative $CO_2H$ radical yields, as estimated from the ESR signal intensity, are 6.3 for CO, 5.2 for $CO+CO_2$, and 2.1 for $CO_2$ (Figure 4). Pre-treatment of silica gel also affects the formation of $CO_2H$. When silica gel was preheated at 450 °C for 6 hr in a vacuum of $10^{-5}$ torr, followed by pumping at room temperature for 16 hr, and then exposed to the mixture of CO and $CO_2$, the yield of $CO_2H$ radicals produced upon UV irradiation was less than half of that obtained when silica gel was preheated at 180°. The decay curve of this radical at −160 °C is shown in Figure 5. Its decay rate became very slow after one hour of standing at this temperature; on warming, however, it started to decay much faster. The ESR signal of $CO_2H$ radicals completely disappeared at about −50 °C and only a weak broad signal remained. This weak broad signal also eventually disappeared at room temperature.

Continued irradiation of all these previously irradiated samples (i.e., CO, $CO_2$, or CO and $CO_2$ on silica gel) at room temperature for 20 hr generated new stable

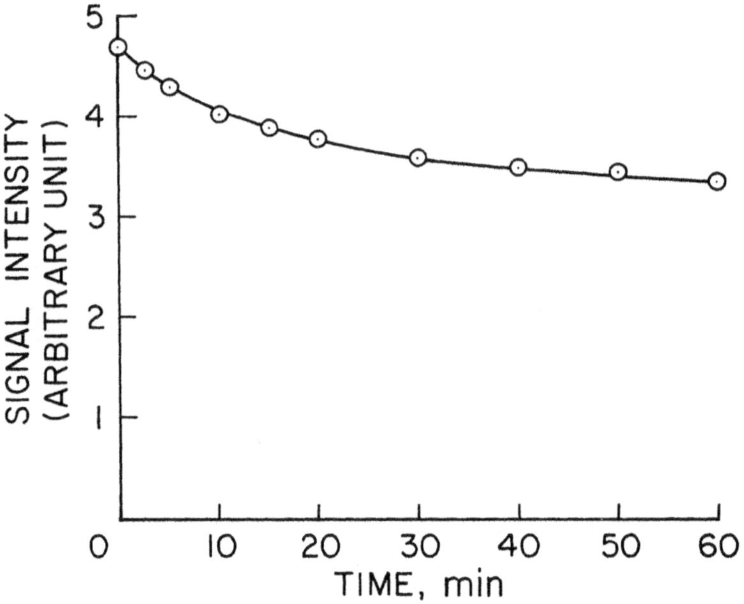


Fig. 5. Decay Curve of $CO_2H$ radicals at −160 °C.

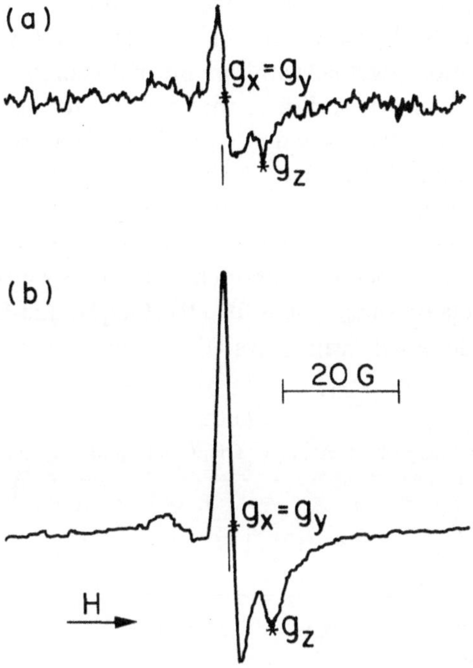


Fig. 6. ESR spectra of $CO_2^-$ radicals produced by UV-irradiation at room temperature. (a) re-irradiation of mixture of CO and $CO_2$ on silica gel pre-irradiated at $-170\,°C$ (See Figure 3); (b) irradiation of formic acid adsorbed on silica gel.

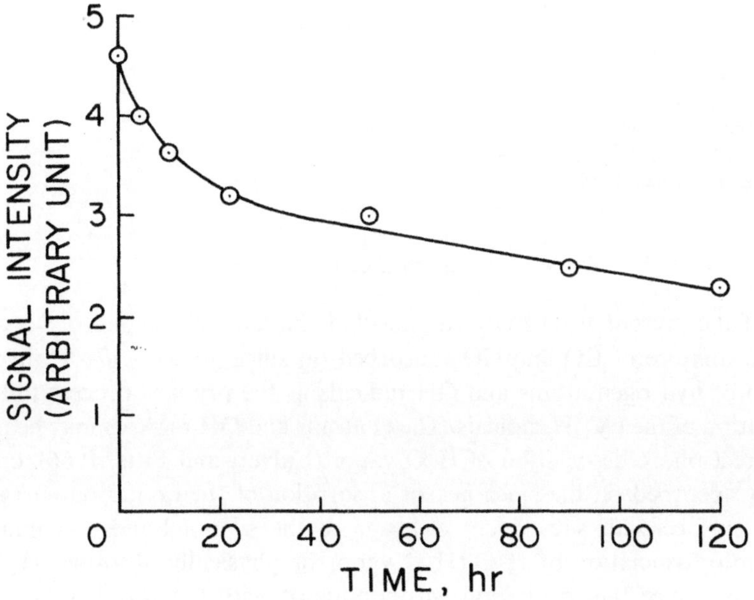


Fig. 7. Decay Curve of $CO_2^-$ radicals at room temperature.

radical species. The ESR spectrum is shown in Figure 6a. The line shape is similar to that of the $CO_2H$ radical observed at $-170\,°C$. However, the line width is smaller ($\simeq 3$ G). This observation together with the unusual stability of the radical at room temperature (Figure 7) clearly indicates that a new radical formed. The $g$ factors for this radical are $g_x = g_y = 2.0024$ and $g_z = 1.9979$. Comparison of the ESR spectrum and $g$ factors of this radical with that information reported in the literature reveals that this is the carbon dioxide anion radical $CO_2^-$ (Miura *et al.*, 1968; Beckwith and Norman, 1969). When formic acid was adsorbed on silica gel and irradiated with UV light at room temperature, an identical ESR spectrum was also obtained (Figure 6b). The results on the $g$ values for OH, $CO_2H$, and $CO_2^-$ radicals are summarized in Table I, along with values available from the literatures for comparison.

TABLE I

Summary of $g$-factors[a] for OH, $CO_2H$, and $CO_2^-$ radicals

| Radical | $g_x$ | $g_y$ | $g_z$ | Isotropic or average $g_0$ |
|---|---|---|---|---|
| OH on silica gel ($-170\,°C$) | | | | 2.0120 |
| OH on $H_2O$ ice[b] | | | | 2.01 |
| $CO_2H$ on silica gel ($-170\,°C$) | 2.0024 | 2.0024 | 1.9971 | 2.0006 |
| $CO_2H$ in $\gamma$-irradiated $KHCO_3$[c] | | | | 2.0005 |
| $CO_2H$ in Ti(III)—$H_2O_2$—HCOOH[d] | | | | 2.0002 |
| $CO_2^-$ on silica gel (25 °C) | 2.0024 | 2.0024 | 1.9979 | 2.0008 |
| $CO_2^-$ on silica gel (25 °C)[e] | 2.0024 | 2.0024 | 1.9978 | 2.0008 |
| $CO_2^-$ in Ti(III)—$H_2O_2$—HCOONa[f] | | | | 2.0003 |

[a] Estimated error: 0.0002
[b] Kroh *et al.*, 1962
[c] Chantry *et al.*, 1962
[d] Norman and West, 1969
[e] Miura *et al.*, 1968
[f] Beckwith and Norman, 1969

## 4. Discussion

Results of the present ESR study suggest that the ultra-violet photolysis of CO or $CO_2$, or a mixture of CO and $CO_2$ adsorbed on silica gel at $-170\,°C$ involves the formation of hydrogen atoms and OH radicals as the primary process followed by the formation of the $CO_2H$ radicals. The H atoms and OH radicals may be produced by the direct photodissociation of $H_2O$ vapor (Calvert and Pitts, 1966), or of $H_2O$ physically adsorbed on the silica gel, or absorption of the excitation energy by the silica gel surface with subsequent cleavage of the silanol bonds ('bound water'). Direct photodissociation of $H_2O$ ($H_2O$ vapor or physically absorbed $H_2O$) seems very unlikely since the irradiation was conducted with the wavelength (253.7 nm) longer than that absorbed by water. Moreover, the ESR signal intensities of hydroxyl radicals appeared to be the same when silica gel alone or silica gel and water vapor (5 torr) were irradiated. It is generally believed that when silica gel surface is heated at 150–200 °C the physically adsorbed water is removed, and only a layer of hydroxyl groups remain on the silica surfaces, i.e., silanol groups (Iler, 1955; Snyder, 1968; Alexander, 1973). Reintroduction of water vapor into the silica gel preheated at 180 °C will only rehydrate the surface and form a physically adsorbed monolayer. If water photolysis is the full story, one would expect an increase in OH radical concentration in the irradiated silica gel containing added water. Absorption of excitation energy by the silica gel surfaces with subsequent cleavage of ≡Si—OH and ≡SiO—H bonds seems a more likely process. Irradiation of silica gel preheated at 450 °C results only in a very weak ESR signal of the OH radicals. Apparently the silanol groups are less available after pretreatment at high temperatures. At 450 °C the major amounts of silanol hydroxyls usually will condense to liberate water and leave surface siloxane groups.

The authors are unaware of any direct evidence of UV-induced photochemical cleavages of ≡Si—OH and ≡SiO—H bonds in the literature. However, Pariiskii *et al.* (1968) recently had reported the ESR detection of deuterium atoms formed by UV irradiation (>253.7 nm) of $D_2$ molecules physically adsorbed on the surface of silica gel. Although an energy transfer process was proposed by them, it is also possible that the ≡SiO—H bonds were cleaved to form hydrogen atoms which then exchanged with $D_2$ to form DH and ·D. Evidence has been provided in a number of reports that hydrogen atoms are released from surface hydroxyl groups of silica gel when irradiated with $\gamma$-rays (Kinell *et al.*, 1969; Kazanskii *et al.*, 1961; Strelko and Suprunenko, 1968). Possible effects of hydroxyl radicals produced by the UV irradiation of silica gel in the presence of some glycine-containing dipeptides also have been suggested by Riuett and Wilshire (1967). The identity of the OH radical observed here was supported by its $g$ factor and by comparison of its ESR spectrum with that obtained from direct photolysis of vapors from dilute aqueous $H_2O_2$ on silica gel. The formation of OH radicals is known to be the photodissociative primary process of $H_2O_2$, and the quantum yield is 0.5 at 2537 Å in aqueous solution (Calvert and Pitts, 1966). Hydrogen atoms were not observed by ESR in the irradiated silica gel, possibly because of their low concentration and their combination to form hydrogen molecules. The size of the H atom presumably makes it more mobile than other radicals and less susceptible to stabilization in trapping sites. Reactions of the hydrogen atom and OH radical with CO can form the HCO radical and $CO_2H$ radical respectively. CO is an efficient scavenger for OH radicals (Buxton and Wilmarth, 1963; Dainton and Hardwick, 1957); therefore, the formation of $CO_2H$ radical is expected to be favored. This $CO_2H$ radical can then react with a hydrogen atom or with a silanol group to form formic acid. $CO_2H$ radical is not necessarily an unstable species. Kinetic study of Buxton and Wilmarth (1963) had provided strong evidence that $CO_2H$ radicals did not decompose to $CO_2$ and H atoms under their experimental conditions. In fact, the silica gel surface does appear to provide some stability for $CO_2H$ radicals. $CO_2H$ radicals also have been prepared in a titanium (III)—$H_2O_2$—HCOOH system and used for study of organic reactions (Norman and West, 1969; Anderson *et al.*, 1971). The CHO radicals will be present in very low concentrations because of more favorable competitive reactions to form $H_2$ and $CO_2H$ radicals; and hence, CHO radicals could not be detected with the ESR conditions used in these experiments. The ESR spectrum of CHO radicals have been obtained by the reaction of CO with hydrogen atoms produced by the photolytic decomposition of HI in solid carbon monoxide (Adrian *et al.*, 1962). When $CO_2$ is adsorbed on silica gel, the photochemically produced hydrogen atoms can react with $CO_2$ and reduce it to $CO_2H$ radicals. This is not an efficient process as is shown in Figure 4, since there is competitive hydrogen atom combination. In the case of mixture of CO and $CO_2$ on silica gel, both gases are adsorbed on the surface, so the sites available for CO absorption will be decreased. Consequently, the total amount of $CO_2H$ radicals formed will be lower than when CO is adsorbed on silica gel alone (Figure 4). This finding is in agreement with the observation of Hubbard *et al.* (1973) that lower yields of organic compounds were obtained when $CO_2$ was added to CO as a diluent gas.

Energy transfer from the silica gel surface to the physically adsorbed $H_2O$ is also possible. However, the fact that no increase of ESR signal due to OH radicals was observed from the irradiation of silica gel containing 5 torr of water vapor suggest that this energy transfer is only a minor process. The mechanism of energy transfer is not known. Transfer of the excitation energy from the surface to CO molecules has been suggested by Hubbard *et al.* (1973). Our results show that irradiation of silica gel bearing $CO_2$ as the only added gas also results in formation of $CO_2H$ radicals. Hubbard *et al.* (1973) also suggested that the adsorption of CO on silica surfaces may have caused a large shift in its absorption spectrum. Although molecules adsorbed on silica surfaces have been shown to exhibit shifts in electronic absorption spectra (Terenin, 1964; Leermakers *et al.*, 1966; Nicholls and Leermakers, 1971), the direction of a spectral shift is dependent on the electronic transition and the solvents involved. For carbonyl compounds the shifts are not large and in some cases the shifts are in the lower wavelength direction. CO absorbs light in a relatively weak band which extends from about 115 nm to 150 nm, but the light used in this study was mainly 253.7 nm with minor components in the longer wavelength region. It is doubtful that the silica surface can cause such a large shift in CO absorption spectrum. However, charge transfer interaction between CO and the surface may induce a new band which then permits absorption of energy to excite the CO molecules. So far such a situation has not been reported. Oxygen was present as an impurity in CO and $CO_2$, its amount was about 30 ppm. The effect of $O_2$ on radical formation will be studied. However, Hubbard *et al.* (1973) reported that in their experiments higher levels of $O_2$ inhibited but did not abolish the conversion of CO to organics.

After $CO_2H$ radicals had disappeared at room temperature, presumably forming formic acid, re-irradiation of the silica gel samples with the same light source at room temperature for 20 hr produced carbon dioxide anion radicals, $CO_2^-$. Irradiation of formic acid adsorbed on silica gel at room temperature for the same period of time gave the identical radical (Figure 6). These radicals were identified by comparison with the ESR spectrum and $g$ factors reported in the literature (Miura *et al.*, 1968; Beckwith and Norman, 1969). These radicals are stabilized on silica gel for days at room temperature (Figure 7). These results demonstrate that formic acid was formed during the decay of $CO_2H$ radicals. The formation of $CO_2^-$ radicals was also observed in the $\gamma$-irradiation of formic acid adsorbed on silica gel (Miura *et al.*, 1968). In view of the strong interaction of the formic acid monomer with the surface by means of hydrogen bonding (Imelik *et al.*, 1959), it may be that both the C—H bond and the O—H bond of the adsorbed acid were ruptured by both UV and $\gamma$-ray irradiations.

## 5. Conclusions

The present study confirms the reports of Hubbard *et al.* (1971, 1973) that formic acid and possibly formaldehyde and other organic compounds were formed during the photolysis of atmospheric gases adsorbed on a simulated Martian surface, thus, suggesting that these compounds may be formed on the surface of Mars as well. An equally important observation in this study is that these organic compounds were formed by the reactions of radicals generated by UV irradiations, the OH radicals playing a very important and initial role. We now know that the atmospheric pressure at the surface of Mars is about 5 mb and that the atmosphere is composed mainly of $CO_2$, with trace amounts of $H_2O$, CO and $O_2$ (Ingersoll and Leovy, 1971; Barker, 1972; Carleton and Traub, 1972). The infrared spectroscopy results of Mariner 9 (Hanel *et al.*, 1972; Hanel *et al.*, 1972) indicate that the fine dust in the Martian atmosphere has an intermediate $SiO_2$ content in the range of $60 \pm 10\%$. Our study shows that water is essential, but not required in large quantity. Thus, photochemical synthesis of organic compounds could occur on Mars provided that the surfaces of $SiO_2$ have enough 'bound water'. In fact, Houck *et al.* (1973) have recently provided the high altitude infrared spectroscopic evidence for bound water on Mars. They estimate that the surface soil layer contains about 1% bound water by weight. The low partial pressure of CO in the Martian atmosphere probably would not contribute serious problems to the organic synthesis, since $CO_2$ which is present in large quantity also could react with hydrogen atoms to form $CO_2H$ radicals, the precursors of formic acid. The effect of $O_2$ on these syntheses probably is not serious. Nevertheless, it will be investigated in detail.

One serious problem that may be encountered is the photodestruction of these organic compounds. Can these organic compounds survive over geological time under UV irradiation? Both formaldehyde and formic acid are known to undergo photodestruction forming $H_2$, CO, and $CO_2$ (Calvert and Pitts, 1966). To answer this question would require more quantitative study on the quantum yields of both photochemical synthesis and photodestruction of these organic compounds. Nevertheless, dust storms similar to the one that was observed in the early months of the Mariner 9 mission to Mars in 1971 (McCauley *et al.*, 1972) or the ozone that has been reported to be present in the polar regions of this planet (Barth *et al.*, 1973) may provide some protection from ultimate UV destruction.

Present findings provide confirmation of the view that organic compounds can be synthesized photochemically on the simulated Martian surface. This observation is very important because it suggests that the synthesis of biologically important organic compounds is also possible in oxidized planetary atmospheres. Our study also demonstrates that photochemically generated free radicals play very important roles in these syntheses. Further study of such heterogeneous systems may provide a mechanistic basis for gauging the potential importance of gas-solid photochemistry for chemical evolution on other extraterrestrial bodies, on the primitive Earth and on dust grains in the interstellar medium.

## Acknowledgements

One of us (S. S. Tseng) wishes to thank National Research Council and National Aeronautics and Space Administration for the support of Research Associateship. Thanks are also due to Mr Wilson Myers for his technical assistance, to Dr Gaylord M. Androes for the helpful discussions, and to Dr Ronald F. Reinisch for the permission to use his ESR spectrometer.

## References

Adrian, F. J., Cochran, E. L., and Bowers, V. A.: 1962, *J. Chem. Phys.* **36**, 1661.
Alexander, G.: 1973, *Silica and Me*, American Chemical Society, Washington, DC.
Anderson, N. H., Dobbs, A. J., Eage, D. J., Norman, R. O. C., and West, P. R.: 1971, *J. Chem. Soc. (B)*, 1004.
Barker, E. S.: 1972, *Nature* **238**, 447.
Barth, C. A., Hord, C. W., Stewart, A. I., Lane, A. L., Dick, M. L., and Anderson, G. P.: 1973, *Science* **179**, 795.
Beckwith, A. L. J. and Norman, R. O. C.: 1969, *J. Chem. Soc. (B)*, 400.
Buxton, G. and Wilmarth, W. U.: 1963, *J. Phys. Chem.* **67**, 2835.
Calvert, J. G. and Pitts, J. N.: 1966, *Photochemistry*, John Wiley and Sons, New York.
Carleton, N. P. and Traub, W. A.: 1972, *Science* **177**, 988.
Chantry, G. W., Horsfield, A., Morton, J. R., and Whiffen, D. H.: 1962, *Molecular Phys.* **5**, 589.
Dainton, F. S. and Hardwick: 1957, *Trans. Faraday Soc.* **53**, 333.
Hanel, R., Conrath, B. J., Hovis, W. A., Kunde, V. G., Lowman, P. D., Maguire, W., Pearl, J. C., Pirraglia, J., Probhakara, C., and Schlachman, B.: 1972, *Icarus* **17**, 423.
Hanel, R. A., Conrath, B. J., Hovis, W. A., Kunde, V. G., Lowman, P. D., Pearl, J. C., Prabhakara, C., and Schlachman, B.: 1972, *Science* **175**, 305.
Houck, J., Pollack, J. B., Schaack, D., Sagan, C., and Decker, J. A., Jr.: 1973, *Icarus* **18**, 470.
Hubbard, J. S., Hardy, J. P., and Horowitz, N. H.: 1971, *Proc. Nat. Acad. Sci. U.S.A.* **68**, 574.
Hubbard, J. S., Hardy, J. P., Voecks, G. E., and Golub, E. E.: 1973, *J. Mol. Evol.* **2**, 149.
Hulett, H. R.: 1969, *J. Theoret. Biol.* **24**, 56.
Iler, R. K.: 1955, *The Colloid Chemistry of Silica and Silicates*, Cornell University Press, New York.
Imelik, B., Rossetti, J. F., and Sigli, P.: 1959, *J. Chim. Phys.* **56**, 1048.
Ingersoll, A. P. and Leovy, C. B.: 1971, *Ann. Rev. Astrophys.* **9**, 147.
Kazanskii, V. B., Pariiskii, G. B., and Voevodskii, V. U.: 1961, *Discussions Faraday Soc.* **31–32**, 203.
Kinell, P. O., Luna, A., and Shimizu, A.: 1969, *J. Phys. Chem.* **73**, 4175.
Kneubuhl, F. K.: 1960, *J. Chem. Phys.* **33**, 1074.
Kroh, J., Green, B. C., and Spinks, J. W. T.: 1961, *J. Am. Chem. Soc.* **83**, 2201.
Kroh, J., Green, B. C., and Spinks, J. W. T.: 1962, *Can. J. Chem.* **40**, 413.

Leermakers, P. A., Thomas, H. T., Weis, L. D., and James, F. C.: 1966, *J. Am. Chem. Soc.* **88**, 5075.
McCauley, J. F., Carr, M. H., Cutts, J. A., Hartmann, W. K., Masursky, H., Milton, D. J., Sharp, R. P., and Wilhelms, D. E.: 1972, *Icarus* **17**, 289.
Miura, M., Hasegawa, A., and Yamada, Y.: 1968, *Bull. Chem. Soc. Japan* **41**, 1320.
Nicholls, C. H. and Leermakers, P. A.: 1971, *Adv. Photochem.* **8**, 315.
Norman, R. O. C. and West, P. R.: 1969, *J. Chem. Soc. (B)*, 389.
Pariiskii, G. B., Mikheikin, I. D., Rez, A. I., and Kazanskii, V. B.: 1968, *Dokl. Akad. Nauk U.S.S.R.* **180**, 123.
Riuett, D. E. and Wilshire, J. F. K.: 1967, *Photochem. Photobiol.* **6**, 867.
Smith, R. C. and Wyard, S. J.: 1961, *Fifth International Symposium on Free Radicals*, Gordon and Breach, Inc., New York, pp. 66–2.
Snyder, L. R.: 1968, *Principles of Adsorption Chromatography*, Marcel Dekker, New York.
Strelko, V. V. and Suprunenko, K. A.: 1968, *Khim Vys. Energ.* **2**, 258.
Swartz, H. M., Bolton, J. R., and Borg, D. C.: 1972, *Biological Application of Electron Spin Resonance*, Wiley-Interscience, New York, p. 23.
Terenin, A.: 1964, *Adv. Catalysis* **15**, 227.
Wertz, J. E. and Bolton, J. R.: 1972, *Electron Spin Resonance Elementary Theory and Practical Applications*, McGraw-Hill, New York, p. 164.

# HIGH FREQUENCY DISCHARGE EXPERIMENT

## I: *Formation of Organic Compounds from Methane and Ammonia*

S. YUASA*

*Dept. of Biology, College of General Education, Osaka University, Toyonaka, Osaka 560, Japan*

and

M. ISHIGAMI

*Biological Laboratory, Jichi Medical College, Yakushiji, Kawachi, Tochigi 329-04, Japan*

**Abstract.** Investigations of the synthesis of organic compounds from methane and ammonia were carried out by using high frequency discharges. The radical of CN was observed in the early phase of the discharge. Therefore it is considered that the CN radical contributes to the successive reaction steps.

An important intermediate, a tetramer of hydrogen cyanide was observed to be formed at room temperature. The amounts of this compound progressively increased with time. It was speculated that the purines which were identified in the refluxed sample were synthesized through the above tetramer. Compounds which are thought to be heterocyclic ($\lambda_{max}$: 250–265 nm) were initially formed in the sample trapped at liquid nitrogen temperature.

After hydrolysis with hydrochloric acid, Gly, Asp, Ala, *etc.* were observed in the sample. Overall yield of organic compounds was estimated at 5–6% based on input carbon of methane.

### 1. Introduction

Ultraviolet light is generally considered (Urey, 1952) to be the most effective energy source for formation of organic compounds among the various energy sources available on the Earth. It is very difficult, however, to use ultraviolet sources when attempting to synthesize large amounts of organic compounds from gaseous mixtures. The syntehsis products adhere cumulatively to the source window and effectively reduce the available energy, thereby stopping any further reaction. We have, therefore, constructed a high frequency electrical discharge apparatus with several excellent characteristics that make it ideal for prebiotic synthesis and chemical evolution studies of simulated primordial gas mixtures. This paper describes this apparatus and generalizes the results of high frequency discharge synthesis of primitive gas mixtures such as methane and ammonia. Detailed analyses of the reaction products are discussed in another paper (Ishigami and Yuasa, 1974).

### 2. Experimental

Only one previous high frequency discharge experiment has been described (Czuchajowski *et al.*, 1968) and our apparatus offers many comparative improvements over this and other energy sources. Figure 1 is a schematic diagram of the system. A pair of copper sleeve electrodes for high frequency discharge were wrapped around the glass tubing, 40 mm in diam, for a distance of 50 mm. High frequency power (42.5 Mc) was supplied by means of a conventional oscillator circuit. The plate-voltage was 1 kV and the current was 200 mA. The output power was about 200 W.

* Present address: Department of Biophysical Sciences, University of Houston, Houston, Tex. 77004, U.S.A.

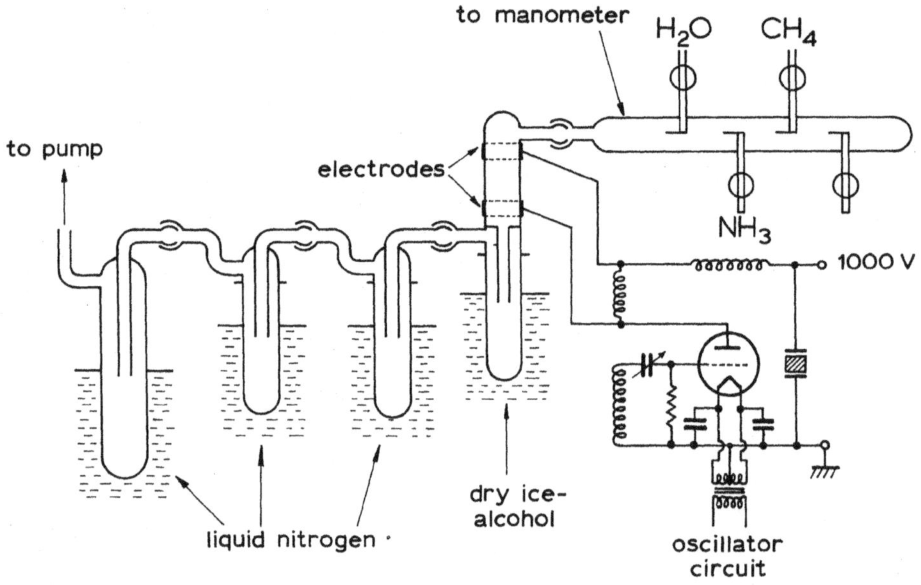


Fig. 1. High frequency discharge apparatus.

Electron temperature and gas temperature in the discharge region were determined to clarify the characteristics of the discharge. Electron temperature was measured by the double probe method (Thonson and Malter, 1950). That is, a floating double probe made from molybdenum wire was inserted into the center of the discharge tube. Gas temperature was determined with a chromel-alumel-thermo-couple fused in a glass rod (4 mm in diam, 10 mm length) inserted into the center of the discharge tube. The values thus obtained were:

| | | |
|---|---|---|
| Electron temperature | $H_2O$ (1.0 Torr) | $8.7 \times 10^4$ K |
| | $NH_3$ (1.5 Torr) | $7.6 \times 10^4$ K |
| Gas temperature | $H_2O$ (1.0 Torr) | 490 K |
| | $NH_3$ (1.5 Torr) | 620 K |

Mean kinetic energy of the electrons was estimated to be $3 \times 10^4$ K which corresponds to that of a 1550 Å photon.

The advantages of this system can be summarized as follows:

(1) The electrodes do not contact the gas mixture because they are wrapped around the outside of the glass-tubing. Accordingly, we can avoid a metal-mediated formation of organic compounds, where internal electrodes might act as an effective catalyst for a recombination of the radicals formed in the discharge region.

(2) The high frequency discharge is not as strong in comparison with a so-called spark discharge from an energetic view point (see the legend in Figure 1). The energy is sufficiently low so that the energetic contribution to the formation of organic compounds would be more synthetic than destructive.

(3) The flow-method (open system) adopted allows us to distinguish the primary products from the subsequent products which are thought to be formed through re-

combination with the primary compounds. In the course of the experiment, we can minimize secondary reaction processes because excited molecules and radicals are immediately transferred away from the discharge region without reexcitation.

(4) Experimental processes are more easily handled than any other discharge. Thus, the apparatus might offer some advantages to another objective besides the study of chemical evolution.

Commercially available gases such as methane (99.9% in purity) and ammonia (99% in purity) were used in the present experiment. Condensable impurities were removed by trapping the former with liquid nitrogen and the latter with a dry ice-alcohol mixture, respectively. Water used in the experiment was distilled after condensing in dry ice-alcohol cooled trap to eliminate impurities. After the entire chamber was well evacuated (less than $10^{-3}$ Torr), the frozen gases in each trap were slowly heated and thus evaporated into a mixing chamber through greaseless cocks and capillary nozzles (2 mm I.D., 40 mm in length). Static pressure of the gas mixture was measured at the chamber with an oil manometer and kept at 1.5 mm Hg.

## 3. Results and Discussion

### 3.1. Molecules and Radicals

The plasmal glow of the discharge region was introduced into a prism-spectrophotometer through a quartz-window. Various kinds of atoms, molecules and radicals were observed and identified by their emission spectra. The results thus obtained are shown in Table I. Arpigny (1965) has observed atomic and molecular glows in a comet head although no discharge is present. For comparative purposes, the emissions from the comet head are also represented additively in Table I. The experiments from #1 to #4 produced close to those observed in the comet. This fact seems to suggest formation of some fundamental organic compounds in the solar nebula.

When mixing ammonia and methane, the radical CN was predominantly observed (exp 4 in Table I). Accordingly, we can expect the radical CN to play an important role in the successive reaction processes when suitable energy is supplied to ammonia and methane mixtures.

### 3.2. Formation of Organic Compounds

#### 3.2.1. *Ammonium Cyanide*

Using methane and ammonia (1.5 Torr), the discharge was carried out for 2 hr. After discharge, each trap was stored at −22 °C for overnight. A glass-like crystal was formed in the trap (mp: *ca.* 40 °C). This crystal was isolated and dissolved in distilled water. Later the solution was measured by means of spectrophotometry ($\lambda_{max}$: 296 nm), and it was suggested that ammonium cyanide ($NH_4CN$) was initially formed in the trap and later converted to the material absorbing at 296 nm. Ponnamperuma and Woeller (1967) also reported the formation of ammonium cyanide in an arc discharge experiment by using only methane and ammonia. This material has received attention in the formation of various organic compounds in chemical evolution experiments.

TABLE I

Comparison of the observed emission spectra with those of comet

| High frequency discharge | | | | | Comet (head) | | |
|---|---|---|---|---|---|---|---|
| Exp | Supplied gas | Emitter | Wavelength[c] ($\lambda$) | Transition | Emitter | Wavelength ($\lambda$) | Transition |
| 1 | $CH_4$ | $C_2$ | 5165.2 | $A^3\Pi g \rightarrow X^3\Pi u$ | $C_2$ | 4350–6200 | $A^3\Pi g \rightarrow X^3\Pi u$ |
| | | CH | 4314.2 | $A^2\Delta \rightarrow X^2\Pi$ | CH | 4280–4350 | $A^2\Delta \rightarrow X^2\Pi$ |
| | | | 3871.3 | $B^2\Sigma \rightarrow X^2\Pi$ | | 3885–3925 | $B^2\Sigma \rightarrow X^2\Pi$ |
| | | | 3869.0 | | | | |
| 2 | $NH_3$ | H[a] | | | | | |
| | | $N_2$ | 3159.3 | $C^3\Pi \rightarrow B^3\Pi$ | | | |
| | | | 3371 | | | | |
| | | | 3577 | | | | |
| | | | 3804.9 | | | | |
| | | NH | 3360 | $A^3\Pi \rightarrow X^3\Sigma$ | NH | 3350–3400 | $A^3\Pi i \rightarrow X^3\Sigma$ |
| 3 | $H_2O$ | H[b] | | | | | |
| | | OH | 2811.3 | $A^2\Sigma^+ \rightarrow X^2\Pi$ | OH | 3070–3160 | $A^2\Sigma^+ \rightarrow X^2\Pi i$ |
| | | | 2829.0 | | | | |
| | | | 3063.6 | | | | |
| 4 | $CH_4$ + $NH_3$ | CN | 3590.4 | $B^2\Sigma \rightarrow X^2\Sigma$ | CN | 3500–4250 | $B^2\Sigma \rightarrow X^2\Sigma$ |
| | | | 3883.4 | | | 7800–11000 | $A^2\Pi \rightarrow X^2\Sigma$ |
| | | | 4152.4 | | | | |
| | | | 4158.1 | | | | |
| | | | 4167.8 | | | | |
| | | | 4181.0 | | | | |
| | | | 4197.2 | | | | |
| | | | 4216.2 | | | | |
| 5 | $CH_4$ + $H_2O$ | CO | 2833.1 | $B^3\Sigma \rightarrow A^3\Pi$ | | | |
| | | | 2977.4 | | | | |
| | | | 3134.4 | | | | |
| | | | 3305.7 | | | | |

[a] Showed H$\alpha$ ($n=3 \rightarrow 2$) and H$\beta$ ($n=4 \rightarrow 2$) Balmer series.

[b] Contained H$\gamma$ ($n=5 \rightarrow 2$) Balmer series in addition to H$\alpha$ and H$\beta$.

[c] The wave length described here showed only strong absorbance-band. In the experiment, medium and weak absorbances were observed. When two kinds of gas were supplied, we described only new emitters such as CN and CO, respectively.

We can expect that this material played an important role in forming the organic compounds noted in our experiments.

### 3.2.2. *Tetramer of Hydrogen Cyanide*

After discharge for 2 hr, volatile and nonvolatile materials were separated, and dissolved into glass-distilled water. The two samples were analyzed by using column chromatography (0.2N—$NH_4Cl$ buffer, pH 10.0; Dowex 1×8, meq. 200, $Cl^-$; Hitachi-034 liquid chromatography). Thus results obtained are shown in Figure 2. ((a) – nonvolatile and (b) – volatile). The material eluted between 40 to 60 min was collected for both samples measured with a UV-spectrophotometer (Shimadzu UV-200). Both showed $\lambda_{max}$: 296 nm. Both were found to have the same profile. This material was identified as diaminomaleonitrile (DAMN, tetramer of HCN) by means of paper chromatography. It was considered that the tetrameter was formed predominantly from ammonium cyanide as already shown in (3.2.1.). It has been reported that the tetramer has been synthesized from HCN and various kinds of bases such as ammonia (Sanchez *et al.*, 1967, 1968; Wakamatsu *et al.*, 1966, 1969; Yuasa *et al.*, 1970), amines (Yuasa and Ishigami, 1972) and alkaline earth metal oxides (Yuasa *et al.*, 1970; Yuasa and Oró, 1974). As the sample shown in Figure 2(b) contained only volatile materials, it seems that the tetramer was formed from HCN and ammonia or amines which were confirmed in the present experiment. These findings would substantiate the formation of some heterocyclic organic compounds in discharge experiments and especially in ours.

We also investigated the material(s) eluted between 12 to 33 min in Figure 2 (a). This material(s) has $\lambda_{max}$: 250 to 265 nm, which was observed in the initial sample after dissolving in water (pH 8.2 to 8.9). Based upon the specific colourimetric analysis (Pauli-diazo reaction) for an imidazole ring, this material(s) might contain some imidazole derivatives. A decade ago Oró (1963) reported that there was material(s) with $\lambda_{max}$: 260 nm when a spark discharged sample was diluted with 0.1 NHCl and directly examined for an ultraviolet-absorbing compound(s). Although our experiment was not identical to his, there is a possibility that similar compounds might be formed if such an imidazole derivative is directly synthesized in a short duration discharge as the results indicate we can expect nucleic acid bases and their related compounds to be formed under our experimental conditions. Accordingly, the above unknown compound(s) should be identified as soon as possible.

### 3.2.3. *Nucleic Acid Bases*

The sample trapped at a temperature of dry ice-alcohol was dissolved into 5 ml of distilled water and refluxed by using a conventional method for 24 hr at 80 °C. After refluxing the sample was dried in order to eliminate most volatile materials. To this

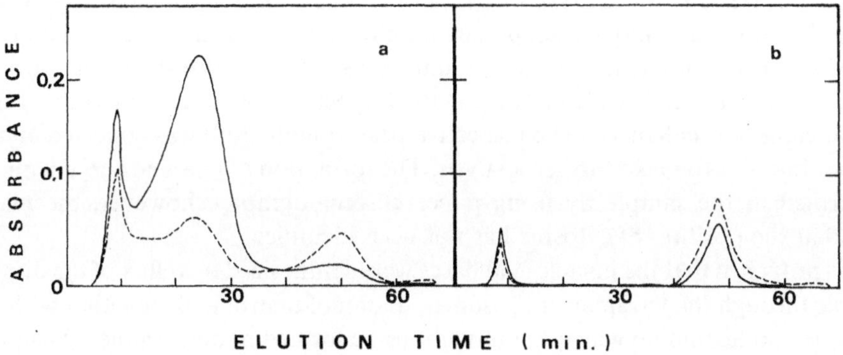


Fig. 2. Column chromatography of nonvolatile and volatile materials. (a) Nonvolatile fraction at an room temperature (25 °C). (b) Volatile fraction ———: 260 nm, ------: 280 nm, respectively.

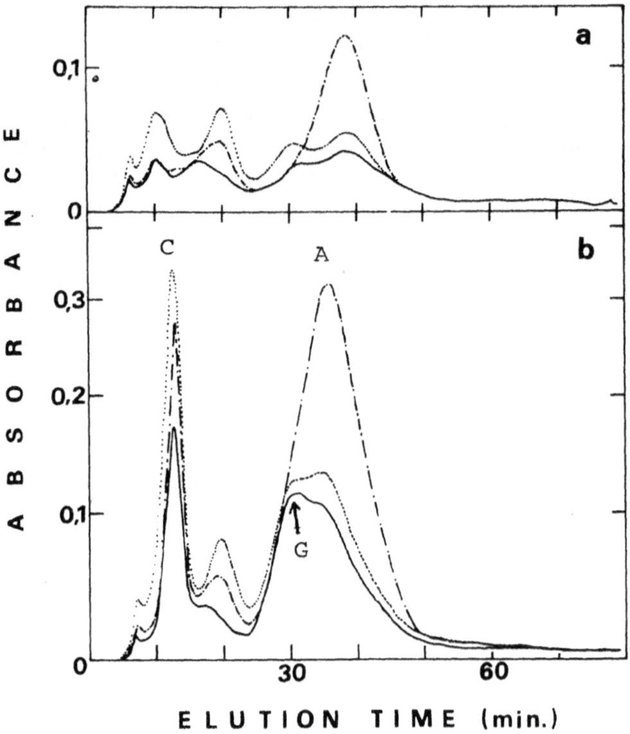


Fig. 3. Analysis of nucleic acid bases. (a) The direct analysis of the sample. (b) To (a) the authentic standard samples 0.1 $\mu$mol of cytosine, guanine and adenine were added and chrlmatographed. -----: 230 nm, —-—-—: 260 nm, ———: 280 nm, respectively.

sample 2 ml of 0.1N HCl was added and an extraction was made at 90 °C for 15 min to obtain nucleic acid bases. The extracted supernatant was collected by centrifuge (15000 g), divided into two fractions then applied to a column (Dowex 1 × 8, $Cl^-$ 0.2N $NH_4Cl$ buffer, pH 10.0).

Figure 3(a) shows the analysis of one of the sample fractions obtained by the above method. Authentic standard samples containing cytosine, guanine and adenine were added to one fraction and analyzed simultaneously. Figure 3(b) shows the chromatogram of this sample. The absorbance at the respective positions of cytosine, guanine and adenine was enhanced. The respective peak (Figure 3(a)) was collected and condensed in order to make further analysis. The formation of guanine and adenine was confirmed in the sample by using paper chromatography, however, the material eluted at the position of cytosine has not been identified.

It is proposed that the nucleic acid bases were synthesized by reflux of the discharge sample through the tetramer or its isomer, diaminofumaronitrile together with formamidine and guanidine which are thought to be formed in our samples. The purines were obtained in yields about 0.1% based on input methane. Since the tetramer of hydrogen cyanide was about 3% yield, the limiting factor synthesizing purines would appear to be a catalyzer such as amidines which might have escaped the reaction system before or during reflux of the sample.

## Acknowledgement

The authors would like to thank Dr K. Imahori, Osaka University, Dr S. Akabori, Protein Research Foundation, Drs J. Oró and D. Flory for their suggestions and discussions to carry out this experiment and present the manuscript.

## References

Arpigny, C.: 1965, *Ann. Rev. Astron. Astrophys.* **3**, 351.

Czuchajowski, L., Francik, H., and Gorecka, W.: 1968, *Ann. Soc. Chim. Polonorum* **42**, 869.

Ishigami, M. and Yuasa, S.: 1974, in preparation.

Oró, J.: 1963, *Nature* **197**, 862.

Ponnamperuma, C. and Woeller, F.: 1967, *Current in Modern Biology* **1**, 156.

Sagan, C. and Khare, B. N.: 1971, *Science* **173**, 417.

Sanchez, R. A., Ferris, J. P., and Orgel, L. E.: 1967, *J. Mol. Biol.* **30**, 223; *ibid.*, **38**, 121 (1968).

Thonson, E. O. and Malter, L.: 1950, *Phys. Rev.* **80**, 58.

Urey, H. C.: 1952, *Proc. Natl. Acad. Sci. U.S.* **38**, 351.

Wakamatsu, H., Yamada, Y., Saito, T., Kumashiro, I., and Takenishi, T.: 1966, *J. Org. Chem.* **31**, 2035; *ibid.*, **42**, 1454 (1969).

Yuasa, S. and Ishigami, M.: 1972, *Tanpakushitsu-Kakusan-Koso*, #200.

Yuasa, S. and Oró, J.: 1974, in J. Oró, S. L. Miller, C. Ponnamperuma, and R. S. Young (eds.), *Cosmochemical Evolution and the Origin of Life*, Vol. II, D. Reidel Publ. Co., Dordrecht-Holland, pp. 295–299.

Yuasa, S., Ishigami, M., Honda, Y., and Imahori, K.: 1970, *Proc. Intl. Symposium Biogeochem. Hydrogeochem.*, Tokyo.

# FORMATION OF PREBIOCHEMICAL COMPOUNDS IN MODELS OF THE PRIMITIVE EARTH'S ATMOSPHERE

## I: $CH_4$—$NH_3$ *and* $CH_4$—$N_2$ *Atmospheres*

G. TOUPANCE, F. RAULIN, and R. BUVET

*Université de Paris – Val de Marne, Laboratoire d'Energétique Biochimique, Creteil 94000, France*

**Abstract.** In order to understand the formation of organic compounds in the primitive atmosphere, the first steps of evolution in models of the primitive atmosphere were investigated. Mixtures containing C—H—N elements were subjected to a low pressure silent electric discharge for several seconds, and the resulting effluents were analysed mainly by gas chromatography, infrared spectrometry and chemical analysis. The formation of hydrocarbons (i.e. ethylene, acetylene, methylacetylene) and of nitrogen containing compounds (i.e. hydrogen cyanide, cyanogen, saturated nitriles, acrylonitrile, cyanoacetylene) is reported. The influence of the initial mixture composition on the amount of compounds formed was systematically studied. The nature of the nitrogen source ($N_2$ or $NH_3$) in the primitive atmosphere has a great influence on the amount and on the very nature of the synthesized products. It is shown that important precursors such as cyanogen and cyanoacetylene are formed only in very rich $N_2$ mediums. These results show the important role played by the nature of the primitive atmosphere in the determination of the chemical evolution pathways.

### 1. Introduction

During the formation of compounds of biological interest under primitive earth conditions, many simple molecules such as hydrogen cyanide, formaldehyde, or cyanoacetylene seem to play important roles. The formation of these molecules under possible primitive conditions has been frequently reported (Miller, 1957; Sanchez *et al.*, 1966). However, no extensive study has yet been performed to determine the range of composition of the primitive atmosphere which allows the formation of the principal key intermediates. We have presented at the Third International Conference on the Origin of Life (Pont à Mousson, 1970) several elements of our study which has the objective of giving a final answer to this problem (Toupance *et al.*, 1971). We now present the results obtained in studies concerning the compounds formed in atmospheres containing the elements C, H, and N. Results obtained from this laboratory (Toupance, 1973) and concerning oxygen-containing atmospheres will be reported elsewhere.

### 2. Experimental

In our experiments, models of primitive atmospheres of known composition enter with a well known flow rate into a reactor in which there is a low pressure (20 torr) corona discharge. The effluents are analysed, and the concentrations of the identified compounds are plotted as a function of the initial atmospheric composition.

The block diagram of our apparatus is given by Raulin and Toupance (1974). Figure 1 gives the diagram of the reactor. The anode is a 10 cm long brass tube,

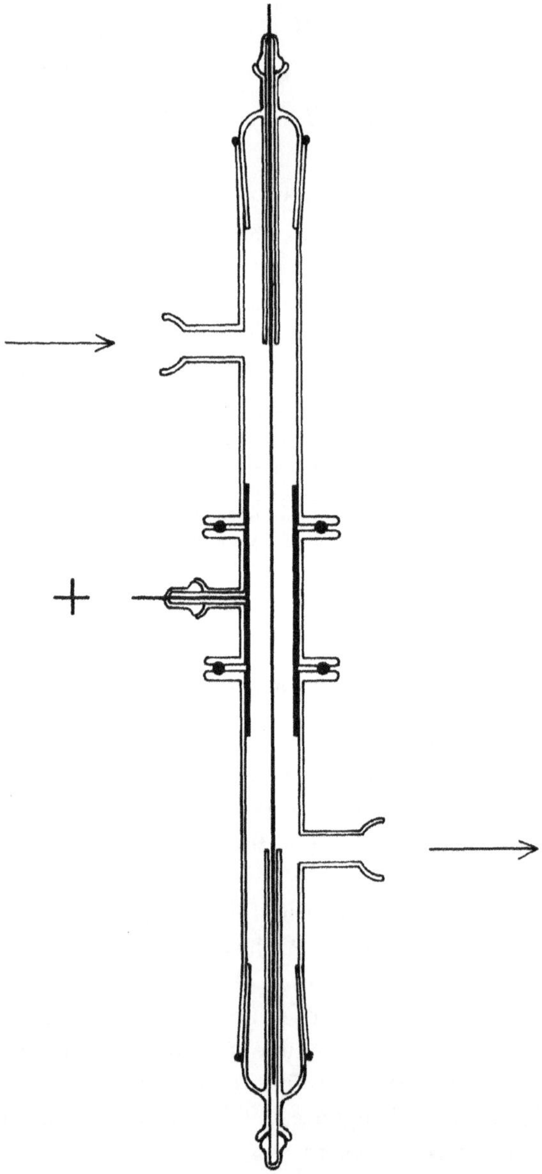


Fig. 1. Discharge tube.

20 mm in diameter. The cathode is a tungsten wire, 1 mm in diameter, positioned in the axis of the cylinder. All the system is enclosed in a glass envelope.

In agreement with previously reported preliminary results (Toupance *et al.*, 1971) all our experiments have been performed under conditions which give a significant synthesis yield simultaneously for many compounds. The pressure in the reactor was maintained at 20 torr in spite of the continuous flow of gases. The residence time of the mixture in the discharge was 3 s and the discharge current was 100 mA. According to the composition of the initial mixtures, the potential drop varied between 450 and 550 V.

Samples of gaseous effluents have been taken directly at the exit of the reactor and submitted to infrared and gas chromatographic analyses without any chemical treatment. Gas chromatographic samples have been recompressed to 1 atmosphere by a very clean pneumatic system in order to increase sensitivity. Gas chromatographic analysis was performed by using a 3 m × 2 mm i.d. stainless steel column packed with Porapak Q, 80–100 mesh (105 °C – He 30 ml/mn – T.C. detector). A detailed description of the experimental system and of the analytical procedure and a discussion of the identifications is reported elsewhere (Toupance, 1973).

An attempt to detect the formation of amines was made by using a 3 m × 2 mm i.d. stainless steel column packed with Chromosorb 103. Because no satisfactory resolution between amines, hydrocarbons, and nitriles was found, the search for amines has not been pursued at this time.

## 3. Results

In a first step, pure methane was subjected to the discharge. Among the compounds formed (Figure 2) hydrogen, ethane, ethylene, acetylene, propane, propene, propyne, and isobutane have been identified. Butane and isobutene have been identified but they have not been resolved from other hydrocarbons containing 4 carbon atoms. It can be seen that the amount of each class of hydrocarbon decreases roughly by

| | % | |
|---|---|---|
| $H_2$ | 41. | |
| $CH_4$ | 46. | |
| $C_2H_6$ | 4.6 | |
| $C_2H_4$ | 2.2 | 10.8 |
| $C_2H_2$ | 4.0 | |
| $C_3H_8$ | 1.2 | |
| $C_3H_6$ | 0.6 | 2.0 |
| $C_3H_4$ | 0.2 | |
| iso-$C_4H_{10}$ | 0.2 | |
| n-$C_4H_{10}$ + t-butene 2 | 0.08 | 0.5 |
| iso-butene+butene 1+butadiene | 0.2 | |

Fig. 2. Molar composition of effluents resulting from subjecting pure $CH_4$ to the discharge.

a factor 5 when the number of carbon atoms in the chain increases by one. Many higher hydrocarbons have been detected, but at very low levels – lower than 0.1% in the total mixture – and they have not been studied. However, the behaviour of these compounds may be predicted, if necessary, from the set of results obtained from the lighter homologous compounds.

Binary mixtures containing $CH_4$—$NH_3$ and $CH_4$—$N_2$ have been made containing various combinations of pure methane and pure ammonia or nitrogen. Figures 3a to 3f give the composition of the effluents as a function of the composition of the initial mixture for the system $CH_4$—$NH_3$.

Figure 3a shows that $CH_4$ and $NH_3$ are similarly decomposed in the discharge. However, the slopes of the curves corresponding to $NH_3$ and $CH_4$ show that ammonia is more stable in low ammonia-containing mixtures than methane in low methane-containing mixtures. A large amount of hydrogen is always formed while nitrogen always appears to be present in relatively low amounts.

Figure 3b shows the synthesis of the two-carbon containing hydrocarbons. It should be noted that there is approximately twice as much $C_2H_2$ formed as $C_2H_4$,

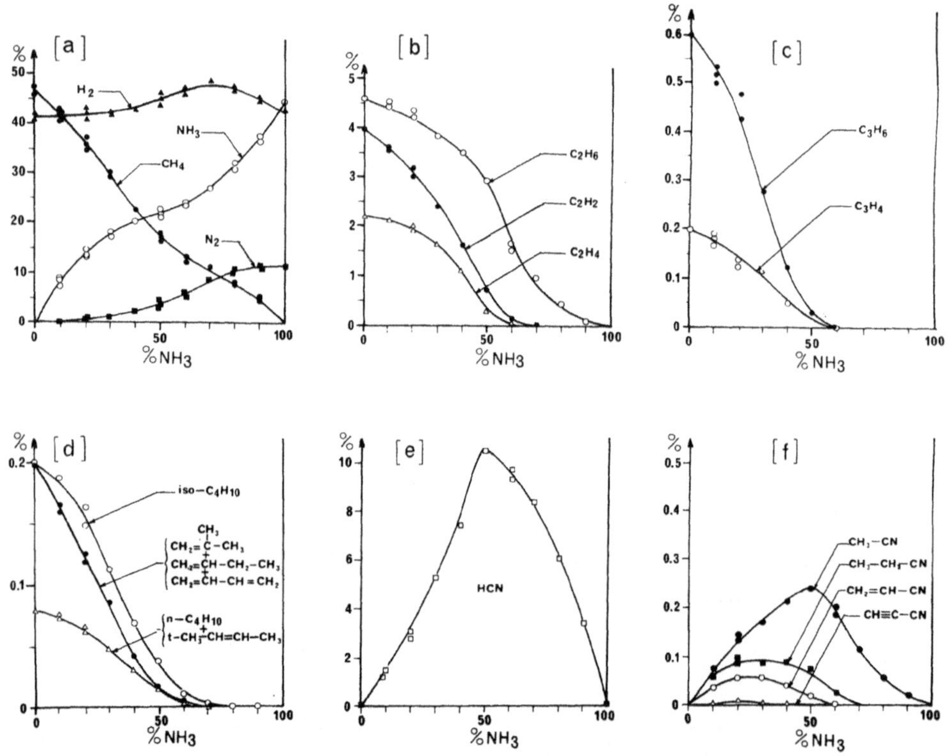


Fig. 3. Molar ratio of several compounds in effluents resulting from subjecting $CH_4$—$NH_3$ mixtures to the discharge, as a function of the $NH_3$ percentage in initial mixtures. 3a, $H_2$, $CH_4$, $NH_3$ and $N_2$; 3b, $C_2$ hydrocarbons; 3c, $C_3$ hydrocarbons; 3d, $C_4$ hydrocarbons; 3e, HCN; 3f, several nitriles.

and that the amount of these unsaturated compounds decreases rapidly with increasing amount of ammonia in the initial mixture. Their concentration becomes very low when the $NH_3$ percentage is higher than 70%. The same behaviour is observed in Figure 3c for the three-carbon containing hydrocarbons. Propane has not been represented because of its interference with HCN on gas chromatograms. In Figure 3d a decrease in the amount of the unsaturated four-carbon containing hydrocarbons can also be observed; the decrease is faster than for the saturated hydrocarbons. Two of the curves represent the area of gas chromatographic peaks corresponding to several unresolved compounds.

Figure 3e represents the synthesis of HCN in $CH_4$—$NH_3$ mixtures. The maximum amount obtained under our standard experimental conditions is about 10% of the effluent mixture. By increasing the discharge current and the residence time, a higher amount of HCN can be obtained, reaching 25%. The yield of HCN, calculated from $CH_4$, is 30% in an equimolecular $CH_4$—$NH_3$ mixture and under our standard conditions.

Figure 3f represents the synthesis of several nitriles. The maximum concentration is about 0.2% of the effluent gases. For acetonitrile, this maximum is reached with an equimolecular mixture, and for the other compounds with mixtures containing 20 to 30% ammonia. Cyanoacetylene has been detected only with a 20% $NH_3$ mixture and at a very low concentration.

The shape of the curves suggests a mechanism of synthesis in which CN radicals react with a previously formed hydrocarbon. Such an hypothesis has been already proposed for HCN synthesis from $CH_4$—$NH_3$ mixtures submitted to the action of radiofrequency waves (Topouzkhanian *et al.*, 1967).

The results obtained from $CH_4$—$N_2$ mixtures are represented on Figures 4a to 4g. As shown on Figure 4a, important differences between $CH_4$—$NH_3$ and $CH_4$—$N_2$ mixtures appear for initial methane-poor mixtures. The amount of hydrogen is here continuously decreasing while it is roughly constant, around 45%, in $CH_4+NH_3$ mixtures. The amount of methane in effluent mixtures decreases at the beginning approximately as it does in $CH_4$—$NH_3$ mixtures, but for initial mixtures, poor in methane, the slopes of the curves are quite different. The amount of remaining methane is here much lower, being a few percent in the range of 70–90% $N_2$, than it is in homologous $CH_4$—$NH_3$ mixtures. It should be noted that ammonia is always formed in low proportions, not more than 4%, even when both nitrogen and hydrogen are present in effluents in significant concentrations.

Figures 4b to 4d show the results for several hydrocarbons products. It can be seen that unsaturated hydrocarbons are synthesized here in noticeable amounts even from initial mixtures poor in methane. In homologous $CH_4$—$NH_3$ mixtures these are not detected and they appear only when the initial amount of methane is higher than 40%. This behaviour may be related to the amount of free hydrogen present in the effluent mixtures.

Figure 4e shows that the maximum amount of hydrogen cyanide formed is around 10%, but this maximum is moved from 50% in the case of a $CH_4$—$NH_3$ mixture

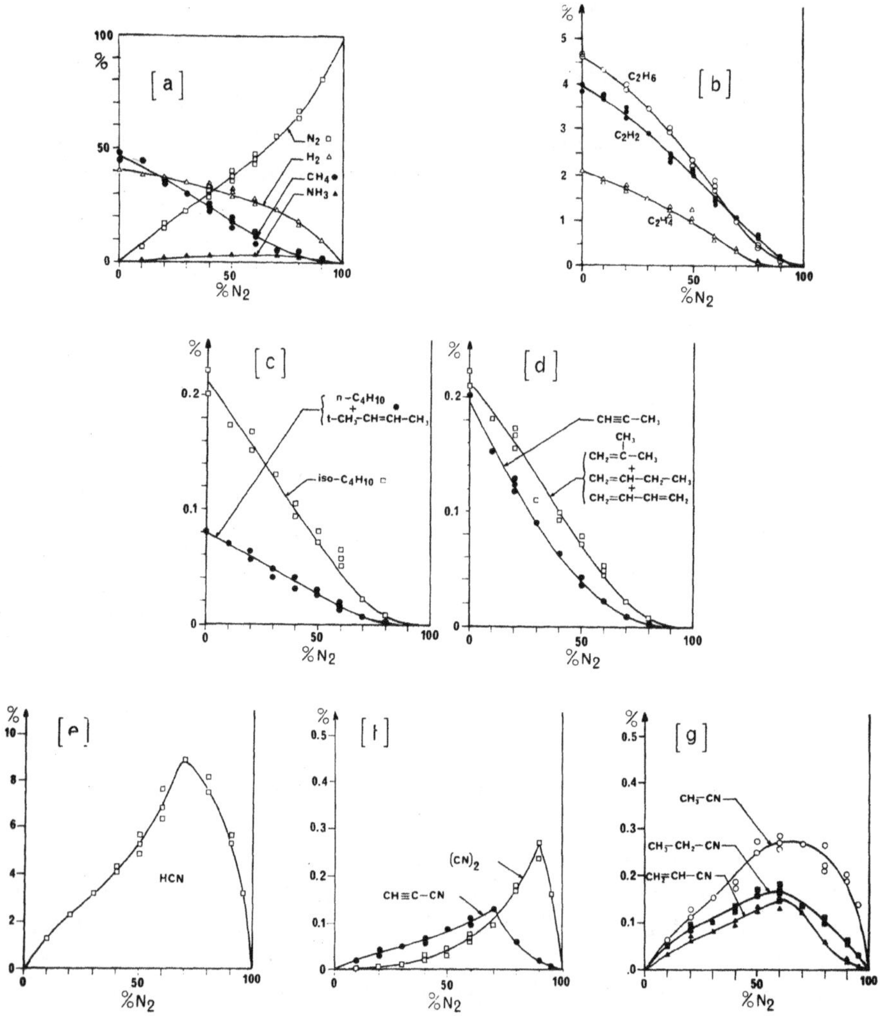


Fig. 4. Molar ratio of several compounds in effluents resulting from subjecting $CH_4$—$N_2$ mixtures to the discharge, as a function of the $N_2$ percentage in initial mixtures. 4a, $N_2$, $CH_4$, $H_2$ and $NH_3$; 4b, $C_2$ hydrocarbons; 4c and 4d, several $C_3$ and $C_4$ hydrocarbons; 4e, HCN; 4f, cyanogen and cyanoacetylene; 4g, several other nitriles.

to 70% in the present case of a $CH_4$—$N_2$ mixture. This result has been confirmed by many control experiments.

In Figures 4f and 4g, the formation of several nitrogen-containing compounds is presented. The synthesis of cyanogen and of noticeable amounts of cyanoacetylene should be noted. Contrary to homologous results obtained from $CH_4$—$NH_3$ atmospheres (Figure 3f) the amount of all these compounds is higher in atmospheres containing large amounts of nitrogen. The maximum amount of formation of cyanogen and cyanoacetylene is reached respectively for 90 and 70% $N_2$. For other nitriles, the maximum synthesis is reached at about 60% $N_2$.

The synthesis of nitrogen-containing compounds from $CH_4$—$N_2$ atmospheres may occur through complex chain reaction involving atomic nitrogen formed in the discharge (Blanton *et al.*, 1966; Safrany and Jaster, 1968). However, we have not yet found a satisfactory explanation for the synthesis of all the compounds.

In order to understand the action of hydrogen during these syntheses, hydrogen has been added to an initial $CH_4$—$N_2$ mixture of fixed composition. It has been found that the amount of saturated compounds decreases more slowly than could be predicted from the diluting action of hydrogen. In contrast, the amount of cyanogen and unsaturated compounds is lowered.

As previously indicated, amines have not been systematically analysed. However, if these compounds are synthetized under our conditions, several results show that they are never formed at a level higher than 0.01%.

## 4. Discussion

Ammonia has been the most regular source of nitrogen used in prebiotic experiments and many interesting results have been obtained. Several experiments also have been carried out using $N_2$. In order to complete these results and to unravel the prebiological pathways, many authors have suggested that several precursors may have played a determining role. Among them, HCN has been most frequently invoked but cyanoacetylene or cyanogen may act as key intermediates in the synthesis of pyrimidines, purines, and in phosphorylation reactions (Miller, 1957; Lohrmann and Orgel, 1968; Degani and Halmann, 1971).

New data which have been reported here show that pathways involving cyanoacetylene or cyanogen are most improbable in $CH_4$—$NH_3$ containing atmospheres. On the contrary, these two compounds may be considered in atmospheres containing a large amount of free nitrogen and a low amount of hydrogen with respect to the amount of methane. Fortunately, this condition corresponds to the present ideas about the nitrogen state in the primitive atmosphere (Bada and Miller, 1968).

The admixtures free of water vapor we have considered in this report are appropriate models of the bulk of the primitive atmosphere since the water percentage decreases rapidly with altitude. On the contrary, in the lower atmosphere, the presence of noticeable amounts of water should be taken into account.

Studies are now in progress in this laboratory in order to establish the role of water vapor in $CH_4$—$NH_3$—$H_2O$ and $CH_4$—$N_2$—$H_2O$ atmospheres. However, important deviations from results reported in this paper are not expected because additional results obtained respectively from $CH_4$—$H_2O$, $CO_2$—$H_2O$, $CO_2$—$N_2$—$H_2O$ and $CO_2$—$N_2$—$H_2$—$H_2O$ mixtures (Toupance, 1973) have shown that water vapor is a very stable compound which does not introduce any important modification in the evolution of mixtures under investigation.

## References

Bada, J. L. and Miller, S. L.: 1968, *Science* **159**, 423.
Blanton, J. R., Grunwald, J., and Gayer, K. H.: 1966, *Rec. Chem. Progr.* **27**, 269.
Degani, C. and Halmann, M.: 1971, in R. Buvet and C. Ponnamperuma (eds.), *Molecular Evolution*, Vol. 1, North Holland, Amsterdam, p. 225.
Lohrmann, R. and Orgel, L. E.: 1968, *Science* **161**, 64.
Miller, S. L.: 1957, *Biochim. Biophys. Acta* **23**, 480.
Raulin, F. and Toupance, G.: 1974, this volume, pp. 91–97.
Safrany, D. R. and Jaster, W.: 1968, *J. Phys. Chem.* **72**, 518.
Sanchez, R. A., Ferris, J. P., and Orgel, L. E.: 1966, *Science* **154**, 784.
Topouzkhanian, A., D'Incan, J., and Janin, J.: 1967, *J. Chim. Phys.* **64**, 506.
Toupance, G.: 1973, Thèse de Doctorat d'Etat – Université Paris.
Toupance, G., Raulin, F., and Buvet, R.: 1971, in R. Buvet and C. Ponnamperuma (eds.), *Molecular Evolution*, Vol. 1, North Holland, Amsterdam, p. 83.

# FORMATION OF PREBIOCHEMICAL COMPOUNDS IN MODELS OF THE PRIMITIVE EARTH'S ATMOSPHERE

## II: $CH_4$—$H_2S$ *Atmospheres*

F. RAULIN and G. TOUPANCE

*Université de Paris – Val de Marne, Laboratoire d'Energétique Biochimique, Creteil 94000, France*

**Abstract.** In order to understand the role of sulfur in the primitive atmosphere, we have studied the action of a silent discharge on mixtures of $CH_4$ and $H_2S$ at low pressure. The nature of the products formed in the gaseous phase, and the influence of several parameters, especially the $H_2S$ percentage, on the yield of the products are reported.

The analysis of the products is carried out by gas liquid chromatography and infrared spectrometry. The formation of sulfur-containing compounds, such as thiols and sulfides, is reported. $CS_2$ is formed in high yield (a few percent) in mixtures containing 40–50% of $H_2S$, while the maximum concentration of thiols (i.e., $CH_3SH$ and $C_2H_5SH$) is reached with lower percentages of $H_2S$. The formation of hydrocarbons decreases rapidly with increasing proportions of $H_2S$. These results show the important inhibitor effect of $H_2S$ on the formation of hydrocarbons and the possibility of occurrence of many sulfur compounds in prebiological evolution.

## 1. Introduction

In order to determine the range of elementary atomic composition for primitive and extraterrestrial atmospheres which permitted the building up of biologically interesting compounds, exhaustive investigation of the primary transformation processes of atmospheres containing C, H, O, and N has already been reported from this laboratory (Toupance, 1973; Toupance *et al.*, 1971 and 1974).

Because of the importance of sulfur in present biochemistry, it was necessary to extend this study to sulfur-containing primitive atmospheres for which only a few results have been published (Lu *et al.*, 1959; Choughuley and Lemmon, 1966; Sagan and Khare, 1971; Van Trump and Miller, 1972). As a first step of this extension, we have studied the action of a silent electric discharge on mixtures of $CH_4$ and $H_2S$ of various compositions and at low pressure.

## 2. Experimental

The block diagram of our apparatus is represented in Figure 1. Most of its parts, particularly the atmosphere generator (A) and the discharge tube (B) have been described elsewhere (Toupance *et al.*, 1971). The atmosphere generator has been slightly modified in order to provide $H_2S$-containing atmospheres of known compositions and flow rates. The gaseous mixture passes continuously through the discharge tube (B) which consists of a cylindrical open metal tube (20 mm I.D. and 10 cm long) acting as the anode and coaxial tungsten wire (1 mm diam) as the cathode. The silent discharge occurs in our mixtures with a voltage between 450 and 800 V and under a pressure of 20 torr.

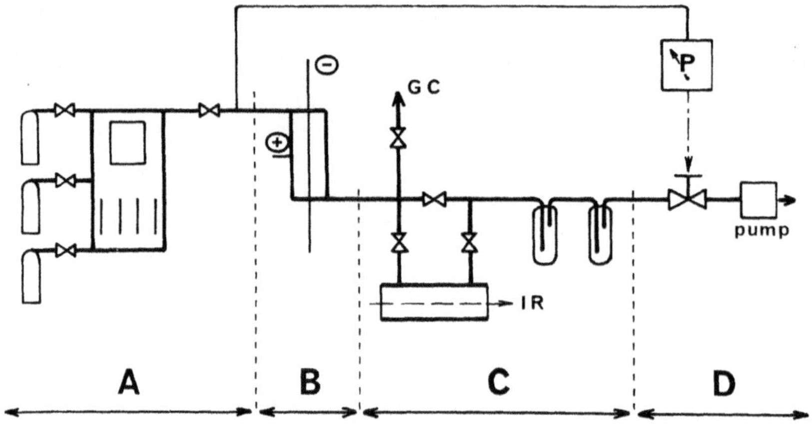


Fig. 1. Block diagram of the apparatus. (A) Atmosphere generator; (B) Discharge tube; (C) Analytical device; (D) Pumping device and pressure regulator.

The gas chromatographic analysis of the effluent gases is accomplished on two different columns. The sulfur-containing compounds and many hydrocarbons are detected and quantitatively analysed on a 2 m long × 2 mm i.d. teflon column packed with Durapak-O.P.N/porasil C (isothermal for 6 min at 25 °C; 3° min$^{-1}$ up to 115 °C) using a flame ionisation detector.

An exhaustive calibration of this column for the thiols and sulfides containing 1 to 3 atoms of carbons and for several additional $C_4$-sulfur-containing compounds has been recently accomplished in this laboratory and is reported elsewhere (Raulin and Toupance, 1974). Nevertheless, $CS_2$, $H_2$, $CH_4$, $C_2H_4$ and $C_2H_6$ are not resolved under the conditions which provide a good separation of the other products. $C_2H_6$, $H_2S$, and $H_2$ are quantitatively analysed by gas-chromatography on a 10 ft long × × 2 mm i.d. stainless-steel column packed with Porapak Q 80–100 mesh (isothermal 6 min at 25 °C: 6° min$^{-1}$ up to 150 °C), using a thermal conductivity detector.

For $CH_4$, $C_2H_2$, $C_2H_4$, and $CS_2$ vapor, quantitative analysis is carried out by infrared spectrometry.

## 3. Results and Discussion

In all experiments, the pressure has been fixed at 20 torr. Preliminary studies have fixed the two parameters of discharge current and residence time of the $CH_4$–$H_2S$ mixture at the value corresponding to the maximum yield of sulfur-containing compounds. By treating fixed composition atmospheres, under a 100 mA discharge current, the maximum amount of formation of compounds such as $CS_2$, $C_2H_4$, and $C_2H_6$ has been obtained for a residence time of 3 s. Under this optimum condition, but under variable discharge currents, the maximum amount of $CS_2$ increases rapidly with current and becomes quite constant for currents greater than 100 mA. All our further experiments have been carried out under a residence time of 3 s and a discharge current of 100 mA. It should be noticed that these values are similar to the values adopted for the study of atmospheres containing C, H, N.

On the basis of this initial study, the nature of the products formed in the gaseous phase, and the influence of the $H_2S$ percentage on the yield of the synthesis have been investigated. First, gas chromatographic identification of the $C_1$–$C_4$ hydrocarbons has been accomplished by injecting simultaneously in the gas chromatograph a mixture of known hydrocarbons and the sample obtained when only $CH_4$ is introduced into the reactor. There is no ambiguity in such an identification, because the expected products are known to be only hydrocarbons, and the behaviour of the column to the homologous series of hydrocarbons had been established. When $H_2S$ is added to $CH_4$, all the new peaks which are evidenced on the gas chromatograms should correspond to sulfur-containing compounds. The identification of these peaks has been achieved by injecting simultaneously a standard mixture of sulfur-containing compounds and the sample to be analysed. This identification is certified by taking into account the behaviour of the Durapak-OPN column to the sulfur-containing homologous series, thiols, and sulfides.

Figure 2 shows a typical gas chromatogram corresponding to a sample obtained by electric discharge in a mixture containing 90% $CH_4$ and 10% $H_2S$. With the

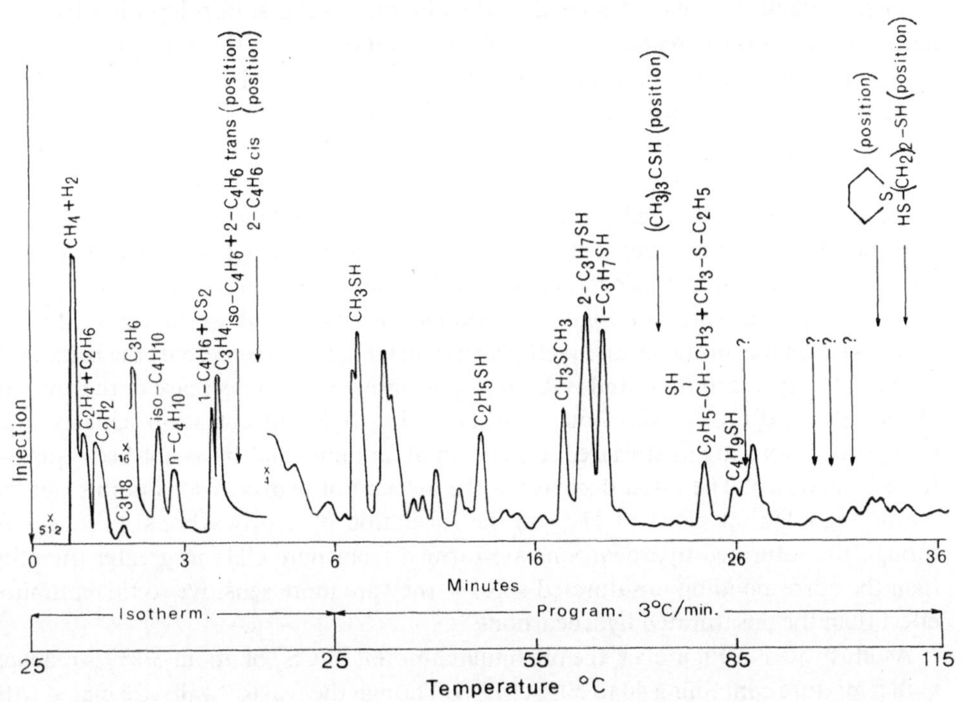


Fig. 2. Gas chromatogram of a sample obtained by the transformation of a 10% $H_2S$ – 90% $CH_4$ mixture. 2 m long × 2 mm i.d. teflon column, packed with Durapak-OPN/Porasil C. Programmed temperature: isothermal for 6 min at 25 °C, 3° min$^{-1}$ up to 115 °C. Injector: 180 °C; detector: F.I.D. 190 °C, carrier gas: $N_2$ 30 ml min$^{-1}$. Sensitivity: $5 \times 10^{-10}$ afs at attenuation × 1.

exception of $CS_2$, there is no interference between the sulfur-containing compounds and the hydrocarbons with less than 5 carbons. Some of the hydrocarbons with more than 4 carbons are not separated from the sulfur-containing products. However, lack of separation is not troublesome because these hydrocarbons are generally in very small amounts (less than $10^{-2}\%$) in our experiments.

There is no doubt of the identification of several volatile thiols; these are methanethiol, ethanethiol, 1-propanethiol, 2-propanethiol, and 1-butanethiol. The identification of the dimethyl sulfide is also secure. The retention times of these sulfur-containing compounds are located in the region where exhaustive gas chromatographic calibrations have been carried out. All these products are formed in the range of $5 \times 10^{-2}$ to $10^{-2}\%$. With our gas chromatographic conditions, 2-butanethiol and methyl ethyl sulfide are coeluted, so we do not know the individual contribution of each of them in the corresponding peak on the chromatogram. Some of other peaks, which correspond to sulfur-containing products with higher boiling points have not been identified with certainty. They are formed in much lower quantity ($<10^{-3}\%$).

The evolution of the formation of several interesting compounds has been studied with respect to the initial composition of the mixture. Figure 3a shows that most of the initial $H_2S$ is destroyed and that simultaneously a large amount of hydrogen is formed. This result must be related to the abundant solid sulfur-deposit which has been observed on the reactor walls. This observation shows the great instability of $H_2S$ in our discharge conditions. The behaviour of $CH_4$ in $H_2S$—$CH_4$ mixtures is roughly similar to its behaviour in $CH_4$—$NH_3$ mixtures (Toupance *et al.*, 1974). Figure 3b represents the variations of the $C_2$-hydrocarbons synthesis with the initial $H_2S$ percentage. The concentration of the three hydrocarbons is in the range of a few percent and decreases rapidly with increasing $H_2S$ percentages. We can notice that the amount of ethane decreases faster than the amount of ethylene and acetylene. In fact, $C_2H_2$ is the only hydrocarbon synthesized in significant amounts in $H_2S$ rich mixtures. As illustrated in Figure 3c, similar results are obtained for $C_3$-hydrocarbons (propane, propene, and methylacetylene). Figure 3d represents the behaviour of methylacetylene, butane and isobutane, and shows the fast decrease of the amount of the saturated $C_4$-hydrocarbons with increasing $H_2S$ concentration. Unsaturated $C_4$-hydrocarbons are synthesized in very small amounts and have not been quantitatively analysed. The rapid decrease of the amount of hydrocarbons points out the important inhibitor effect of $H_2S$ on the formation of hydrocarbons. In addition, though the saturated hydrocarbons are formed from pure $CH_4$ in greater quantity than the corresponding unsaturated species, they are more sensitive to this inhibitor effect than the unsaturated hydrocarbons.

As illustrated in Figure 4a, the maximum amount of $CS_2$ of about 3.2% is reached with a mixture containing 40 to 50% of $H_2S$. Though the reactor walls are plated with a carbon and sulfur deposit, no $CS_2$ is formed when only $CH_4$ is present in the reactor. On the contrary, a significant amount (1%) of $CS_2$ is obtained when only $H_2S$ is used. This fact suggests that a fraction of $CS_2$ may be formed by direct reaction between carbon and $H_2S$.

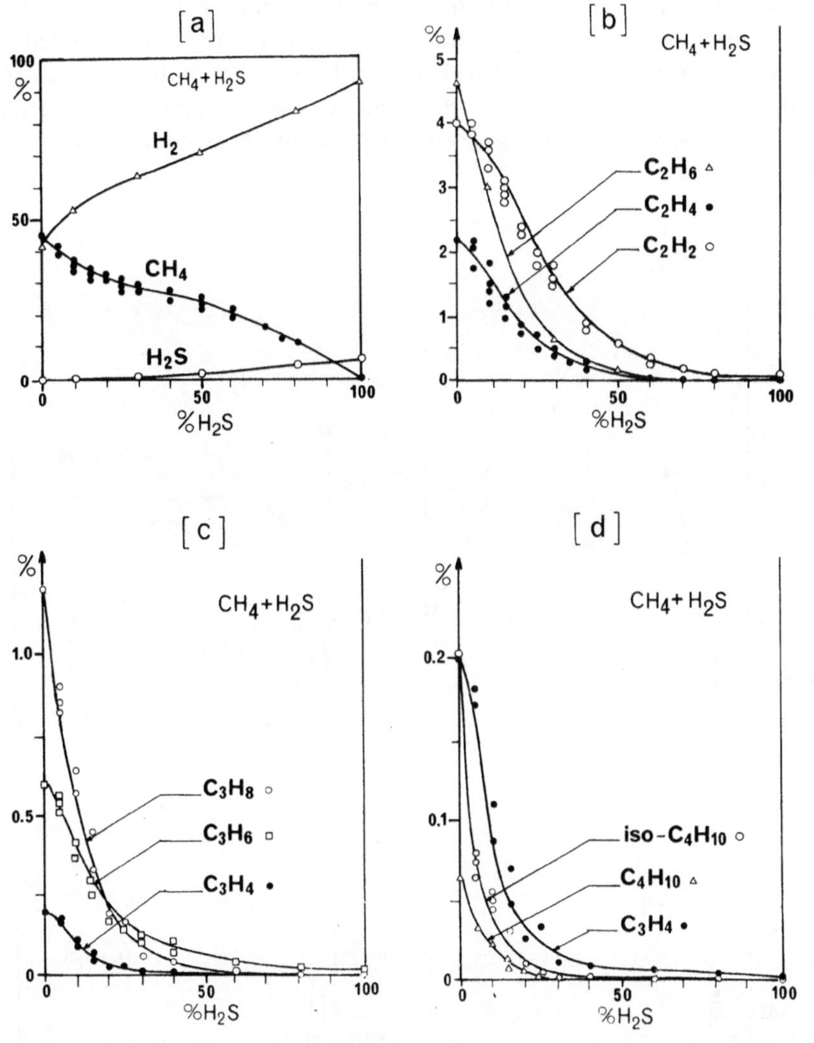


Fig. 3. Ratio of several compounds in effluents as a function of the $H_2S$ percentage in the initial $H_2S$—$CH_4$ mixture. (A) $H_2$, $CH_4$ and $H_2S$. (B) $C_2$-hydrocarbons. (C) Propane, propene and methylacetylene. (D) butane, isobutane and methylacetylene.

In Figure 4b it appears that the maximum amount of methanethiol formed (about 0.04%) corresponds to a mixture containing 30 to 40% $H_2S$. The curve corresponding to $C_2H_5SH$ presents a maximum of 0.014% for an initial mixture containing about 5% $H_2S$. The percentage of ethanethiol is much smaller than the percentage of methanethiol, and it is practically zero for mixtures containing more than 40% $H_2S$. This strong difference suggests a possible mechanism of formation for these sulfur-containing compounds, involving a reaction between $H_2S$, or sulfur-containing radicals such as HS˙ and the preformed hydrocarbon skeletons. Because $CH_4$ is always present in great quantity in the reactor, methanethiol can be formed in significant amounts from mixtures relatively rich in $H_2S$. On the contrary, the decrease of the concentration of other hydrocarbons moves the maximum of formation of the corresponding sulfur-containing products to lower $H_2S$ concentration around 5 to 10%.

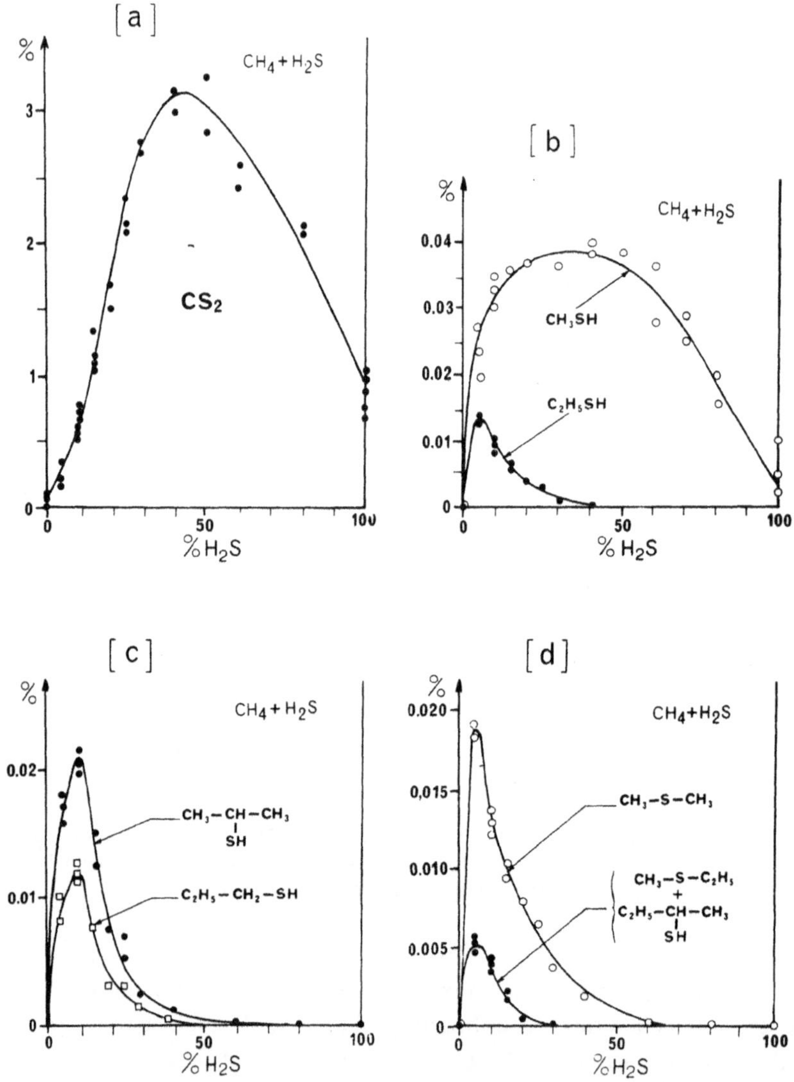


Fig. 4. Ratio of sulfur-containing compounds in gaseous effluents as a function of the $H_2S$ percentage in the initial $H_2S$—$CH_4$ mixture. (A) carbon disulfide. Residual $CS_2$ is obtained from pure $H_2S$ because of carbon deposits in the reactor due to previous experiments. (B) methanethiol and ethanethiol. (C) 1-propanethiol and 2-propanethiol. (D) dimethyl sulfide and (methyl ethyl sulfide + 2-butanethiol).

In Figures 4c and 4d, similar results are shown for the other thiol and sulfide products.

## 4. Conclusion

Though all conditions of excitation have not been investigated, $H_2S$ seems quite unstable in models of primeval environments. According to this result, it may be supposed that the primitive atmosphere was always poor in $H_2S$. Nevertheless, a noticeable amount of sulfur-containing compounds may have been synthesized because of the low $H_2S$ concentrations needed for their synthesis. Studies are in progress in this laboratory in order to simulate the prebiological role of these sulfur containing products.

Prebiological pathways, involving precursors such as HCN, HCHO, and $CH{\equiv}C{-}CN$ have already been proposed to explain the evolution of C, H, N, and O containing atmospheres (Miller, 1957; Sanchez *et al.*, 1967). The intervention of the same pathways in $H_2S$-containing atmospheres is questionable. In order to solve this problem, we are going to study the formation of the most important precursors with respect to the proportion of $H_2S$ in different models of the primitive atmosphere.

## References

Choughuley, A. S. U. and Lemmon, R. M.: 1966, *Nature* **210**, 628.

Lu, H. K., Chou, T. K., Wang, C. W., Yang, C. K., Ku, T. J., and Li, W. L.: 1959, *K'O Hsu ëh Chi Lu* **3**, 150.

Miller, S. L.: 1957, *Biochim. Biophys. Acta* **23**, 480.

Raulin, F. and Toupance, G.: 1974, *J. Chromatogr.*, in press.

Sagan, C. and Khare, B. N.: 1971, *Science* **173**, 4–17.

Sanchez, R. A., Ferris, J. P., and Orgel, L. E.: 1967, *J. Mol. Biol.* **30**, 223.

Toupance, G.: 1973, Doctoral Thesis, Paris.

Toupance, G., Raulin, F., and Buvet, R.: 1971, in R. Buvet and C. Ponnamperuma (eds.), *Molecular Evolution*, Vol. 1, North Holland, Amsterdam, p. 83.

Toupance, G., Raulin, F., and Buvet, R.: 1974, this volume, pp. 83–90.

Van Trump, J. E. and Miller, S. L.: 1972, *Science* **178**, 859.

# ORGANIC SYNTHESIS BY QUENCH REACTIONS

W. K. PARK*, A. R. HOCHSTIM

*Research Institute for Engineering Sciences, Wayne State University, Detroit, Mich. 48202, U.S.A.*

and

C. PONNAMPERUMA

*Laboratory of Chemical Evolution, Dept. of Chemistry, University of Maryland, College Park, Md. 20742, U.S.A.*

**Abstract.** The effects of chemical quench reactions on the formation of organic compounds at a water surface under simulated primordial earth conditions were investigated for the study of chemical evolution.

A mixture of gaseous methane and ammonia over a water surface was exposed to an arc discharge between an electrode and the water surface. This discharge served as a source of dissociated, ionized and excited atomic and molecular species.

Various organic molecules were formed in the gaseous, aqueous, and solid states by a subsequent quenching of these reactive species on the water surface. The effects of these water-surface quench reactions were assessed by comparing the amounts of synthesized molecules to the amounts which formed during the discharge of an arc above the water level.

The results showed that: (1) the water-surface quench reaction permitted faster rates of formation of an insoluble solid and (2) the quench discharge yielded twice as much amino acids and 17 times more insoluble solids by weight than the other discharge. The highest yield of amino acids with the quench reaction was $9 \times 10^7$ molecules per erg of input energy.

These observations indicate that quench reactions on the oceans, rain, and clouds that would have followed excitation by lightning and shock waves may have played an important role in the prebiotic milieu. Furthermore, the possibility exists that quench reactions can be exploited for the synthesis of organic compounds on a larger scale from simple starting materials.

## 1. Introduction

In an effort to study the chemical events and reaction systems of the primordial Earth, many investigators have performed experiments on the formation of amino acids under simulated conditions (Park, 1973; Gabel and Ponnamperuma, 1971). Various primary and secondary sources of natural energies were applied to the presumed primeval, atmospheric gas-mixture at several stages of its development resulting in the production of biologically relevant molecules. However, it is only recently that attention has been given to the roles of chemical and physical parameters in the reaction processes. Specifically, quenching has been recognized as a selection mechanism in the recombination process. Faster rates of formation, larger yields, and larger molecules were predicted for quench conditions (Hochstim, 1963; Park *et al.*, 1973), which would indicate that quenching may have been an important process in the abiotic synthesis of organic molecules.

* Present address: Laboratory of Chemical Evolution, Department of Chemistry, University of Maryland, College Park, Md. 20742, U.S.A.

Gaseous quench processes can be classified by the physical event which causes the temperature to drop: (1) gas expansion, (2) gas mixing, (3) contact with a cool solid surface, and (4) contact with a cool liquid surface. Among these, quenches on surfaces may have been the most significant in the natural processes of chemical evolution because the molecules produced near or at the surfaces were protected from further reaction and decomposition from re-exposure to the energetic sources.

In the design of an investigation of quenches on surfaces, the liquid-surface quench (water-surface quench in this work) was thought to be more germane, because of the possibility of rapid mass transfer from the surface to the inner layers of the water, subsequent shielding, and reactions in the water phase.

The quench discharge (wet discharge) was carried out (Park, 1973) by forming a vertical arc column between an electrode and the water surface; the non-quench discharge (dry discharge) was done by forming an arc column along the line of the electrodes in the gas phase. In the wet discharge, quenching is achieved due to the contact of the hot arc column with the cold water surface. This causes the temperature of the heated gaseous species to decrease rapidly on the water surface.

The experiments were performed at NASA, Ames Research Center in Moffett Field, California and at the University of Maryland at College Park, Maryland.

## 2. Experimental

A 2 liter pyrex flask, cleaned with dichromic acid, was fitted with a water reservoir and electrodes (Figure 1). For each run the whole system was evacuated and maintained at a pressure of 5 $\mu$ for 12 hr for further out-gassing. Matheson research grade $CH_4$ was used to flush the flask and its connecting parts 6 times (each time evacuating to a pressure of 30 $\mu$). The reaction vessel was then filled with research grade $NH_3$ and $CH_4$ (300 mm Hg each). One hundred ml of distilled water was bubbled with filtered $N_2$ in the water reservoir for 20 min to remove dissolved $O_2$, and was then transferred to the vessel by a pressure differential caused by opening the valve to the water reservoir, which was at atmospheric pressure, just long enough to bring in the water.

In the wet discharge a tungsten electrode was located 1 cm above the water which itself served as a second electrode. A gold wire place within the water was used to complete the circuit. In the dry discharge two tungsten electrodes with a 1 cm gap were placed 9 cm above the water level.

As soon as the water was introduced into the reactor, the ammonia began dissolving with concomitant reduction in the pressure of the system. Twelve minutes after the water had been added, approximate phase equilibrium for ammonia was reached resulting in a total pressure of 357 mm Hg. This process was expedited by agitation of the water with a magnetic stirrer.

Two power supplies, transformer and Tesla coil, were used for the discharges. Their characteristics were determined as follows:

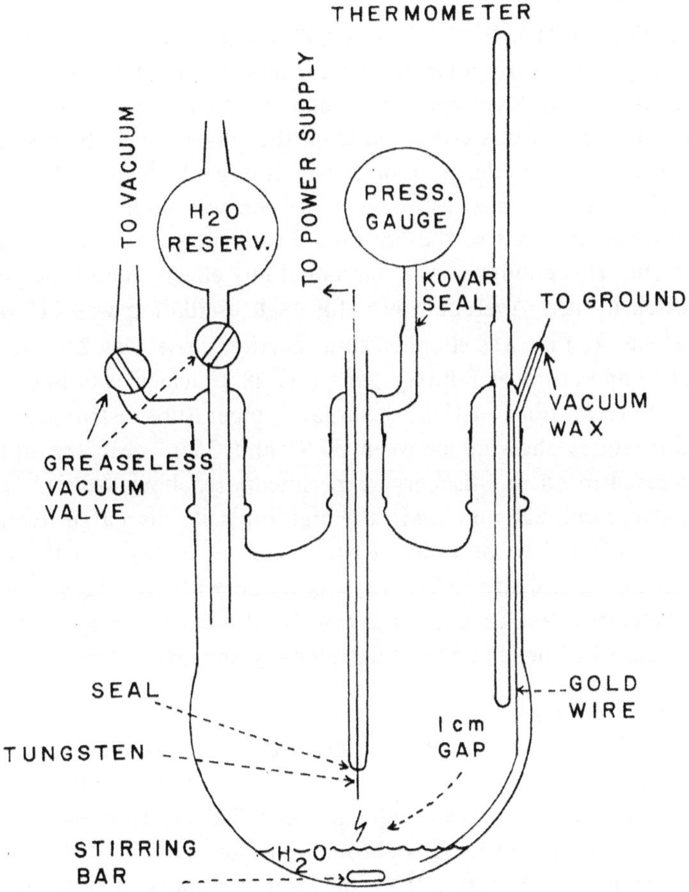


Fig. 1. Wet discharge vessel.

## 2.1. Arc discharges and energy measurements

### 2.1.1. *Low Intensity Arc*

A high voltage transformer (Peschel Instruments Model No. T20-0.6AR), whose one secondary terminal was grounded, was controlled by an autotransformer on the primary side to supply the necessary low intensity power. In order to limit the current and, at the same time, to measure voltage drops across known resistances, resistors were lined up in series. The gold wire in the wet discharge vessel was grounded to draw the necessary current. Voltage measurements of the secondary terminals were made with a Tektronix (1000 × attenuation) high voltage probe connected to a Tektronix 531A oscilloscope. The power (5.1 W) and the total energy ($2.2 \times 10^5$ J) for the reaction were calculated from the measured rms voltage (3 kV) and rms current (1.7 mA) and discharge time (12 hr).

### 2.1.2. *High Intensity Arc*

A Tesla coil (Sargent-Welch Cat. No. S30978) was used as a source for the high intensity arc. It produced irregular decaying sinusoidal oscillations with random time intervals. The average peak voltage for the load was 8 kV with a frequency of 250 kHz. The rms voltage (2.3 kV) was estimated from the ringing wave by assuming a sinusoidal wave for each half cycle. By monitoring voltage drops across known resistors, rms current (50 mA) was also estimated. For the wet discharge vessel, a piece of aluminum foil was attached to the bottom of the reactor to draw a high frequency transient current. This configuration eliminated any effects due to the gold wire used in the low intensity arc. Average power for each oscillation was 115 watts and the effective total energy that was supplied to a reaction vessel was $2.9 \times 10^5$ J.

The initial temperature of both systems was 28 °C which was the ambient room temperature. After an hour of the discharges, temperature distributions in the flask reached steady states showing between 30 °C and 35 °C near the flask wall. The pressure increased in all the discharge experiments reaching 490 mm Hg in the low intensity discharge and 820 mm Hg in the high intensity discharge. Even though the spark gap was adjusted to be about 1 cm, a carbon deposit on the electrode tips decreased the gap, especially in the dry discharge with the transformer. Frequent tapping on the electrodes prevented the growth of a carbon deposit. In general, the discharge pressure was dependent on arc intensity and spark gap.

### 2.2. Chemical Analysis

The resultant solution (97 ml), after 12 hr discharge, was filtered and reduced to 11 ml by partial evaporation on a rotary evaporator at 50 °C. Of this solution 6 ml was removed for analysis by the computer-controlled Durrum amino acid analyzer and the remainder was used for gas chromatography. One half of each sample was then hydrolyzed in 6 N HCl at 110 °C for 12 hr. Aliquots of both hydrolyzed and unhydrolyzed samples for the amino acid analyzer were freeze-dried and dissolved in 0.01 N HCl solution before injection.

The hydrolyzed and unhydrolyzed solutions which had been set aside for gas chromatography, were derivatized to form volatile N-trifluoroacetyl amino acid n-butyl esters (Roach and Gehrke, 1969). These were chromatographed on a 3% OV-17 column in a Perkin-Elmer 990 gas chromatograph (Figure 3). Amino acids were tentatively identified by their retention times in the amino acid analyzer and gas chromatograph. Mass spectra of each resolved peak were later obtained from a Dupont 21-492 mass spectrometer in tandem with the gas chromatograph. The fragmentation patterns of the spectra were then interpreted and matched with the patterns of standard compounds to confirm the identities of the amino acids.

Yields of amino acids were calculated on the basis of carbon. The total carbon present in the 36 millimoles of $CH_4$ with which the system had been charged was 430 mg. The yields of each amino acid were expressed as the per cent of the carbon-content of the amino acid compared to 430 mg of total carbon. Yields of solid material were similarly calculated from the elemental composition of the solid.

Using the same spark gap in identical reactors and feeding the same amount of reactants for both dry and wet discharges, the electrical parameters were characterized by current, voltage, and frequency. In this investigation, peak voltage, rms voltage and rms current were controlled for a given input power supply of finite frequency so as to apply qualitatively and quantitatively the same energy to both discharge systems. Accordingly, a difference in product-distribution would be attributable to chemical reactions in the two different discharges that took different forms in relating chemical kinetics of fragmentation and recombination, transportation properties, and arc variables.

## 3. Results

### 3.1. Insoluble solids

A characteristic difference between the dry and wet discharges of this study was made vividly apparent by the solid particles which separated from the liquid phase. The wet discharge produced the particles within one minute after the onset of arcing. The quantity increased continuously during the discharge, producing an orange-colored solution with orange colored solids in suspension. On the other hand, in the dry discharge, black carbon particles, which dropped from the electrode, were apparent after 30 min, and particles due to reaction products were noticed only after one hour. The particles in both discharges are known to have been formed on the water surface (rather than in solution) from the observation that they were generated as a surface slick when the stirring bar was stopped during the discharge process.

The immediate formation of the solid particles in the wet discharge means that the solid could be formed before the system variables (e.g., temperature, pressure, and possibly concentrations) changed very much and that these changes of the system variables did not affect appreciably their continuous formation. In fact, solid particles were formed as long as the gas phase contained an excess of carbon in relation to nitrogen.

Since this insoluble solid is generally believed to be a cross-linked high molecular weight polymer (Wilson, 1960; Oro, 1963; Henrey, 1969), a rationalization of the aforementioned observations permits the assertion that the solid was produced by a rapid process which could combine a large number of fragmented species together into a finite structure within a short period of time. Such fast reactions are considered to have occurred in the formation of many other organic compounds in the wet discharge. Nonetheless, this process was insignificant in the dry discharge in which most of the solid formed slowly as a film on the flask wall, possibly through a wall recombination of reactive species – partial solid surface quench.

### 3.2. Amino acids

As a result of the faster reactions, yields were larger in the wet discharge than in the dry discharge. Amino acids (see Figure 2) were more abundant in the wet discharge products as free amino acids (before hydrolysis) and especially in the form

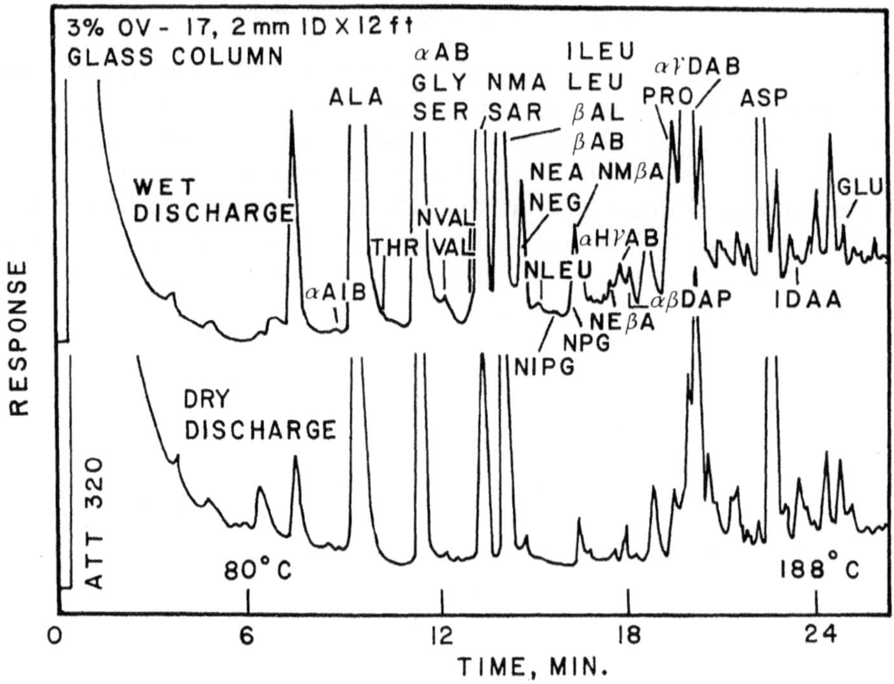


Fig. 2. Gas chromatogram of amino acids from high intensity arc discharge of $CH_4$, $NH_3$ and $H_2O$. NMA is N-Methyl Alanine. NEG is N-Ethyl Glycine. NIPG is N-Isopropyl Glycine. αHγAB is α-Hydroxy-γ-Aminobutyric Acid. αβDAP is α,β-Diaminopropionic Acid. IDAA is Iminodiacetic Acid.

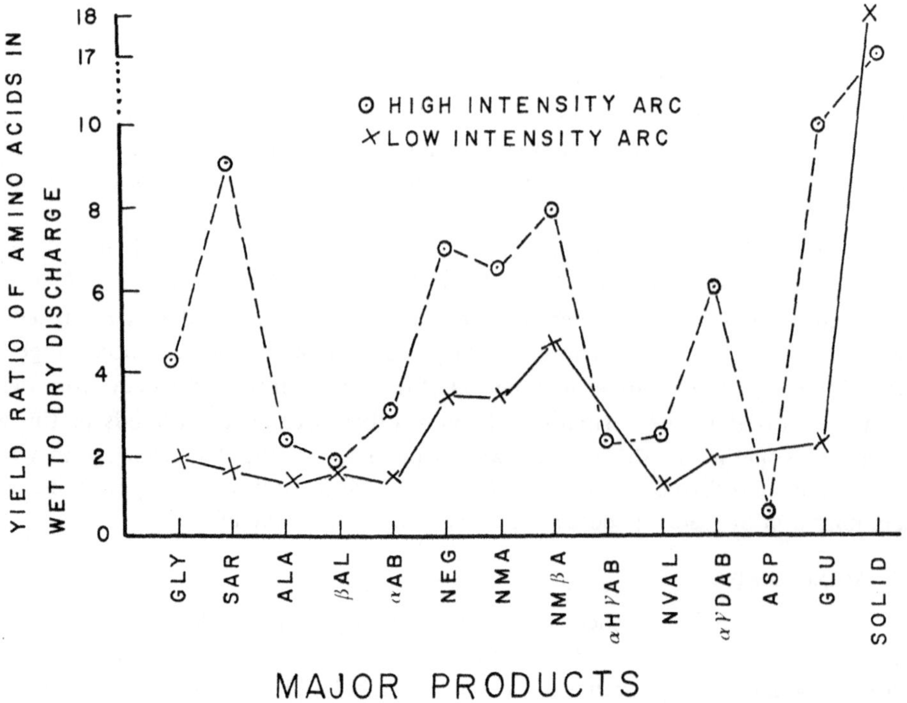


Fig. 3. Yield ratios of major products in wet discharge to dry discharge.

of their precursors which yielded the amino acids after hydrolysis. Except for aspartic acid, yields of all other amino acids were greatest in the wet discharge. The ratio of yields of the wet to the dry discharge for major products are shown in Figure 3. The figure indicates that especially high ratios were associated with the formation of N-alkyl amino acids and the insoluble solid. This fact might be related to the similarly higher yield of N-alkyl amino acids in the low intensity arc discharge than in the high intensity arc discharge.

TABLE I

Comparison of yields of major products for initial energy input

| | | Amino acids | | Solid | | Total % |
|---|---|---|---|---|---|---|
| | | $\mu$M J$^{-1}$ | % | mg J$^{-1}$ | % | |
| Low intensity Arc discharge | Dry | $8.2\times10^{-4}$ | 1.55 | $1.4\times10^{-5}$ | 0.55 | 2.10 |
| | Wet | $14.1\times10^{-4}$ | 2.77 | $25\times10^{-5}$ | 10.16 | 12.93 |
| High intensity Arc discharge | Dry | $1.4\times10^{-4}$ | 0.39 | $1.7\times10^{-5}$ | 0.83 | 1.22 |
| | Wet | $3.8\times10^{-4}$ | 0.94 | $29\times10^{-5}$ | 14.20 | 15.44 |

TABLE II

Absolute yields of major amino acids after hydrolysis (high intensity arc)

| | Dry discharge | | Wet discharge | |
|---|---|---|---|---|
| | $\mu$M ($\mu$M M$^{-1}$ $CH_4$) | %[a] | $\mu$M ($\mu$M M$^{-1}$ $CH_4$) | % |
| Aspartic Acid | 9.4 (260) | 0.10 | 4.6 (128) | 0.05 |
| Sarcosine | 0.9 (25) | 0.01 | 8.0 (222) | 0.07 |
| Glutamic Acid | Trace | | 0.8 (22) | 0.01 |
| Glycine | 7.1 (196) | 0.04 | 29.6 (820) | 0.16 |
| Alanine | 11.4 (3.7) | 0.10 | 26.4 (732) | 0.22 |
| $\alpha$-Amino Butyric Acid | 0.7 (18) | 0.01 | 2.2 (60) | 0.02 |
| Norvaline | 0.4 (10) | 0.01 | 1.0 (27) | 0.01 |
| $\alpha$-Hydroxy $\gamma$-ABA | 2.0 (56) | 0.03 | 4.3 (118) | 0.07 |
| $\beta$-Alanine | 7.2 (199) | 0.06 | 12.8 (354) | 0.11 |
| $\alpha,\gamma$-Diamino Butyric Acid | 2.6 (72) | 0.03 | 15.7 (436) | 0.18 |
| N-Ethyl Glycine | Trace | | 2.1 (58) | 0.02 |
| N-Methyl Alanine | Trace | | 1.3 (37) | 0.01 |
| N-Methyl $\beta$-Alanine | Trace | | 0.8 (22) | 0.01 |
| Total | 41.7 (1153) | 0.39 | 109.6 (3036) | 0.94 |

[a] Percentage based on mass of carbon atom.

The low intensity wet discharge yielded 1.8 times more amino acids than the low intensity dry discharge and the high intensity wet discharge yielded 2.7 times more than the high intensity dry discharge as shown in Tables I and II.

### 3.3. Gases

The most abundant gas in the discharges was hydrogen. The generation of hydrogen accounted for a pressure build-up in the reaction vessel during the discharges. In a typical experiment 51% of the total gases after 12 hr discharge was hydrogen. Other major gases included 34% unreacted methane and $C_2$–$C_4$ hydrocarbons. A quantitative comparison of the gaseous products from the wet and dry discharges was not made.

## 4. Conclusion

For both amino acids and insoluble solid, the ratio of yields of the low intensity wet discharge to the low intensity dry discharge was 6 and the ratio of yields of the high intensity wet discharge to the high intensity dry discharge was 12. Summarizing the quench effect, it is concluded that the quench process led to (1) faster rates of reactions, (2) higher molecular weight organic compounds, and (3) one order of magnitude larger yields.

As described in the introduction, any cold solid surface can also affect the reaction sequences, i.e., solid surface quench. In many experiments conducted so far, with all sorts of reaction vessel configurations designed for synthesis of amino acids, the wall surfaces could have played a significant role in the formation of micromolecular species through adsorption and recombination on the wall. Similarly, processes for the formation of micromolecular species near or at the earth's surface might have been more important than those processes which took place in the atmosphere where the yield would be smaller and where there would be more likely subsequent decomposition of the products. This proposition is further emphasized by the results of our experiments on the water-surface quench, which could have occurred at the atmospheric-hydrospheric interface of the earth and on water droplets of rain and clouds.

Among the major products formed by using the direct discharge on the water surface was an insoluble solid. Even though its detailed structure is not known, it can be postulated that it could be a good reactive surface for macromolecular condensations because of its inert unsaturated structure.

We may conclude that lightning and shock waves, interacting with water via quench reactions, might have played an important role in the primordial milieu leading to the chemical evolution of organic molecules.

## Acknowledgements

This work was partially supported by NASA Grant NGR 05-007-215 to the University of California at Los Angeles, through the courtesy of Prof. W. F. Libby to Prof. A. R. Hochstim, and partially by the Research Institute for Engineering Sciences at Wayne State University.

## References

Gabel, N. W. and Ponnamperuma, C.: 1971, in C. Ponnamperuma (ed.), *Exobiology*, North-Holland, Amsterdam.

Henley, R. R.: 1969, Ph.D. Thesis, Stanford University, Stanford.

Hochstim, A. R.: 1963, *Proc. Nat. Acad. Sci.* **50**, 200.

Oró, J.: 1963, *Nature* **197**, 862.
Park, W. K.: 1973, Ph.D. Thesis, Wayne State University, Detroit.
Park, W. K., Hochstim, A. R., and Ponnamperuma, C.: 1973, *Am. Chem. Soc. Regional Meeting*, Wash., D.C., U.S.A.
Roach, D. and Gehrke, C. W.: 1969, *J. Chromatog*. **44**, 269.
Wilson, A. T.: 1960, *Nature* **188**, 1007.

# SHOCK SYNTHESIS OF AMINO ACIDS II

A. BAR-NUN

*Dept. of Physical Chemistry, The Hebrew University, Jerusalem, Israel*

**Abstract.** The synthesis of amino acids behind shock waves in methane, ethane, ammonia, and water vapor was further investigated. Aldehydes and HCN are formed separately in the non-homogeneous gas during the high temperature period and recombine with ammonia during the thermal quench period, to form α-amino nitriles. The α-amino nitriles are either hydrolyzed by excess water vapor during the quench period or emerge as such after the reaction is completed. A combined gas chromatograph-mass spectrometer analysis of the reaction product showed identical amounts of D and L amino acids, thus confirming the absence of contaminants. Thunder shock waves were shown to be a suitable source of energy for the production of amino acids.

## 1. Introduction

The synthesis of amino acids behind high temperature shock waves in an atmosphere consisting of methane, ethane, ammonia, and water vapor was recently demonstrated by Bar-Nun *et al.* (1970). The energy-efficiency of this synthesis was some 4 orders of magnitude higher than the efficiency of the unsensitized UV photoproduction reaction (Sagan and Khare, 1971). In addition, the yields were very high. Up to 36% of the ammonia present in the reaction mixture was incorporated into the amino acids. This high yield came as a surprise since reactions behind shock waves occur in the gas phase without any catalysis and the processes are governed by thermo-chemical and kinetic parameters only. Such high yields will require a highly selective set of sequential reactions in the gas phase, leading ultimately to amino acids. This study was undertaken in order to confirm the previously obtained results, to elucidate the mechanism of the reaction, and to account for the high yield.

## 2. Experimental

The single pulse shock tube technique was described previously by Bar-Nun *et al.* (1970). This study was carried out on two different shock tubes of similar designs. One (Bar-Nun and Lifshitz, 1967) was made of Pyrex and brass tubing of 50 mm I.D. The second, 35 mm I.D. made of stainless steel, is shown schematically in Figure 1. The driven sections of both shock tubes could be heated to 80 °C and pumped to below $10^{-4}$ torr, with leak rates of 1–5 mtorr min$^{-1}$. Two reaction mixtures were prepared at high pressure in stainless steel cylinders. One mixture (A), consisting of 3.3% $CH_4$, 11.0% $C_2H_6$, 5.6% $NH_3$ (Matheson 'pure' grades, used without further purification), and 80.1% Matheson ultra pure argon. The second mixture (B), consisted of 3.9% $CH_4$, 10.0% $C_2H_6$, 20.0% $NH_3$, and 66.1% argon.

Two different methods were employed in introducing the water vapor and reaction mixture into the driven section: (1) After the shock tube was evacuated, water vapor was introduced from a glass bulb containing degassed liquid water, into the entire

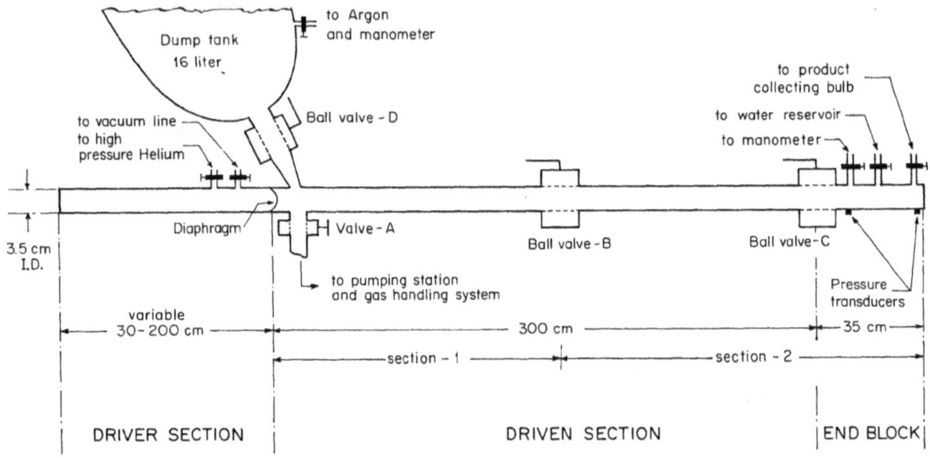


Fig. 1. A schematic diagram of the 35 mm I.D. single-pulse shock tube.

driven section through a valve located at the end block. The reaction mixture was then introduced into the entire driven section from the opposite direction through the pump port located near the diaphragm. Less than half a minute was allowed for partial mixing while the helium pressure in the driver section was increased until the diaphragm burst and the shock wave was initiated. (2) Alternatively, valve-B was closed after the evacuation, water vapor was introduced only to section 2 of the driven section, and the reaction mixture was introduced to Section 1. Valve B was then opened and the shock wave was initiated in less than half a minute. In this method the distribution of water vapor and reaction mixture along the driven section was controlled to a greater extent than in the first method. After quenching, the products were collected in an evacuated glass bulb containing 10 or 20 ml of 0.1 N HCl or ammonia solution.

The solutions of products were either analyzed without any further treatment or, since it was suspected that the end products also contained α-amino nitrites, some of the solutions were hydrolyzed in 6 N HCl at 100 °C for 24 hr. Various methods of analysis were employed; (a) Paper chromatography, with n-butanol, methanol, water, and concentrated ammonia solution in the ratio 10:10:5:2, respectively. (b) Column chromatography in a Beckman-Spinco amino acid analyzer. (c) Gas chromatography of the n-propyl, N-acetyl esters (Coulter and Hann, 1968) combined with a mass spectrometer (GC-MS), (d) GC-MS of the N-trifluoroacetyl sec-butyl esters, to separate and identify diastereoisomers resulting from the D and L amino acids.

## 3. Results

The experimental conditions of the shocks run on the 50 mm I.D. shock tube are summarized in Table I and the results of analysis in Table II. In all shocks the water vapor and gas mixture were introduced into the entire driven section in the manner described earlier (1). Dwell times at the high temperature ranged from 200 to 500 $\mu$s. Analyses of the products from this series were carried out directly on the samples from the collecting bulbs, without any further treatment.

TABLE I

Experimental conditions for shock synthesis in the 50 mm I.D. shock tube

| Run | Shock No. | Initial pressure $P_{H_2O}$ (torr) | $P_{mixt.}$ (A) (torr) | Incident shock speed (mm $\mu s^{-1}$) | Approximate[a] reflected shock temperature (K) mixt. | mixt. $+H_2O$ |
|---|---|---|---|---|---|---|
| 1[b] | 1 | 19 | 31 | 1.305 | 3830 | 2170 |
| | 2 | 18 | 32 | 1.405 | 4400 | 2430 |
| | 3 | 18 | 32 | 1.316 | 3900 | 2190 |
| | 4 | 18 | 32 | 1.319 | 3910 | 2200 |
| 2 | 5 | 20 | 30 | – | – | – |
| | 6 | 20 | 30 | 0.635 | 790 | 550 |
| | 7 | 20 | 30 | 0.652 | 820 | 570 |
| 3 | 8 | 20 | 30 | 1.230 | 2220 | 1290 |
| | 9 | 20 | 30 | 1.230 | 2220 | 1290 |
| | 10 | 20 | 30 | 1.223 | 2210 | 1280 |
| | 11 | 20 | 30 | 1.263 | 2320 | 1340 |
| 4[c] | 12 | 20 | 30 | – | – | – |
| | 13 | 20 | 30 | – | – | – |
| 5 | 14 | 14 | 35 | 1.330 | 2530 | 1440 |
| | 15 | 14 | 35 | 1.330 | 2530 | 1440 |
| | 16 | 14 | 35 | 1.340 | 2560 | 1460 |

[a] Reflected shock temperature calculated from the incident shock speed for two cases: (1) Reflected shock heating of the reaction mixture alone. (2) Reflected shock heating of the reaction mixture +40% water. In both cases thermicity of chemical reactions in the gas was not included in the temperature calculation.

[b] Mixture consisting of argon only.

[c] No shocks were run. Reactants remained in the heated driven section for 5 min and pushed into the collecting bulb by a stream of helium.

TABLE II

Summary of products analysis, $\mu$moles amino acid[a]

| Run | Glycine | Alanine | Valine | Leucine | Isoleucine[b] |
|---|---|---|---|---|---|
| 2 | 0.98 | 0.08 | 0.06 | 0.22 | – |
| 3 | 50.50 | 27.50 | 10.20 | 9.50 | <0.01 |
| 5 | 121.30 | 60.10 | 4.40 | 12.60 | 0.30 |

[a] Calculated from the results of column chromatography and gas chromatography.

[b] Isoleucine was identified only by its elution time in column chromatography. In the gas chromatography and paper chromatography it was not separated from the leucine.

Control runs 1 and 4 did not contain any amino acids, showing that the shock tube was clean and that amino acids were formed only upon shock heating and quenching of the reaction mixture with water vapor. The yields are hard to calculate accurately in these experiments since not all the reacted gas was collected in the sampling bulb. However, using an estimated 50% collection efficiency, the yields (based on the ammonia) are 0.08%, 8.70% and 20.00% for runs 2, 3 and 5 respectively. The yields in runs 3 and 5 are in good agreement with the results reported by Bar-Nun *et al.* (1970).

A GC-MS analysis of the n-propyl, N-acetyl esters confirmed identification of glycine, alanine, valine, and leucine. A GC-MS analysis of the N-trifluoroacetyl sec-butyl esters, from the combined runs 3+5, gave double peaks of identical intensities, of the D and L isomers, for alanine and valine, while leucine gave peaks too small for positive identification.

It can be concluded that the amino acids collected after the shock in the sampling bulb are indeed produced in the shock tube and do not result from contamination.

Another set of experiments was run in the 35 mm I.D. shock tube. The water vapor and reaction mixture were introduced separately into Sections 2 and 1 of the driven section, as described earlier (2), thus ensuring better control over their distribution along the driven section. In this series the products were collected in 10 ml 0.1 N ammonia solution and the collecting bulb was frozen in liquid air after

TABLE III

Experimental conditions and results for shock synthesis in the 35 mm I.D. shock tube

| Run | Initial pressure | | Incident shock speed (mm $\mu s^{-1}$) | Approximate[b] reflected shock temperature (K) | | Dwell time at the high temperature ($\mu s$) | Products ($\mu$mole) | |
|---|---|---|---|---|---|---|---|---|
| | $P_{H_2O}$[a] (torr) | $P_{mixt.}$ (B) (torr) | | mixt. | mixt. $+H_2O$ | | Glycine | Alanine |
| 49 | 19 | 50 | 1.310 | 1750 | 1410 | 200 | 0.51 | 1.29 |
| | 19 | 50 | 1.230 | 1600 | 1290 | 200 | | |
| 50 | 19 | 50 | 1.440 | 2010 | 1610 | 300 | 1.20 | 2.50 |
| | 19 | 50 | 1.425 | 1970 | 1590 | 300 | | |
| 51 | 20 | 50 | 1.335 | 1790 | 1450 | 400 | 0.84 | 1.83 |
| | 20 | 50 | 1.345 | 1810 | 1460 | 400 | | |
| 52 | 30 | 50 | 1.405 | 1930 | 1560 | 300 | 0.60 | 1.43 |
| | 50 | 50 | 1.397 | 1920 | 1540 | 300 | | |
| 53 | 20 | 50 | 1.600 | 2350 | 1870 | 600 | 2.70 | 2.52 |
| | 20 | 50 | 1.704 | 2590 | 2050 | 800 | | |
| 54 | 20 | 50 | 1.515 | 2160 | 1730 | 2700 | 0.95 | 1.66 |
| | 20 | 50 | 1.465 | 2060 | 1650 | 2700 | | |

[a] In runs 49, 50, 53 and 54 water vapor was introduced into Section 2 of the driven section. In run 51 water vapor was introduced only to the end block. In run 52 liquid water was introduced only to the end block.

[b] The reflected shock temperature is calculated from the incident shock speed for two cases: (1) reflected shock heating of the reaction mixture alone, (2) reflected shock heating of the reaction mixture with the appropriate percentage of water. In both cases the thermicity of chemical reactions in the gas was not included in the temperature calculation.

each run. The solution was analyzed both before and after hydrolysis. The yields after hydrolysis were considerably higher. The ratios of yields before and after hydrolysis, however, varied from run to run and only yields after hydrolysis are reported. The considerable increase in the yield of amino acids after hydrolysis suggests that α-amino nitriles are formed as intermediates in the reaction.

Table III summarizes the experimental condition for this series and the results.

Several additional shocks were run without water vapor, covering a wide range of temperatures, compositions, and dwell times at the high temperature. The quantities of amino acids obtained after hydrolysis were less than 10 n moles, which could well be a result of contamination. It was thus demonstrated that water vapor was essential for the production of the α-amino nitriles, and was not required only for their hydrolysis to amino acids.

In another series of experiments, using reaction mixtures without ammonia, the oxidation of hydrocarbons by water vapor was investigated. A homogeneous reaction mixture of 2.5% methane, 10.0% water vapor, and 87.5% argon was shock heated to 1550K. About 1.4% of the methane was found, by gas chromatography, to be converted to formaldehyde. Shock heating of the same reaction mixture to 2000K resulted in the conversion of 0.15% of the methane to formaldehyde and 50% to carbon monoxide. In another set of experiments a mixture of 10.0% ethane in argon was introduced to Section 1 of the driven section and water vapor was introduced to Section 2, as in the experiments summarized in Table III. After shock heating to 1550K, 0.16 and 0.26% of the ethane were converted to formaldehyde and acetaldehyde respectively. In an experiment which was carried out under conditions identical to those of run 50 in Table III only without ammonia, 2.7 $\mu$moles of acetaldehyde were formed, while in run 50 2.5 $\mu$moles of alanine were found. At temperatures above 2000K the aldehydes vanished and the products were mainly carbon monoxide, carbon dioxide, and acetylene. In shocks run above 2000K with reaction mixtures containing ammonia large quantities of HCN were formed as well. The gas chromatographic analysis of the aldehydes were inaccurate because of their condensation on the walls and the results quoted here are only approximate. The presence of aldehydes was also confirmed by precipitate formation with 2,4-dinitrophenylhydrazine.

## 4. Discussion

The increase in yield upon hydrolysis, in the series summarized in Table III, indicates that α-amino nitriles are part of the end products of the reaction, although their presence was not confirmed by analysis. Their hydrolysis in the shock tube, to give the corresponding amino acids, can be partial and depends upon the conditions in the shock tube. If the temperature after their formation is still high enough, they will be hydrolyzed by excess water vapor, whereas if they are formed at a later stage of the quenching period they will emerge from the shock tube as nitriles. The negligible yields of amino acids or nitriles in the absence of water vapor lead to the conclusion that some intermediate containing oxygen is essential for the reaction. Such an intermediate is most likely an aldehyde. Aldehydes were shown to be the products of the reaction preceding ignition in hydrocarbon-air mixtures compression heated to around 700 K (Martinengo *et al.*, 1965). They were also shown to be intermediates in the slow oxidation of hydrocarbons by air and in cool flames (Yang and Gray, 1969). Water vapor can oxidize hydrocarbons at elevated temperatures just like air, as was demonstrated in the last series of experiments and by Glick (1959).

If the temperature is high enough, the aldehydes oxidize to carbon monoxide. However, keeping the temperature low, enables the isolation of aldehydes as we have indeed demonstrated. The aldehydes are one of the starting materials in the Strecker reaction for the formation of amino acids. The other two are HCN and ammonia. Ammonia is one of the components of the reaction mixture and HCN is well known as one of the main products of high temperature reactions of hydrocarbons and any nitrogen bearing compound. HCN is not formed, however, in large amounts at temperatures below 2000 K, and at these temperature a large fraction of the aldehydes is already oxidized to carbon monoxide. This leads to the inevitable conclusion that the shock synthesis of amino acids occurs in an essentially non-homogeneous medium, where the temperature is non-uniform along the reacting mixture.

The series of experiments summarized in Table III was performed under such inhomogeneous conditions. When valve B, which separates the water vapor from the reaction mixture is opened, only partial mixing results during the elapse of less than half a minute before the onset of the shock. Thus, the shock wave passes through three regions in the driven section: the first region containing the reaction mixture only; the second, the reaction mixture and water vapor; and the third, water vapor only. For a certain shock speed the temperature behind the shock front depends upon the heat capacity of the gas. The reaction mixture has the lowest heat capacity and hence reaches the highest temperature. The reaction mixture with water vapor has a higher heat capacity and reaches a much lower temperature. This can be clearly seen from the reflected shock temperatures in Tables I and III. There is a temperature gradient along the reacting gas, ranging from high near the diaphragm to low towards the end of the driven section. The temperatures of the reaction mixture together with water vapor were even lower than indicated in Tables I and III. This occurs because the gas is enriched with water vapor towards the end of the driven section, whereas the temperatures in the tables are calculated for a homogeneous mixture of water vapor and the reaction mixture according to their initial pressures prior to mixing.

The conditions in the first series of shocks, summarized in Table I, are also non-homogeneous, for two reasons: (a) water vapor is introduced from one end and the reaction mixture from the opposite end of the driven section; (b) only half a minute elapses before the onset of the shock. However, such mixing is non-reproducible and hard to control, whereas the conditions in the series summarized in Table III are reproducible.

As mentioned earlier shocks run at very high temperatures did not yield any amino acids, but produced large quantities of HCN and CO, while low temperature shocks in hydrocarbon-water-argon mixtures yielded aldehydes. Thus, it can be concluded that HCN is formed in the high temperature region while aldehydes are formed in the low temperature region. Ammonia is present from the beginning and the Strecker reaction can take place. This mechanism explains how such a high percentage of the ammonia in the reaction mixture was incorporated into amino acids: HCN is the main product of the high temperature shocks through the methane-ethane-ammonia-argon mixture while sufficient unreacted ammonia remains in the low temperature region.

With regard to the synthesis of amino acids in thunder shock waves, two processes could have taken place. One is the formation of amino acids or the corresponding α-amino nitriles during thunder-storms, where the rain droplets evaporating as a result of the high temperature shock will form a gradient in the water vapor concentration and consequently a gradient in the temperature. The second is the separate formation of HCN and aldehydes in different temperature regions of the decaying thunder shock wave (Bar-Nun and Tauber, 1972) and their subsequent reaction together with the dissolved ammonia in the ocean to form α-amino nitriles and ultimately, amino acids. This last process has the advantage of not requiring ammonia in the atmosphere. HCN is formed as well in large quantities from high temperature shocks through hydrocarbons and molecular nitrogen.

## Acknowledgements

The author wishes to express his gratitude to Dr Y. Wolman for separating and identifying the diastereoisomers of D and L amino acids on the GC-MS, and to Mrs P. Gairon and Mr E. Frimovitch for performing the analyses on the amino acid analyzer.

## References

Bar-Nun, A. and Lifshitz, A.: 1967, *J. Chem. Phys.* **47**, 2878.
Bar-Nun, A. and Tauber, M. E.: 1972, *Space Life Sci.* **3**, 254.
Bar-Nun, A., Bar-Nun, N., Bauer, S. H., and Sagan, C.: 1970, *Science* **168**, 470.
Coulter, J. R. and Hann, C. S.: 1968, *J. Chromatog.* **36**, 42.
Glick, H. S.: 1959, *7th Symposium (International) on Combustion (London)*, p. 98.
Martinengo, A., Melczer, J., and Schlimme, E.: 1965, *10th Symposium (International) on Combustion*, p. 323.
Sagan, C. and Khare, B. N.: 1971, *Science* **173**, 417.
Yang, C. H. and Gray, B. F.: 1969, *J. Phys. Chem.* **73**, 3395.

# QUANTUM CHEMICAL STUDY OF THE THERMODYNAMICS, KINETICS OF FORMATION AND BONDING OF $H_2CN$: RELEVANCE TO PREBIOTIC CHEMISTRY

G. H. LOEW

*Dept. of Genetics, Stanford University Medical Center, Stanford, Calif. 94305, U.S.A.*

and

S. CHANG

*Chemical Evolution Branch, Ames Research Center, NASA, Moffett Field, Calif. 94035, U.S.A.*

**Abstract.** Using the Iterative Extended Huckel Theory (IEHT), energy-conformation studies have been carried out for $H_2CN$ (I), *trans*-HCNH (IIA), and *cis*-HCNH (IIB), three possible isomers formed by addition of a hydrogen atom to hydrogen cyanide. Calculations show that the order of decreasing thermodynamic stability is I$\gg$IIA$>$IIB. Additionally, from calculated energies along simulated reaction pathways, the formation of I from HCN+H appears to be kinetically favored over IIA. Calculated properties of the minimum energy conformers of I and IIA are described and the potential role of $H_2CN$ (I) as a reactive intermediate in prebiotic organic synthesis and its possible relevance to interstellar organic chemistry are discussed.

## 1. Introduction

Hydrogen cyanide in the gaseous state is an initial product and an important intermediate in most typical chemical evolution experiments designed to study the transformation of simple primitive atmospheric gases to more complex organic compounds (Abelson, 1955–56; Abelson, 1966; Fox and Harada, 1965; Miller, 1955; Miller, 1957; Palm and Calvin, 1962; Calvin, 1969; Lemmon, 1970; Kenyon and Steinman, 1969). Furthermore, hydrogen cyanide is one of the more abundant molecules found in the interstellar medium (Townes, 1971; Rank *et al.*, 1971) H atoms are also abundant in both prebiotic atmosphere experiments and interstellar space. Thus, it seems clear that the reactions of H atoms with hydrogen cyanide may have played an important role in both terrestrial and extraterrestrial chemical evolution.

Yet little is known of the contribution of free radical species derived from hydrogen cyanide to such processes. We have begun consideration of the possibly relevant free radical chemistry of hydrogen cyanide with a quantum chemical study of the $H_2CN$ radical, a compound formed by the simple addition of a hydrogen atom to hydrogen cyanide.

This study of $H_2CN$ consists of five parts: (1) energy-conformation calculations to determine the minimum energy tautomeric form and conformer for this species; (2) total energy calculations at various points along a specified reaction pathway for the formation of each tautomer from HCN and H atoms; (3) calculation of a number of properties such as bond densities, net atomic charges, dipole moments, and ionization potential for the molecule; (4) description of the nature of the bonding in the molecule; and (5) a discussion of the potential roles of this species as a reactive intermediate in prebiotic organic synthesis experiments and as an interstellar molecule.

All calculations were based on the so-called Iterative Extended Huckel Theory (IEHT). The formulation used is essentially that of Zerner and Gouterman and has been well documented in the literature (Zerner and Gouterman, 1966; Zerner *et al.*, 1966). It has been used for many studies including HCN (Loew and Chang, 1971), HCN dimer (Loew and Chang, 1971) and tetramer (Loew and Chang, 1972). Moffat and Tang have applied another semi-empirical method called CNDO to similar studies of two nitriles (Moffat and Tang, 1973).

The species $H_2CN$ is an odd-electron radical with 11 valence electrons and 10 valence atomic orbitals which form 10 molecular orbitals. Each molecular orbital (MO) represents the spatial distribution of a single electron or two electrons with opposite spins. Specifying occupancy of the MO's specifies a valence electron configuration for the molecule. It has been shown that, in general, the valence configuration energy though only a small part of the total molecular energy should vary with nuclear geometry as the total energy (Pan and Allen, 1967; Allen and Russell, 1967). Therefore, it can be used with some degree of confidence to determine minimum energy conformers and other energy variations with geometry, including changes along a reaction pathway.

## 2. Energy Conformation Studies: Relative Thermodynamic Stability of $H_2CN$ (I), *trans*-HCNH (IIA) and *cis*-HCNH (IIB)

Although the formation of a radical species from HCN and H atoms has been reported (Cochran *et al.*, 1962), very little is known of its characteristics. Three possible isomers are shown below:

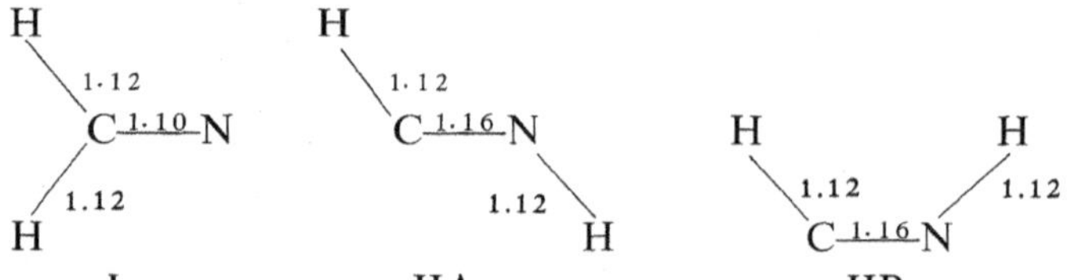


I IIA IIB

One of the first aims of our study was to ascertain by energy-conformation calculations which is the most stable in the gas phase. In the calculations reported here, we have fixed the bond angles at 120° for the three isomers of $H_2CN$, and varied the C—N bond length in each from 1.10 to 1.32 Å and the CH and NH bond lengths from 0.96 to 1.12 Å. From 16 such calculations for $H_2CN$ (I), and 22 for the trans (IIA) and cis (IIB) forms of HCNH, minimum energy conformers were obtained with bond lengths shown above. The results indicated that the $H_2CN$ species (I) is the most stable by 2.53 eV (58 kcal mole$^{-1}$) while the trans (IIA) is more stable than the cis (IIB) by 0.3 eV (7 kcal mole$^{-1}$). From our results we conclude that in the gaseous state, as well as in a solid matrix (Cochran *et al.*, 1962) the formation of $H_2CN$ is thermodynamically favored over HCNH.

## 3. Kinetic Studies: Formation of $H_2CN$ (I) and *trans*-HCNH (IIA)

If, under conditions where I and/or IIA may be found, equilibrium is not reached before further reactions take place, then the kinetically favored species could be the relevant one in understanding the properties of this molecule. Thus we sought to determine the relative energy of activation for the formation of each isomer by simulating reaction pathways from HCN and H which can lead to both forms of the radical. If we assume that the entropy of activation for each path is the same, then we can obtain a measure of the relative rates of formation of each isomer from

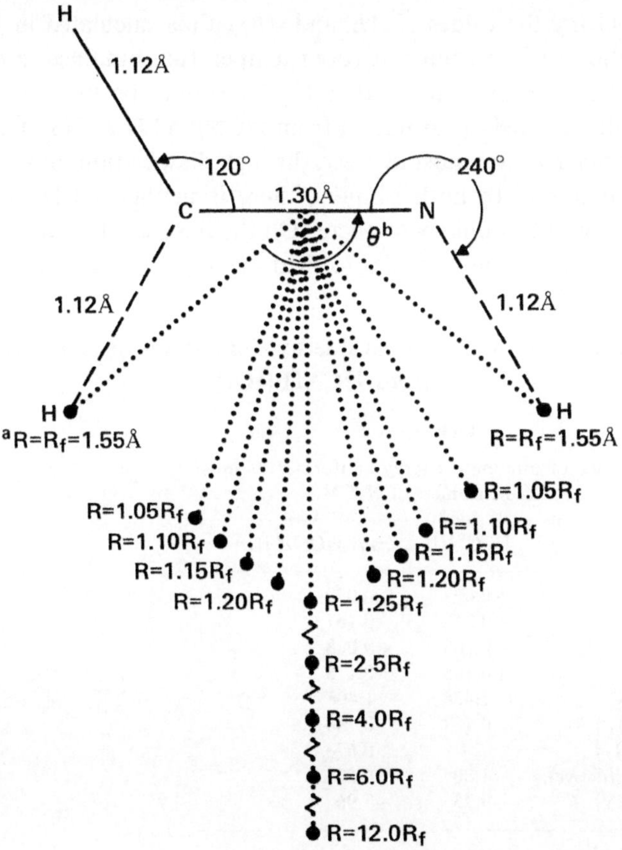


Fig. 1. Reaction pathway for formation of $H_2CN$ (I) and HCNH (IIA) from HCN and H·. (a) $R$= distance from mid-point of C—N bond to H· atom at various separations. $R_f = R = 1.55$ Å in product molecules. (*b*) $\theta$ = angle made by intersection of various $R$ at midpoint of C—N bond. $\theta_f = \theta$ in product molecules: $\theta_f = 140°$ for I, 40° for IIA.

the Arrhenius equation:

$$k = Ae^{-\Delta E/kT}$$

$$\frac{k_{\mathrm{I}}}{k_{\mathrm{IIA}}} = \frac{e^{-\Delta E_{\mathrm{I}}/kT}}{e^{-\Delta E_{\mathrm{IIA}}/kT}}$$

$$\ln(k_{\mathrm{I}}/k_{\mathrm{IIA}}) = (\Delta E_{\mathrm{IIA}} - \Delta E_{\mathrm{I}})/kT \quad (1)$$

where $k_{\mathrm{I}}$ is the rate of formation of $H_2CN$, $\Delta E_{\mathrm{I}}$ is the energy of activation for this formation, and $k_{\mathrm{IIA}}$ and $\Delta E_{\mathrm{IIA}}$ are the corresponding quantities for the HCNH isomer.

At a number of points indicated along the reaction pathway depicted in Figure 1 corresponding to different values of $R$, the energy was minimized with respect to $\theta$ for two separate regions of $\theta$: $140 \geqslant \theta \geqslant 90$, corresponding to the formation of $H_2CN$ (I); and $40 \leqslant \theta \leqslant 90$, corresponding to the formation of HCNH (IIA). In addition, at each point, the net charges on each atom and the bond densities were also monitored. Using the values of $\Delta E_{\mathrm{I}}$ and $\Delta E_{\mathrm{IIA}}$ thus calculated in Equation (1), the ratio of reaction rate constants at room temperature obtained is $k_{\mathrm{I}}/k_{\mathrm{IIA}} = 10^{47}$. Thus, our calculation clearly shows that $H_2CN$ should be kinetically as well as thermodynamically favored in formation from HCN and H atoms. The results also indicate that tautomerization must proceed through dissociation. Since at $2.5R_f$ and $\theta = 90°$, a point common to both isomers, there is neither a CH nor NH bond, tautomerization would be unlikely to occur. On the basis of all of our results then it appears that $H_2CN$ is the only species formed in the addition of a hydrogen atom to HCN.

## 4. Properties of the Minimum Energy Conformers of $H_2CN$ (I) and *trans*-HCNH (IIA)

TABLE I

Properties of minimum energy conformer of most stable isomers of $H_2CN$

| | $H_2CN$(I) | *trans*-HCNH (IIA) |
|---|---|---|
| C[a] | +0.007 | −0.079 |
| N | −0.218 | −0.167 |
| $H_1$[a, b] | +0.105 | +0.068 |
| $H_2$[a, c] | +0.105 | +0.178 |
| C—N[d] | 1.438 | 1.408 |
| C—$H_1$[d] | 0.758 | 0.788 |
| N—$H_2$[d] | – | 0.678 |
| D.M. (debye) | 1.80 | 0.62 |
| I.P. (eV) | −9.75 | −7.96 |

[a] Net charge on atom in units of electrons; consistency = ±0.008.
[b] $H_1$ = H on C atom.
[c] $H_2$ = H on N atom.
[d] electron density between atoms in units of electrons.

The electron distribution in the filled molecular orbitals allows the calculation of the net atomic charges and bond densities by using Mulliken population analysis (Mulliken, 1955). Table I contains these quantities for the most stable conformer of the $H_2CN$ (I) and trans-HCNH (IIA). Results in Table I show that the CN bond is more polar in the $H_2CN$ than in HCNH. The three-fold difference in calculated dipole moment and the substantial difference in ionization potential shown in this table could be of use in the identification of the isomers in a gaseous state.

## 5. Nature of the Bonding in $H_2CN$ (I) and *trans*-HCNH (IIA)

From a careful analysis of the nature of each filled molecular orbital, it is possible to obtain a description of the way in which each of the atoms bonds in the molecule, which is formally like that of the valence bond picture, i.e. the 'hybridization' and number of electrons used in each bond. The results of this analysis are given in Table II for $H_2CN$, accompanied by a figure depicting a composite picture of the bonding in this molecule. A number of interesting features emerges from a consideration of this table and figure. In particular while there is one full pi bond between the C and N, there is only the smallest residue of the third, in-plane pi nitrile bond, ($p_y$) originally present in HCN. For the C atom for which this is a partial 5th bond,

TABLE II

Bonding in $H_2CN$

| Atom | Bond | Contrib. MO's | Bond character | Approx. hybrid. | No. of electrons in bond |
|---|---|---|---|---|---|
| Carbon | $C\overset{\sigma}{-}H$ | 1, 2, 3, 5, 6 | $0.43s+0.73p_x+0.48p_y$ | $sp^3$ | 1.64 |
| | $C\overset{\sigma}{-}H$ | 1, 2, 5 | $0.60s+0.43p_x$ | $s^2p$ | 1.03 |
| | $C\overset{\pi'}{-}N$ | 3 | $0.26p_y$ | $p_y$ | 0.26 |
| | $C\overset{\pi}{-}N$ | 4 | $1.01p_z$ | $p_z$ | 1.01 |
| Nitrogen | $N\overset{\sigma}{-}C$ | 1, 2, 5 | $1.36s+0.23p_x$ | $s^4p$ | 1.59 |
| | $N\overset{\pi'}{-}C$ | 3 | $0.26p_y$ | $p$ | 0.26 |
| | $N\overset{\pi}{-}C$ | 4 | $0.99p_z$ | $p$ | 0.99 |
| | $N_{non\text{-}bond}$ | 2, 5, 6 | $0.7s+1.46p_x+0.84p_y$ | $p$ | 2.37 |

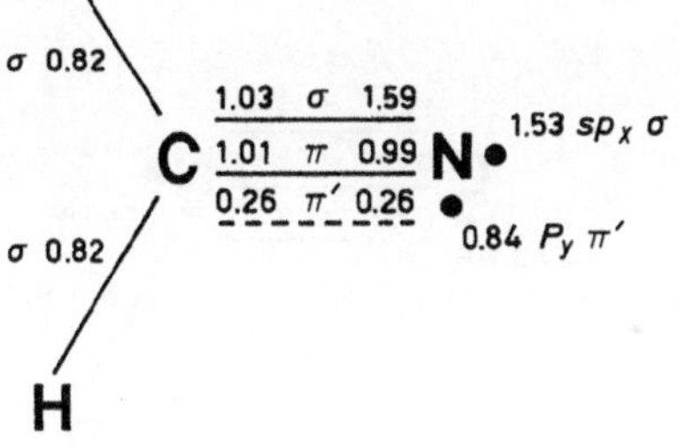

the 0.26 electrons come from the electron-deficient sigma bond to H. For the N atom the remainder of this normal third bond is in the form of an additional non-bonding electron fraction in the $p_y$ orbital.

From this bonding picture of $H_2CN$, it seems clear that the reactivity of this molecule would center on the electron-donating ability of the N atom with its substantial negative charge and added unpaired non-bonding electron ($p_y$) which is in the highest energy half-filled MO. The partial 5th bond and the neutral charge on the C would mitigate against its role as an initiator of electrophilic reactions.

Table III and the accompanying figure give the analogous bonding description for HCNH. In this molecule the triple bond character of the CN bond is retained only slightly more intact than in $H_2CN$. The carbon atom, bonded to only two atoms, appears to have at least 0.43 non-bonding electrons in the half-filled MO, three 'normal bonds', and a partial 4th bond. The nitrogen seems to have some of the features of a quarternary nitrogen with solid sigma bonds to the C and H atoms, an electron-rich pi bond with 1.22 electrons, and a partial bond with 0.40 $p_y$ ($\pi'$) electrons, leaving only one unpaired electron mainly in a $p_x$ sigma orbital and partially in a $p_y$ orbital.

TABLE III

Bonding in HCNH (IIA)

| Atom | Bond | Contrib. MO | Bond character | Approx. hybrid. | No. of electrons in bond |
|---|---|---|---|---|---|
| Carbon | $C\overset{\sigma}{-}H$ | 1, 2, 3, 4, 6 | $0.56s+0.35p_x+0.31p_y$ | $sp$ | 1.17 |
| | $C\overset{\sigma}{-}N$ | 1, 3, 6 | $0.59s+0.61p_x$ | $sp$ | 1.40 |
| | $C\overset{\pi'}{-}N$ | 2, 4 | $0.44p_y$ | $p_y$ | 0.44 |
| | $C\overset{\pi}{-}N$ | 5 | $0.77p_z$ | $p_z$ | 0.77 |
| | $C_{non\text{-}bond}$ | 6 | $0.03s+0.40p_y$ | $p_y$ | 0.43 |
| Nitrogen | N—C | 1, 2, 3, 6 | $1.03s+0.34p_x$ | $s^3p$ | 1.36 |
| | N—C | 4 | $0.40p_y$ | $p_y$ | 0.40 |
| | N—C | 5 | $1.22p_z$ | $p_z$ | 1.22 |
| | N—H | 1, 2, 3 | $0.28x+0.25p_x+0.63p_y$ | $sp^3$ | 1.16 |
| | $N_{non\text{-}bond}$ | 4, 6 | $0.02s+0.68p_x+0.26p_y$ | $p$ | 0.96 |

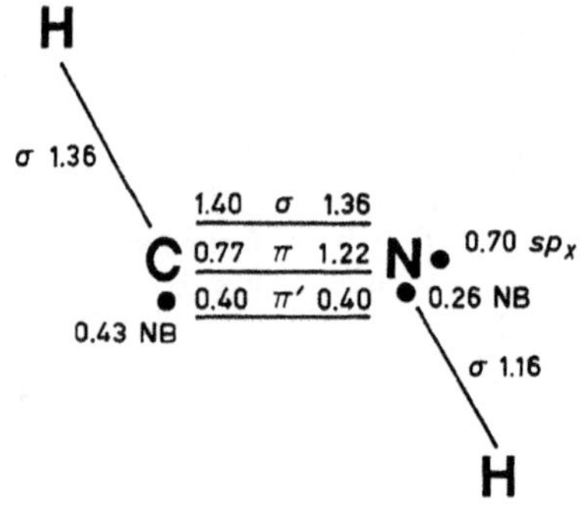

While it is more dfficult to predict the type of reactivity of this molecule, it would appear that the nitrogen atom with a reduced negative charge and a paucity of non-bonding electrons, is not a strong nucleophilic agent, but that perhaps the electron-deficient pi and pi′ orbitals and the unusual non-bonding electrons of the C atom serve as an electrophilic center for free radical reactions.

## 6. Possible Involvement of $H_2CN$ in Prebiotic Organic Chemistry

The calculations presented support the view that the reaction of hydrogen atoms with hydrogen cyanide produces $H_2CN$ (I) as the only thermodynamically and kinetically probable primary molecule product; therefore, we shall focus attention on this species. From the bonding description (Table II) and distribution of unpaired spin in the highest occupied molecular orbital of I, clearly the reactivity of this species should be that of an imine free radical. Predominant localization of unpaired spin on the N atom means that hydrogen abstraction, radical combination and addition to unsaturated bonds and dimerization should occur at N affording formaldimine, N-substituted formaldimines and formaldazine.

A product having an empirical formula corresponding to that of formaldimine was obtained by Geib and Harteck (1933) at liquid air temperature. The $H_3CN$ was formed by allowing hydrogen atoms to react with hydrogen cyanide, presumably in stepwise fashion with $H_2CN$ as an intermediate. In addition, amino acid precursors, $\alpha$-aminoacetonitriles, can arise directly from formaldimines, under conditions where formaldimine may be quickly transported to an aqueous phase. Solvolysis of formaldazine is a potential source of hydrazine in primordial atmosphere experiments.

Oligomerization of formaldimine may occur in the gas or solution phase in a fashion typical of aliphatic aldimines (Smith, 1965).

In the gas phase, combination of $H_2CN$ with other plausible free radicals (e.g. OH, CN, $CH_3$, $NH_2$, $H_2CN$, etc.) or addition of $H_2CN$ to unsaturated bonds of other molecules can produce a wide variety of N-substituted formaldimines, whose organic chemistry in the gas and solution phases is expected to be very rich and complicated providing solvolytic oligomerization and condensation routes to a variety of simple and more complicated compounds (Smith, 1965; Loew *et al.*, 1972). Significantly, Chadha *et al.* (1971) reported that a homologous series of low molecular weight imine-like compounds were produced during electrical discharges through methane-ammonia mixtures. Apparently, oligomerization of imines may contribute substantially to the polymeric material produced in prebiotic syntheses (Abelson, 1955–1956; Miller, 1955; Grossenbacher and Knight, 1965) which has previously been attributed to ionic polymerization of hydrogen cyanide.

Direct photolysis of hydrogen cyanide in the gas phase yielded methane, ammonia, ethane, hydrazine, methylamine, cyanogen, and hydrogen along with a reddish polymer; the primary reaction reportedly involved addition of a hydrogen atom to hydrogen cyanide (Mizutani and Mikuni, 1972) to form $H_2CN$. Although it is not clear what mechanisms were involved in the production of amino acids (Abelson, 1955–1956) and adenine (Ponnamperuma and Mariner, 1963; Ponnamperuma, 1965) during ultraviolet irradiation of hydrogen cyanide solutions, a similar free radical reaction may have initiated processes which were completed through ionic mechanisms. Such combined free radical-ionic mechanisms have been presented to explain the radiolysis of aqueous hydrogen cyanide solutions (Ogura, 1967; Ogura and Kondo, 1967; Ogura, 1968; Ogura *et al.*, 1972). Products of aqueous hydrogen cyanide radiolysis include formaldehyde, ammonia, diaminomaleonitrile, and other oligomers of hydrogen cyanide which can serve as precursors of amino acids (Ferris *et al.*, 1972; Sweeney *et al.*, 1974). The proposed reactions (Ogura *et al.*, 1972) involve $H_2CN$ as an intermediate in a way which is compatible with our calculated properties.

In interstellar clouds where hydrogen cyanide has been detected, reaction of hydrogen cyanide with hydrogen atoms should be plausible both in the gas phase and on the surface of interstellar grains (Brecher and Arrhenius, 1971; Salpeter, 1971). Our kinetic and thermodynamic calculations indicate that the simplest such addition product would be $H_2CN$ rather than HCNH. Further, our analysis of the reactivity of $H_2CN$ led us to the prediction that formaldimine would be the product formed from it by the addition of another H hydrogen atom. Within clouds where destructive ultraviolet radiation is attenuated and where the kinetic temperature may range from 10 to 100 K, both $H_2CN$ and formaldimine, once formed, may be stable, particularly if they are adsorbed on siliceous grain surfaces. Preliminary evidence for the possible stability of at least $H_2CN$ in interstellar space under the latter condition, has recently been obtained in our laboratory. Irradiation with a low pressure mercury lamp of hydrogen cyanide absorbed on porous Vycor glass yields $H_2CN$ radicals which can be detected by ESR and which have half-lives of hours at room temperature (Chang and Dalan, unpublished work). A search for these molecules in cloud regions containing hydrogen cyanide may be fruitful provided that sufficient theoretical data are available to predict their microwave spectra. Remarkably, since this work was completed, one of our predicted products of HCN + H addition, formaldimine, has indeed been detected in the interstellar medium (Turner, 1973).

It is noteworthy that hexamethylenetetramine (HMT) has been suggested (Sagan and Khare, 1970) as a possible constituent of interstellar grains. Interstellar HMT would be consistent with the existence of interstellar ammonia and formaldehyde in the same cloud region. On the other hand, interstellar HMT may also be derived from hexamerization of formaldimine $6CH_2{=}NH \rightarrow (CH_2)_6N_4 + 2NH_3$. If it occurs, the polymerization of formaldimine or formaldehyde and ammonia to HMT is expected to occur on the surface of interstellar grains where the monomers can accumulate.

It seems clear from the foregoing discussion that the free radical chemistry of hydrogen cyanide extending from $H_2CN$ can lead to a rich variety of products in prebiotic type experiments and may provide pathways for formation of a number of other interstellar molecules, some of which may be amenable to detection by radioastronomy. It is heartening indeed that one of our predictions, the existence of $H_2CNH$ in interstellar space, based on the calculated properties of (HCN+H) has already come to pass.

## References


Abelson, P. H.: 1955–1956, *Carnegie Inst. of Wash. Yearbook* **53**, 171.
Abelson, P. H.: 1966, *Proc. Natl. Acad. Sci. (U.S.)* **55**, 1365.
Allen, L. C. and Russell, J. D.: 1967, *J. Chem. Phys.* **46**, 1029.
Brecher, A. and Arrhenius, G.: 1971, *Nature, Phys. Sci.* **230**, 107.
Calvin, M.: 1969, *Chemical Evolution*, Oxford University Press, New York.
Chadha, M. S., Flores, J., Lawless, J. G., and Ponnamperuma, C.: 1971, *Icarus* **15**, 39, and unpublished results.
Chang, S. and Dolan, E.: unpublished results.
Cochran, E. L., Adrian, F. J., and Bowers, V. A.: 1962, *J. Chem. Phys.* **36**, 1938.
Ferris, J. P., Donner, D. B., and Lotz, W.: 1972, *J. Am. Chem. Soc.* **94**, 6968.
Fox, S. W. and Harada, H.: 1965, in S. W. Fox (ed.), *Origins of Prebiological Systems and of Their Molecular Matrices*, Academic Press, New York.
Geib, K. H. and Harteck, P.: 1933, *Berichte* **66 B**, 1815.
Grossenbacher, K. A. and Knight, C. A.: 1965, in S. W. Fox (ed.), *Origins of Prebiological Systems and of Their Molecular Matrices*, Academic Press, New York.
Kenyon, D. H. and Steinman, G.: 1969, *Biochemical Predestination*, McGraw-Hill, New York.
Lemmon, R.: 1970, *Chem. Rev.* **70**, 95.
Loew, G. H. and Chang, S.: 1971, *Tetrahedron* **27**, 3069.
Loew, G. H. and Chang, S.: 1971, *Tetrahedron* **27**, 2989.
Loew, G. H. and Chang, S.: 1972, *Theor. Chim. Acta* **27**, 273.
Loew, G. H., Chadha, M. S., and Chang, S.: 1972, *J. Theoret. Biol.* **35**, 359.
Miller, S. L.: 1955, *J. Am. Chem. Soc.* **77**, 2351.
Miller, S. L.: 1957, *Biochim. Biophys. Acta* **23**, 488.
Mizutani, H. and Mikuni, H.: 1972, *Chem. Letters* **1972**, 573.
Moffat, J. B. and Tang, K. F.: 1973, *Tetrahedron* **29**, 3111.
Mulliken, R. S.: 1955, *J. Chem. Phys.* **23**, 1833, 2338, 2343.
Ogura, H.: 1967, *J. Rad. Res.* **8**, 93.
Ogura, H.: 1968, *Bull. Chem. Soc., Japan* **41**, 2871.
Ogura, H. and Kondo, M.: 1967, *Bull. Chem. Soc., Japan* **40**, 2448.
Ogura, H., Fujimura, T., Murozono, S., Hirano, K., and Kondo, M.: 1972, *J. Nucl. Sci. Tech.* **9**, 339.
Palm, C. and Calvin, M.: 1962, *J. Am. Chem. Soc.* **84**, 2115.
Pan, D. C. and Allen, L. A.: 1967, *J. Chem. Phys.* **46**, 1797.
Ponnamperuma, C.: 1965, in S. W. Fox (ed.), *Origins of Prebiological Systems and of Their Molecular Matrices*, Academic Press, New York.
Ponnamperuma, C. and Mariner, R.: 1963, *Rad. Res.* **19**, 183.
Rank, D. M., Townes, C. H., and Welch, W. J.: 1971, *Science* **174**, 1083.
Sagan, C. and Khare, D. N.: 1970, *Bull. Amer. Astron. Soc.* **2**, 340.
Salpeter, E. E.: 1971, in *Highlights of Astronomy, Trans. Intern. Astron. Union* **14B**, D. Reidel Publ. Co., Dordrecht, Holland.
Smith, P. A. S.: 1965, *The Chemistry of Open Chain Organic Nitrogen Compounds*, Chapter 7, W. A. Benjamin, Inc., New York.
Sweeney, M. A., Toste, A. P., and Ponnamperuma, C.: 1973, unpublished.
Townes, C.: 1971, in *Highlights of Astronomy* **2**, *Trans. Intern. Astron. Union*, D. Reidel, Dordrecht, p. 359.
Turner, B. E.: 1973, *Sci. Amer.* **228**, 50.
Zerner, M. and Gouterman, M.: 1966, *Theor. Chim. Acta* **4**, 44.
Zerner, M., Gouterman, M., and Kobayashi, H.: 1966, *Theor. Chim. Acta* **6**, 363.

# AMINONITRILES: POSSIBLE ROLE IN CHEMICAL EVOLUTION

MOHINDRA S. CHADHA
*Bio-Organic Division, Bhabha Atomic Research Center, Bombay, India*

and

PETER M. MOLTON and CYRIL PONNAMPERUMA
*Laboratory of Chemical Evolution, Dept. of Chemistry, University of Maryland, College Park, Md. 20742, U.S.A.*

**Abstract.** The formation of HCN, ammonium cyanide, alkylnitriles, aminoacetonitrile and its C- and N-methyl homologs was demonstrated earlier in a simulated Jovian atmosphere. The polymeric material resulting in these experiments was shown to give glycine, alanine, sarcosine, aspartic acid and some imino dibasic acids on acid hydrolysis suggesting thereby the participation of the monomeric nitriles into the formation of the polymeric product(s). Further examination of products resulting from semi-corona and arc discharge through a mixture of methane and ammonia has provided evidence for the formation of alkylaminopropionitriles as a complex mixture and also some pyridyl and pyrimidyl type heterocyclic compounds. A GC-MS examination of the heterocyclics showed resemblance with those found in some carbonaceous chondrites. The significance of these findings in relation to chemical evolution will be discussed.

## 1. Introduction

The formation of nitriles in a variety of primitive Earth experiments has led to several interesting developments in the area of chemical evolution (Miller, 1957; Ponnamperuma and Woeller, 1967; Woeller and Ponnamperuna, 1969; Chadha *et al.*, 1971a, b; Noda and Ponnamperuma, 1971). The simplest of the nitriles, hydrogen cyanide has been shown to give rise to condensation products which can lead to the formation of amino acids and nucleic acid bases (Ponnamperuma and Gabel, 1968; Lemmon, 1970). Recently, the presence of HCN has been found in the interstellar medium (Snyder and Buhl, 1971) which further supports the importance of this simple yet reactive molecule as a building block in organic synthesis during chemical evolution. A quantum chemical study of HCN has shown that the alkylation of $CN^-$ is possible under prebiotic conditions and the organic cyanides thus formed could undergo further reactions to enrich the pool of organic matter in the primitive oceans (Loew and Chang, 1971).

Investigations have been carried out on a simulated Jovian atmosphere consisting of methane and ammonia. Gaseous and volatile products resulting from semicorona (0.3 mA) and electric discharge (0.5 and 5 mA) have been analyzed. Depending upon the experimental conditions, quantitative differences in the formation of gaseous products were observed (Ponnamperuma and Woeller, 1967; Woeller and Ponnamperuma, 1969). HCN was formed to a higher extent when the current through the system was 5 mA. In a typical 0.5 mA arc experiment, the products in the volatile fraction were shown to be ammonium cyanide, acetonitrile, propionitrile, aminoacetonitrile, N-methyl-aminoacetonitrile and 2-aminopropionitrile (Ponnamperuma and Woeller 1967). The formation of some alkylamines viz. methyl, ethyl and propylamine, etc., was observed in a 0.5 mA arc experiment (Chadha and Ponnamperuma, 1970). The nonvolatile colored product was acid hydrolyzed and examined by ion exchange chromatography. The hydrolysate was derivatized and the resulting *n*-butylated N-trifluoroacylated amino acids were examined by combined gas chromatography – mass spectrometry, when glycine, alanine, sarcosine, iminodiacetic acid, iminoacetic-2-propionic acid, iminoacetic-3-propionic acid and aspartic acid were identified (Chadha *et al.*, 1971a, b).

The formation of aminoacetonitrile as a constituent of the volatile fraction and the formation of glycine on the hydrolysis of the colored material clearly leads one to the conclusion that aminoacetonitrile is one of the primary and significant products of $CH_4$–$NH_3$ discharge experiments. The chemistry of aminoacetonitrile has therefore been a subject of primary interest. Even before aminoacetonitrile was characterized from the $CH_4$—$NH_3$ discharge experiments, Akabori (1955) had proposed that peptides could result by the polymerization of α-aminoacetonitrile (AAN) on a solid phase, followed by the hydrolysis of the polymer to polyglycine and ammonia. Subsequently Hanafusa and Akabori (1959) reported the formation of diglycine and triglycine when aminoacetonitrile sulfate was heated on clay to 120–140° for 3–5 hr. Moser and Matthews (1968) examined the products resulting from acidic, neutral and basic hydrolysis of AAN and reported the formation of a number of amino acids including aspartic acid and traces of alanine, serine, threonine and lysine. Since the characterizations were carried out only by the use of ion exchange chromatography, these findings needed further confirmation. In a subsequent study, the chemistry of aminoacetonitrile at pH 9.0, in aqueous solution was investigated to determine whether intermolecular reactions resulting in the formation of amide bonds could occur. It was observed (Chadha *et al.*, 1971c) that hydrolysis was the predominant reaction and not condensation, and this was borne out on the basis of the molecular orbital calculations also (Loew *et al.*, 1972). The major products formed were glycine, iminodiacetic acid and nitrilotriacetic acid. However, the presence of phosphate in the reation mixture led to the formation of a small amount of a product which had the elution time of gly-gly on an ion exchange column (Chadha *et al.*, 1971c). The solution which contained AAN and alanine showed the formation of a few elution peaks in the dipeptide region. The ion-exchange chromatographic pattern for the AAN+alanine+Pi reaction mixture was similar to the solution where Pi was not added. Of the several peaks from AAN and alanine, two were found to coincide with the elution times of gly-gly and gly-ala representing 0.3% and 0.05% incorporation of AAN. It would therefore appear that in solution, aminoacetonitrile undergoes hydrolysis in preference to condensation reactions. However, under anhydrous conditions, polymerization of 2-aminopropionitrile was shown to be facile at 80° and is reported to yield polymers resembling polyalanine (Morimoto and Ponnamperuma, 1970). The colored product that is formed during the electric discharge through anhydrous $CH_4$ and $NH_3$ was shown to consist of polymerized material, with the maximum molecular weight being a few thousand (Noda and Ponnamperuma, 1971). For the formation of the polymeric material, perhaps the cyano and amino groups of the aminonitriles could have interacted to give amidine linkages, and other monomers could have also participated, but the true nature of this complex mixture of products has been difficult to elicit. The monomeric compounds which could have participated in a variety of condensation reactions are many, viz., HCN, the alkylnitriles, alkylamines, and aminoalkylnitriles.

The present paper deals with further attempts to characterize the products resulting from the semicorona or spark discharge of mixtures of methane and ammonia and their significance in chemical evolution. The products appear to be 3-alkylaminopropionitriles and some heterocyclic compounds.

## 2. Materials and Methods

### 2.1. Preparation

3-Aminopropionitrile was prepared by the reaction of aqueous ammonia with acrylonitrile under pressure (Buc, 1955). The 3-alkylaminopropionitriles were prepared by the method of Cook and Reed (1945) wherein acrylonitrile is treated with appropriate amines. The dialkylaminopropionitriles were prepared likewise from the corresponding dialkylamines, except for the propyl-, and butyl-methyl and ethylaminopropionitriles where the procedure was slightly modified (Molton and Ponnamperuma 1973).

### 2.2. Discharge experiments

A 1 l flask evacuated to 20 $\mu$ pressure and filled to 40 cm with 20 cm each of research grade anhydrous methane and ammonia. A 0.5 mA spark (6 keV) or 0.3 mA semicorona was applied for 24 hr or 4 days respectively. The flask was fitted with a cold finger maintained at $-80°$ with solid $CO_2$/acetone to trap low boiling products and remove them from further reaction for 1 set of experiments. In another set of experiments, the products were allowed to collect on the walls of the flask at 20°. In one experiment in a flask evacuated to 20 $\mu$, 7 cm of acetonitrile and 20 cm of $NH_3$ were introduced and subjected to 0.5 mA spark for 24 hr.

The products were removed from the flask with 20 ml of pure anhydrous methanol which was then evaporated under reduced pressure. A portion of the products were derivatized with bis-(trimethylsilyl) acetamide (BSA) in pyridine in an acid-cleaned vial sealable with a teflon cap, which was then heated at 120° for 30 min. The reaction products and the derivatized materials were analyzed on an Aerograph 1520 gas chromatograph which has a FID detector and a 10 ft $\times$ 0.25 in., 3% OV–17 coated Celite 545 stainless steel column. The column was programmed from 80–250° at 4° $min^{-1}$ and held at the initial and final temperature for 10 min. The combined gas chromatography – mass spectrometry (GC–MS) was carried out on a Loenco gas chromatograph attached to a CEC-21-491 mass spectrometer equipped with a membrane separator.

## 3. Results and Discussion

The products which were collected in a spark discharge experiment utilizing a cold-finger were subjected to GC-MS without derivatization, the spectra indicated the presence of alkylaminopropionitriles. One of the products which was obtained from the methane/ammonia experiment produced a spectrum similar to that of 3-ethylaminopropionitrile (Figure 1). The principal peaks in the two spectra are the same, viz., at m/e 30, 42, 56, 83 and 98. The difference in relative intensities between the plot of the standard sample and of the unknown and the presence of some additional peaks in the unknown is possibly due to the lack of separation of closely related isomeric compounds. Under the conditions used, it was found difficult to separate diethyl- and methyl, propylaminopropionitriles. It is likely to be the case with other isomeric compounds also. So far, it has not been possible to assign structures to each of the compounds resolved by GC but a careful consideration of the GC-MS

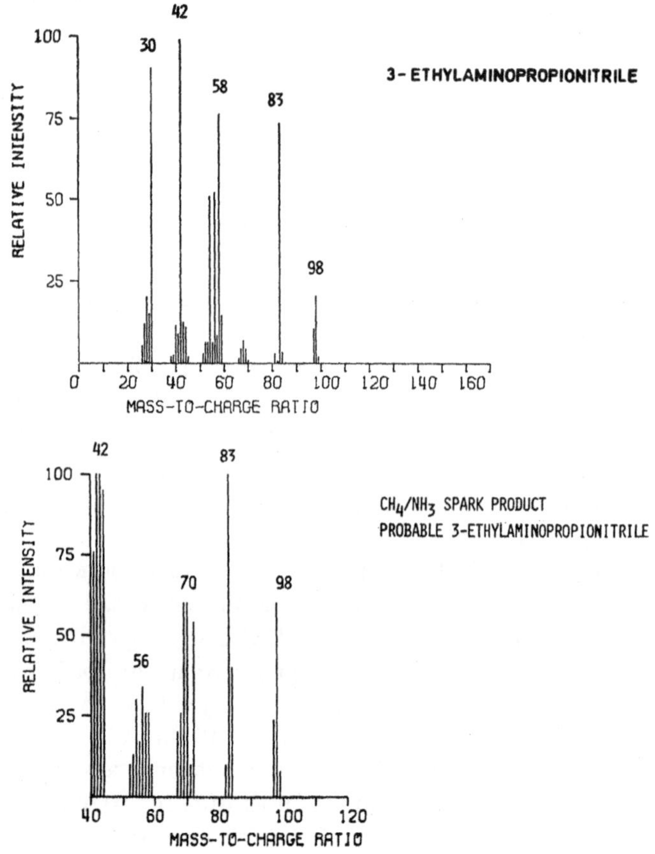


Fig. 1. Comparison of mass spectra from 3-ethylaminopropionitrile and an unknown present in the $CH_4/NH_3$ spark discharge product.

data leads us to conclude that they represented alkylaminopropionitriles with the general formula ($R_1$, $R_2$) $N.CH_2CH_2CN$; where $R_1$, $R_2$ are alkyl groups. The sum of the two alkyl ($R_1+R_2$) groups, followed by the number of separate GC-MS peaks in each case is indicated below:

$$CH_4\,(1);\ C_2H_6\,(2);\ C_3H_8\,(4);\ C_5H_{12}\,(2);\ C_6H_{14}\,(3);$$
$$C_7H_{16}\,(1);\ C_8H_{18}\,(2);\ C_9H_{20}\,(3);\quad \text{and}\quad C_{10}H_{22}\,(1),$$

which represent 19 peaks showing mass spectra corresponding to 3-alkyl amino-propionitriles.

The formation of aminoacetonitrile, N-methylaminoacetonitrile and 2-amino-propionitrile, which were characterized earlier, could have resulted from the direct combination of radicals. Likewise, the formation of 3-alkylaminopropionitriles can be rationalized on the basis of direct combination of radicals or acetylene, HCN and alkylamines. The formation of acetylene and HCN in the electric discharge experiments has already been demonstrated and these can give rise to acrylonitrile. The alkylamines resulting from electric discharge experiments can on addition to acryl-onitrile, be expected to give 3-alkylaminopropionitriles.

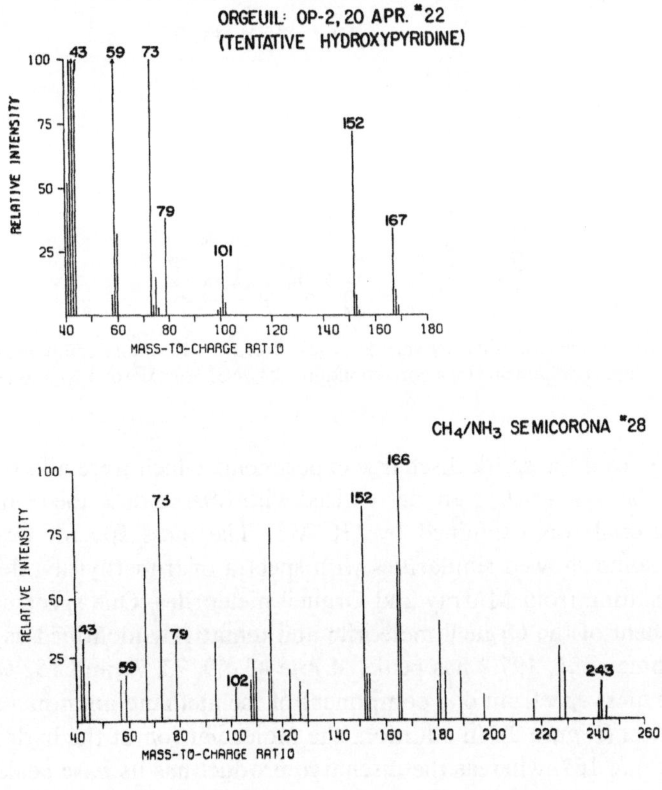


Fig. 2. Mass spectra of trimethylsilyl derivatives of (a) Orgeuil, OP:2 sample component; (b) a corresponding $CH_4/NH_3$ semicorona sample component.

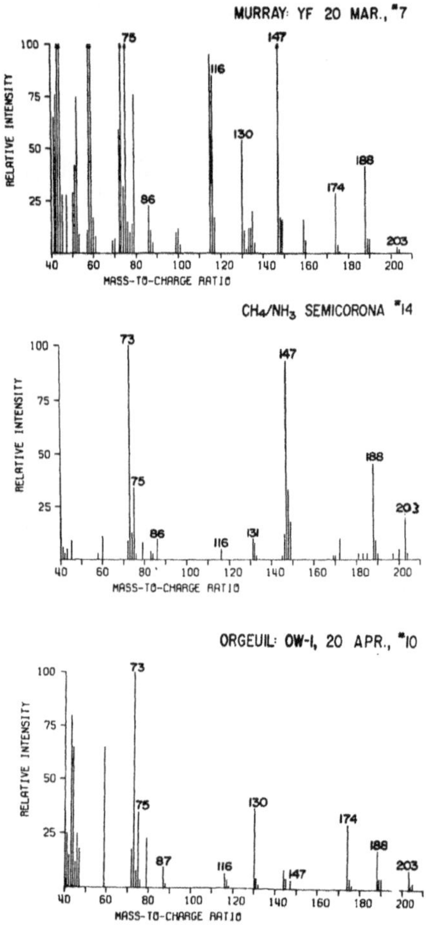


Fig. 3. Mass spectra of trimethylsilyl derivatives of (a) Murray: YF sample component; (b) Orgeuil: OW-1 sample component; (c) a corresponding $CH_4/NH_3$ semicorona component.

The products from the spark discharge experiments which were allowed to collect on the walls of the flask at 20° were derivatized with BSA as described earlier and the derivatized material was examined by GC-MS. The mass spectra were carefully examined and some showed similarities with spectra of trimethylsilyl derivatives of compounds resulting from Murray and Orgueil meteorites. One spectrum obtained from a component of the Orgueil meteorite and tentatively identified as a hydroxypyridine (Folsome *et al.*, 1973) has peaks at m/e 43, 59, 73, 79 and 152 which correspond with the mass spectrum of a component of the methane/ammonia semicorona discharge product (Figure 2). In addition, the molecular ion of the hydroxypyridine derivative is at m/e 167, whereas the discharge product has its base peak at m/e 166 and a number of peaks at higher m/e values, suggesting that it could be a hydroxypyridine with a substituent which is easily cleaved.

A component of the semicorona discharge showed peaks at m/e 45, 73, 75, 86, 116, 131, 147, 172, 188 and 203 and similarities were observed when compared to spectra from the Murray meteorite having peaks at m/e 45, 73, 75, 86, 116, 130, 147, 174, 188 and 203 and from an Orgueil component, with m/e 73, 75, 116, 130, 147, 174, 188 and 203. All three spectra had the same base and molecular ion peaks (Figure 3). Furthermore, a component from the semicorona discharge showed peaks at m/e 58, 75, 79, 115, 130, 135, 141, 174, 188, 204 and 221, which had similarities with a component of the Murray meteorite having peaks at m/e 58, 75, 79, 116, 129, 135, 147, 173, 188, 204 and 221 (Figure 4).

The closest correspondence was, however, observed between the mass spectrum of a component of the acetonitrile/ammonia semicorona discharge experiment and a component of the Murray meteorite. The peaks at m/e 43, 74, 89, 100, 104, 116, 147, 188, 204 and 219 were observed in both cases (Figure 5). Interestingly enough, there was no detectable correspondence between any spectra of the acetonitrile/ammonia

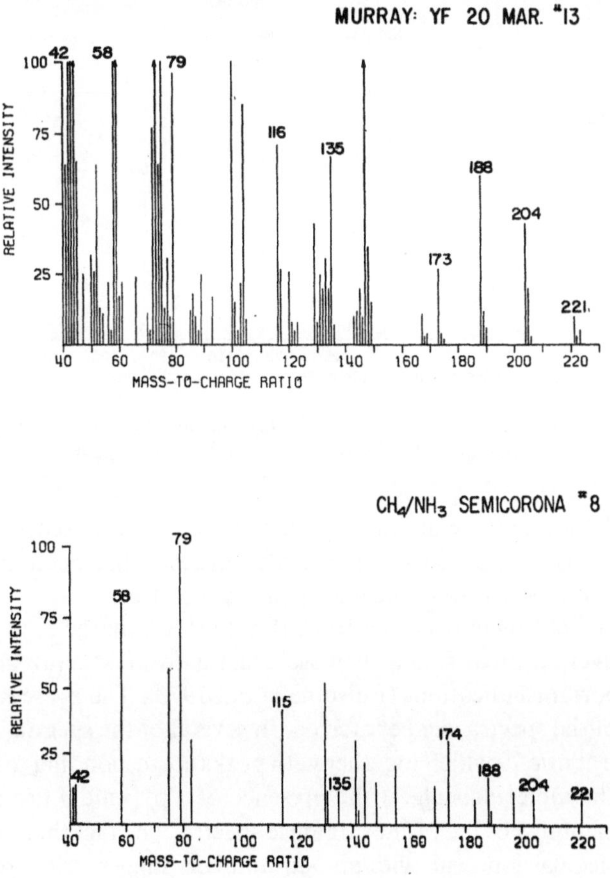


Fig. 4. Mass spectra of trimethylsilyl derivatives of (a) Murray: YF sample component; (b) a corresponding $CH_4/NH_3$ semicorona component.

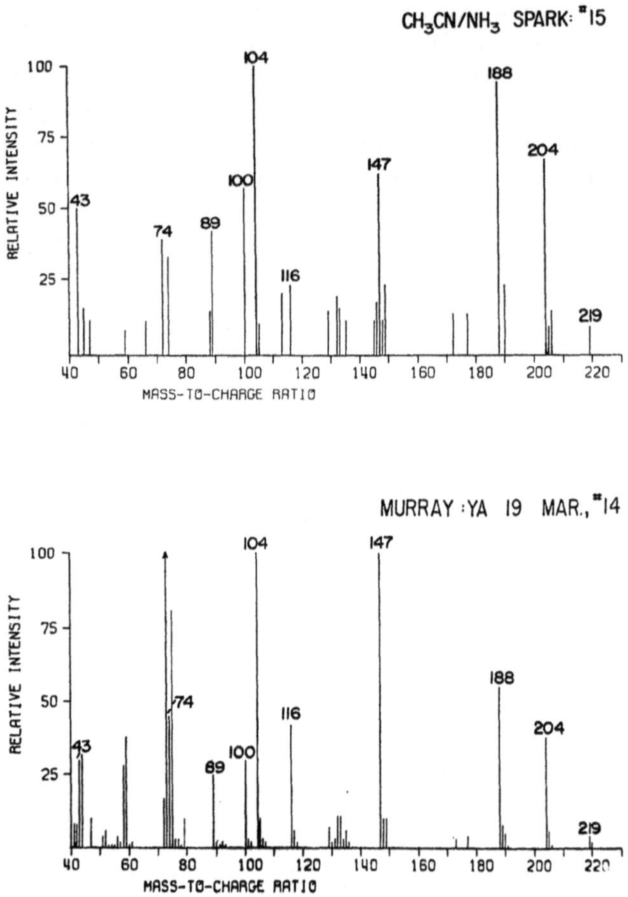


Fig. 5. Mass spectra of trimethylsilyl derivatives of (a) Murray: YA sample component; (b) a corresponding $CH_3CN/NH_3$ spark sample component.

experiment and the methane/ammonia experiments. The presence of acetonitrile-derived compounds in the Murray meteorite suggests that acetonitrile may have played a part in organic synthesis during chemical evolution.

The mixture of products resulting from the spark discharge is undoubtedly very complex. However, comparison with mass spectra from Murray and Orgueil has given some important indications (Folsome *et al.*, 1973). The presence of a pyridine nucleus in one of the spectra can be inferred. In several of the spectra, the major peak was shown to be at m/e 79 which together with peaks corresponding to the elimination of HCN from the molecule suggest the presence of a pyrimidyl group. The peak at m/e 79 is prominent in 5 of the 21 most intense spectra. The fact that most of the spectra have the molecular ion peak with an odd number, suggests the presence of $2n+1$ nitrogens (where $n=0$, 1, 2 etc.) in the system. The presence of one nitrogen can be attributed to the ring nitrogen in pyridine and three nitrogens could be accounted for by two in the ring and the third in the side chain. It is thus possible to rationalize the formation of pyrimidines in the electric discharge experiments. Evidence for the possible formation of 3-alkylaminopropionitriles has been discussed in this paper and a schematic presentation of their possible transformation to pyrimidine derivatives is shown in Figure 6. Recently some heterocyclic compounds were characterized from the Murchison meteorite (Folsoms *et al.*, 1971, 1973).

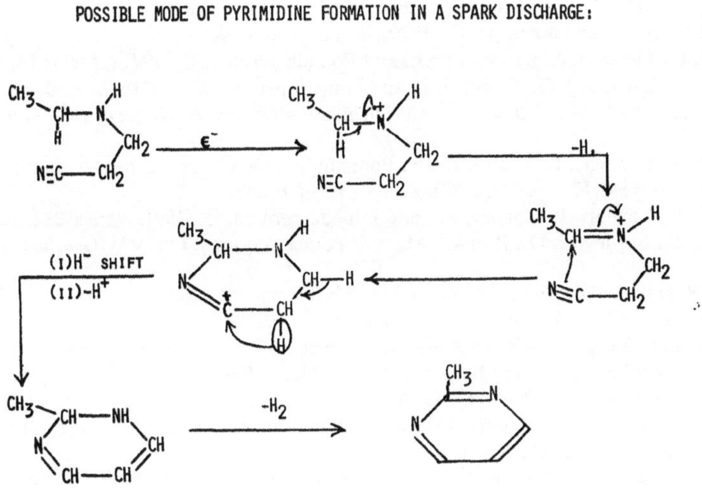


Fig. 6. Possible mechanism for the formation of pyrimidines from alkylaminopropionitriles.

Needless to say that an unequivocal characterization of the alkylaminopropionitriles, the pyridines and pyrimidines still remains to be carried out and this should be possible when better separation of the very complex mixture is possible.

The evidence so far obtained on the formation of aminoacetonitriles, their hydrolysis to some amino and imino acids and the presently discussed evidence for the possible formation of alkylaminopropionitriles and some pyridyl and pyrimidyl derivatives further focuses one's attention on the variety of organic compounds which could have resulted from a primitive earth atmosphere or a contemporary Jovian atmosphere and the intermediary role the aminonitriles could have played in the formation of more complex organic compounds.

It is anticipated that equally interesting results will be forthcoming when the products resulting from $CH_4/N_2$, $CH_4/N_2/NH_3$ and other suitably modified systems are examined for products like alkylaminopropionitriles and for some heterocyclics.

## Acknowledgements

Thanks are due to Dr Clair Folsome, University of Hawaii for mass spectra of the meteorite samples. The low temperature discharge experiments were performed by Mr Louis Fielder at the Ames Research Center. His assistance and the provision of facilities by NASA is acknowledged. Supported in part by NASA grant on Chemical Studies on the Origin of Life, NGR 21-002-317.

## References

Akabori, S.: 1955, *Kagaku, Tokyo* **25**, 54.

Buc, S. R.: 1955, *Organic Syntheses*, Collective vol. III, J. Wiley and Sons, p. 93.

Chadha, M. S. and Ponnamperuma, C.: 1970, unpublished work.

Chadha, M. S., Flores, J. J., Lawless, J. G., and Ponnamperuma, C.: 1971a, *Icarus* **15**, 39.

Chadha, M. S., Lawless, J. G., Flores, J. J., and Ponnamperuma, C.: 1971b, in R. Buvet and C. Ponnamperuma (eds.), *Molecular Evolution I – Chemical Evolution and the Origin of Life*, North-Holland Publ. Co., p. 143.

Chadha, M. S., Replogle, L., Flores, J., and Ponnamperuma, C.: 1971c, *Bio-Organic Chemistry* **1**, 269.

Cook, A. H. and Reed, K. J.: 1945, *J. Chem. Soc., London*, 399.

Folsome, C. E., Lawless, J., Romiez, M., and Ponnamperuma, C.: 1971, *Nature* **232**, 108.

Folsome, C. E., Lawless, J. G., Romiez, M., and Ponnamperuma, C.: 1973, *Geochim. Cosmochim. Acta* **37**, 455.

Hanafusa, H. and Akabori, S.: 1959, *Bull. Chem. Soc., Japan* **32**, 626.

Lemmon, R. M.: 1970, *Chemical Rev.* **70**, 95 and references therein.

Loew, G. H. and Chang, S.: 1971, *Tetrahedron* **27**, 3669.

Loew, G. H., Chadha, M. S., and Chang, S.: 1972, *J. Theor. Biol.* **35**, 359.

Miller, S. L.: 1957, *Biochim. Biophys. Acta* **23**, 480.

Molton, P. M. and Ponnamperuma, C.: 1973, Paper presented at the Middle Atlantic Regional ACS Meeting, Washington, D.C., Jan. 17.

Morimoto, S. and Ponnamperuma, C.: 1970, Paper presented at the 160th ACS meeting, Chicago, Sept. 13–18.

Moser, R. E. and Matthews, C. M.: 1968, *Experimentia* **24**, 658.

Noda, H. and Ponnamperuma, C.: 1971, in R. Buvet and C. Ponnamperuma (eds.), *Molecular Evolution I – Chemical Evolution and the Origin of Life*, North-Holland Publ. Co., p. 236.

Ponnamperuma, C. and Gabel, N. W.: 1968, *Space Life Sci.* **1**, 64 and references therein.

Ponnamperuma, C. and Woeller, F.: 1967, *Curr. Mod. Biol.* **1**, 156.

Snyder, L. E. and Buhl, D.: 1971, *Astrophys. J.* **163**, L 47.

Woeller, F. and Ponnamperuma, C.: 1969, *Icarus* **10**, 386.

# PART III

# ABIOTIC ORGANIC SYNTHESIS AND INTERACTIONS: POLYMERS

# MACROMOLECULES AND THE ORIGIN OF LIFE

HARUHIKO NODA, HIROSHI MIZUTANI and HIROYUKI OKIHANA

*Dept. of Biophysics and Biochemistry, Faculty of Science, University of Tokyo, Tokyo, Japan*

**Abstract.** From our knowledge of present day organisms, it is hard to imagine a living assembly, even at its most primitive stage, without macromolecules.

In order to look for the macromolecules which possibly participated in the assembly of the primitive organisms, the reaction and formation of polymers in HCN under irradiation of ultraviolet ray of 184.9 nm. was studied.

As an example of a simple way of producing an assembly of macromolecules, the mechanism of coacervation was studied by using gelatin as the material.

## 1. Introduction

Although it is almost impossible to give a complete definition of life, a tentative definition from the chemical point of view may be "Life is a system of coordinated chemical reactions." In the present organisms, chemical reactions are interconnected because the products of a reaction are the reactants of another reaction, and in addition there are cross-linking connections.

Whenever a covalent bond is made or broken in the living system, there is always an enzyme. An enzyme helps the system to attain equilibrium, but in the absence of an enzyme or when the enzyme is inactive, the particular reaction virtually stands still. Thus an enzyme functions as a valve in the pipeline system. Therefore, the chemical reactions in an organism are controlled by coordinated manipulation of the enzymes.

Enzymes in the present organisms are macromolecules. Enzymes are catalysts, and it seems that there is no reason for an enzyme to be a macromolecule, because the active site of an enzyme usually consists of a few chemical functional groups. A chemically functional group is composed of a few atoms and is on the average of a size of less than 10 Å. However, if these chemically functional groups are to be placed in proper orientation with each other in order to make an active site with a proper specificity, a large number of residues are necessary to support these functional groups. One reason for the great number is the fact that almost all biological molecules are single chain molecules, because their structure must be coded on the deoxyribonucleic acid. Another reason is that the main chain structure of a biomolecule is limited in its flexibility and a large amount of detour is necessary when a supporting structure is designed for two groups, the distance and relative orientation between which is fixed by the specificity requirement of the active site. An enzyme often has more than a single active site, and this would require a larger molecular size.

Biomolecules other than enzymes also have similar active sites, because all biological functions are chemical when viewed at the level of atoms and molecules. Thus, most of the molecules with some function in the living system are macromolecules by the same reason as that for enzymes.

By these reasons, macromolecules are necessary constituents of present day organisms. When we think of the first primitive organism on the Earth, there is no logical necessity that it was composed of macromolecules, but it is much more difficult to imagine one which was made only of smaller and simpler compounds.

If the formation of some macromolecules preceded the origin of life, what kind should we expect? In simulated primitive atmosphere experiments in which carbon and nitrogen were included in the components, HCN is always formed (cf. Gol'danskii *et al.*, 1971; Raye *et al.*, 1971; Mizutani *et al.*, 1972). Moreover, HCN is stable under heat and is very reactive because of the triple bond between carbon and nitrogen and will readily polymerize at a low temperature with generation of heat. In order to study a possible formation of polymer molecules in a simple and defined condition, we studied the reaction of HCN in the gas phase under excitation by 184.9 nm line of a mercury lamp.

Once macromolecules are formed, they have to be assembled in a limited space in order for them to have any kind of interaction. For such limited space, various kinds of phenomena have been proposed, such as droplets of coacervates, adsorptive surface of clay mineral crystals, etc. Especially, the droplets of coacervates are interesting because they look like cells under the microscope (Oparin, 1966; Bungenberg de Jong, 1949). The process of complex coacervation of gelatins has been studied.

## 2. Methods

A low-pressure mercury lamp was used for light source. The light intensity of the lamp at 184.9 nm was determined by $N_2O$ actinometry, and was estimated to be about $5.6 \times 10^{16}$ quanta $s^{-1}$. As hydrogen cyanide has no absorption above 191.5 nm, we did not use any filter to exclude 253.7 nm light. The analysis of the reaction products was made by gas chromatography, spot tests and mass spectrometry.

An acid-precursor gelatin (A) and an alkali-precursor gelatin (B) were used as the materials for coacervation experiments. Coacervation was restricted to the symmetrical mixtures, namely the mixtures of equal amounts of A and B of the same concentrations. Each of gelatins was deionized through the mixed-bed ion exchange column. Solutions of A and B were mixed in glass tubes at 40 °C. After mixing, the mixture was centrifuged for 10 min at 3000 rpm, and then the volume of coacervates was measured. The total volume of mixed solution was 10 ml in every case.

## 3. Results and Discussion

### Part I

Cyanogen and hydrogen were the major products, and smaller amounts of methane, ethane, ammonia, hydrazine and methylamine were also formed from HCN by UV irradiation. After some period of irradiation, solid material of reddish brown color deposited on the inner surface of the reaction cell. It is known that polymerization of HCN is accelerated by a base, and in this case the presence of ammonia and methylamine seemed to have accelerated the polymerization. In fact, the formation of colored material on the cell surface continued slowly after termination of ultraviolet irradiation.

From various preliminary measurements on the deposit on the cell surface, the polymerized substance is very similar to the substance (Noda and Ponnamperuma, 1971) formed by electric discharge in a gas mixture of methane and ammonia. The molecular weight seems to be a little larger, the maximum being in the order of $10^4$ or more.

Figure 1 shows the dependence of quantum yield of hydrogen, cyanogen and methane on the initial pressure of hydrogen cyanide. The quantum yield of methane shows a maximum at 180 torr of the initial pressure of HCN, while that for cyanogen stays constant and that for hydrogen decreases rapidly as the initial pressure of HCN increases. The addition of NO which is a scavenger of radicals remarkably reduced the quantum yield of hydrogen and methane, indicating that hydrogen atoms and methyl radicals are involved in the reaction.

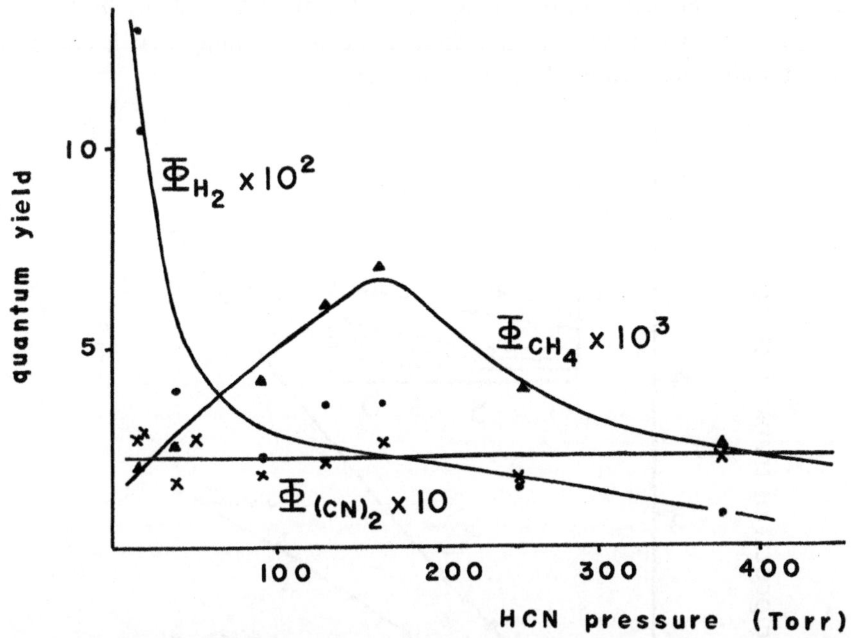


Fig. 1. Dependence of the quantum yield of $H_2$, $(CN)_2$ and $CH_4$ on the initial pressure of HCN. Irradiation was performed at the room temperature for 500 s.

From these findings, it is assumed that such reactions as shown below are taking place in the hydrogen cyanide gas when irradiated by 184.9 nm line of mercury. In these reactions, hydrogen atoms and CN radicals recombine and hydrogen atoms are added to the triple bond of HCN. The results shown in Figure 1 can be interpreted by the reactions in Table I, except the fact that the quantum yield of methane decreases

TABLE I

| | |
|---|---|
| HCN | $\rightarrow \cdot H + \cdot CN$ |
| $\cdot H + \cdot H$ | $\rightarrow H_2$ |
| $\cdot CN + \cdot CN$ | $\rightarrow (CN)_2$ |
| $\cdot H + HCN$ | $\rightarrow H_2CN$ (radical) |
| $\cdot H + H_2CN$ | $\rightarrow (H_2C{=}NH)$ |
| $\cdot H + (H_2C{=}NH)$ | $\rightarrow H_4CN$ (radical) |
| $\cdot H + H_4CN$ | $\rightarrow \cdot CH_3 + \cdot NH_2$ |
| $\cdot H + \cdot CH_3$ | $\rightarrow CH_4$ |
| $\cdot H + \cdot NH_2$ | $\rightarrow NH_3$ |
| $\cdot CH_3 + \cdot CH_3$ | $\rightarrow C_2H_6$ |
| $\cdot NH_2 + \cdot NH_2$ | $\rightarrow N_2H_4$ |
| $\cdot CH_3 + \cdot NH_2$ | $\rightarrow CH_3NH_2$ |

for higher initial pressure of HCN. This may be related to the formation of polymers of HCN, because the yield of the polymer increases as the initial pressure of HCN increases. Figure 2 indicates that the colored material is not the primary product of the reaction. The absorption spectrum of the colored material is featureless, and decreased monotonously towards longer wavelength. We wish to look into the mechanism of formation of polymers from radicals, assuming that it could be an example of abiotic formation of macromolecules.

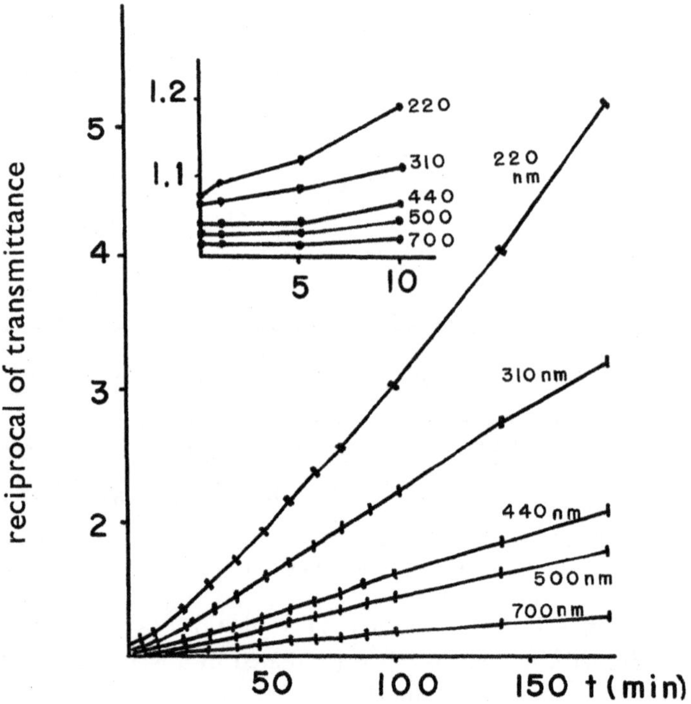


Fig. 2. Ultraviolet and visible absorption of polymeric substance developed on the cell surface when HCN was irradiated by ultraviolet rays of 184.9 nm at a pressure of 115 torr.

PART II

When measurements were made on complex coacervation of gelatin solutions (Figure 3), the concentration (Cc) of gelatin in the phase of coacervates was constant irrespective of the initial concentration (Ci) of gelatin and agrees with the expectation from a formula which was proposed in order to understand and interpret the phenomenon of complex coacervation (Overbeek and Voorn, 1957; Veis, 1961). However, the concentration (Ce) of gelatin in the external phase which was in equilibrium with the coacervates varied with the variation of the initial concentration (Ci) of gelatin.

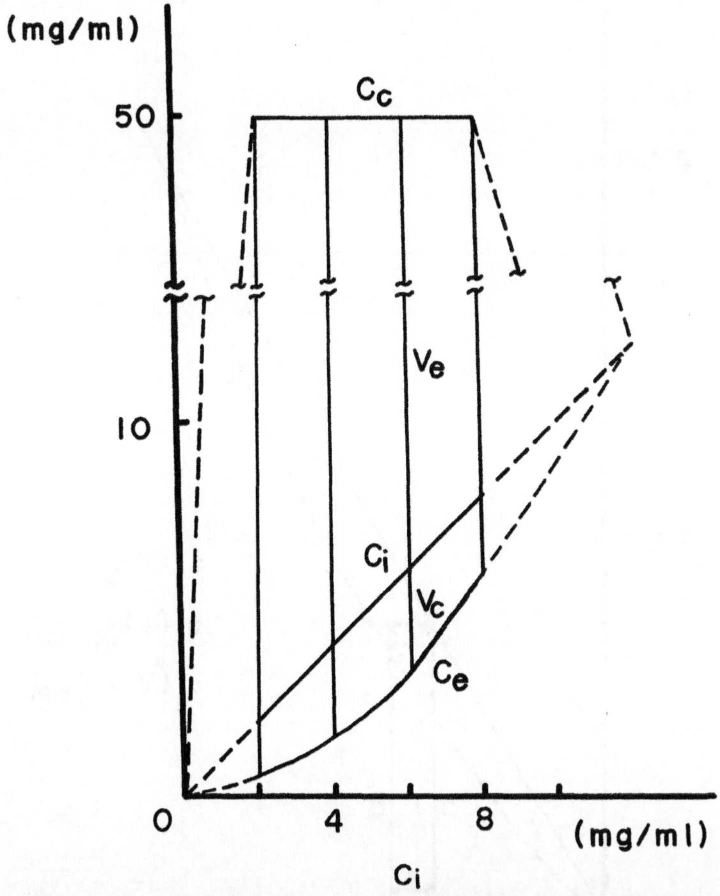


Fig. 3. Complex coacervation of gelatin. Cc is the concentration of gelatin in the coacervate phase. Ce is the concentration of gelatin in the solution outside of the coacervate phase. Ci is the concentration of gelatin in the initial solution. *Vc* and *Ve* stand for volume of the coacervate phase and external phase. Scale for *Vc* is magnified 10 times.

If the conditions postulated in the derivation of the foregoing formula are valid, the latter concentration (Ce) should be constant and independent of the initial concentration (Ci). The result obtained by Veis (1963) shows the same dependence of the external concentration on the initial concentration. Nakajima and Sato (1972) obtained similar results working on synthetic polymers. These results suggest that the phenomenon of complex coacervation is complicated and various other factors should be taken into consideration.

As one of the explanations that Ce is not constant on varying Ci, it is supposed that gelatins were inhomogeneous with regard to their ability to form coacervates. Some fractions of gelatins that have larger chain lengths have the ability to form coacervates, and others smaller ones do not. The fraction which has no ability tends to reduce the amount of coacervates because such a fraction acts as a coexisting salt.

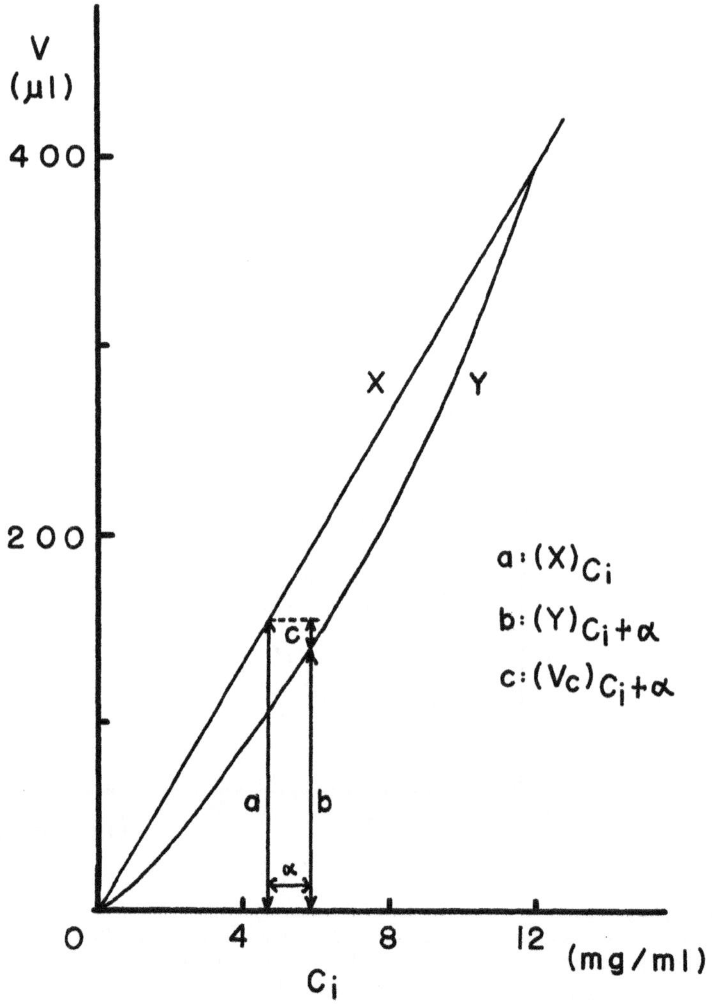


Fig. 4. The volume of coacervates without inhibition ($X$) and the suppression by electrostatic effect ($Y$). $(Vc)_{Ci+\alpha}=(X)_{Ci}-(Y)_{Ci+\alpha}$ where $(Vc)_{Ci+\alpha}$ is the volume of coacervates when the concentration of gelatin is $Ci+\alpha$. $(X)_{Ci}$ is the value of the curve $X$ when the concentration of gelatin is Ci. $(Y)_{Ci+\alpha}$ is the value of the curve $Y$ when the concentration of gelatin is $Ci+\alpha$.

In order to check up this assumption, the curve of the volume of coacervates ($Vc$) was resolved into a curve $X$ and a curve $Y$ (Figure 4). For the curve $X$, the data by using the coacervated gelatins (Okihana, 1973) was used. This curve should represent the volume of coacervates when only the fraction of high ability to form coacervates was used as the material. The curve $Y$ was obtained by subtracting the curve $Vc$ from the curve $X$. This curve will represent the extent to which coacervation is suppressed due to charges. The addition of a salt is equal to the increase of the value of the curve $Y$ because the fraction having no ability to form coacervates acts as a coexisting salt. So, by the following equation we can predict the volume of coacervates when a salt is coexistent. $(Vc)_{Ci+\alpha}=(X)_{Ci}-(Y)_{Ci+\alpha}$. By using NaCl as a representative salt, the effect of salt was measured as seen in Figure 5. The relation found agrees well with the expected one of the above equation.

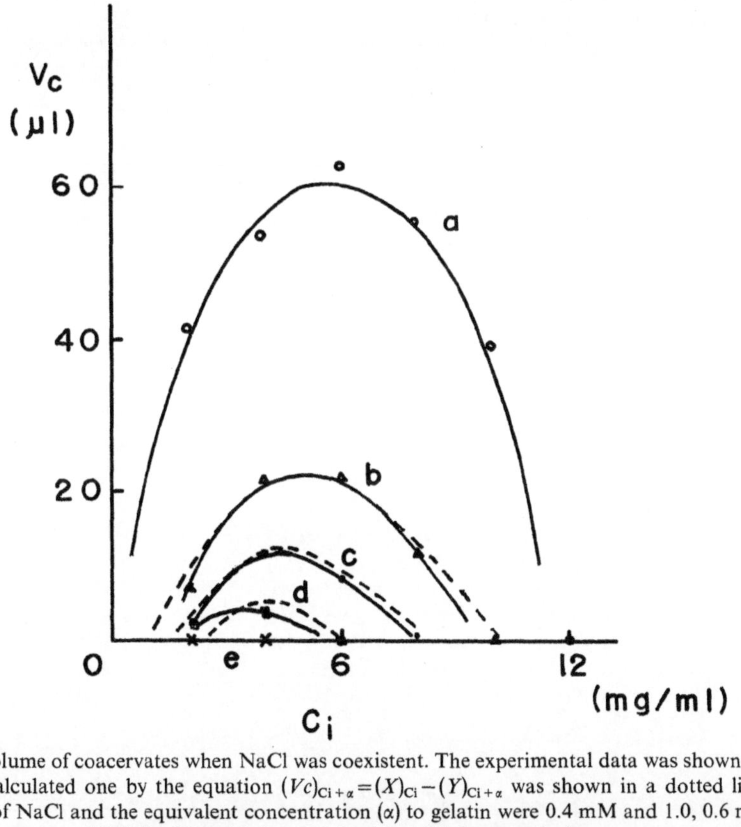


Fig. 5. The volume of coacervates when NaCl was coexistent. The experimental data was shown as solid line, and the calculated one by the equation $(Vc)_{Ci+\alpha}=(X)_{Ci}-(Y)_{Ci+\alpha}$ was shown in a dotted line. The concentration of NaCl and the equivalent concentration ($\alpha$) to gelatin were 0.4 mM and 1.0, 0.6 mM and 1.3, 0.8 mM and 1.5, and 1.0 mM and 1.8, respectively.

From our knowledge of modern biology, macromolecules and their interactions seemed to be the prerequiste for the primitive life. However, the kind of macromolecule which took part in the formation of the first life and the function of those molecules are totally unknown and even imagination cannot go far. We cannot sit still and wait. We must look for the material and the condition for the formation of a system which might lead to the creation of the primitive life.

## References

Bungenberg de Jong, H. G.: 1949, in H. R. Kruyt (ed.), *Colloid Science*, Vol. II, Elsevier, New York.
Gol'danski, V. I., Dzantiev, B. G., Popov, V. N., and Gavrilov, K. A.: 1971, *Atomic Energ.* **30**, 262.
Mizutani, H., Mikuni, H., and Takahashi, M.: 1972, *Chem. Letters* **7**, 573.
Nakajima, A. and Sato, H.: 1972, *Biopolymers* **10**, 1345.
Noda, H. and Ponnamperuma, C.: 1971, in R. Buvet and C. Ponnamperuma (eds.), *Chemical Evolution and the Origin of Life*, North-Holland Publ. Co., p. 236.
Okihana, H.: 1973, Thesis, Tokyo University.
Oparin, A. I.: 1966, *The Appearance and Early Development of Life*, Medizina Printing House, Moscow.
Overbeek, J. T. G. and Voorn, M. J.: 1957, *J. Cellular Comp. Physiol.* **49**, Suppl. 1, 7.
Raye, T. D., Juul-Dam, T., and Brockmeier, F. N.: 1971, *Chem. Eng. Progr. Sym. Ser.* **67**, 12.
Veis, A.: 1961, *J. Phys. Chem.* **65**, 1798.
Veis, A.: 1963, *J. Phys. Chem.* **67**, 1960.

# NMR STUDIES OF PREBIOTIC POLYPEPTIDES

SALVATORE ANDINI, ETTORE BENEDETTI, LUCIANO FERRARA, LIVIO PAOLILLO, and PIERO ANDREA TEMUSSI

*Instituto Chimico, Università di Napoli, 80134 Naples, Italy*

**Abstract.** Several polypeptides prepared by means of pyrocondensation have been the subject of structural investigations. Attention has been focused on the constitutional characterization of homo- and co-polymers containing Asp and Glu residues, whose role is essential for the formation of the so-called proteinoids. Contrary to the literature data based on chemical degradation, nmr studies show conclusively that in thermal poly-aspartic acid only $\beta$-peptide linkages are present. This result casts serious doubt on the role thermal condensation might have played in prebiotic polypeptide syntheses.

## 1. Introduction

Peptides have been synthesized in the last few years according to several possible prebiotic methods. Nearly all investigations on prebiotic syntheses of polypeptides have been performed mainly to test the possibility of spontaneous polymerization of amino acids and simple derivatives or to determine the relative reactivity of these biomonomers. Very little effort, however, has been devoted to structural investigations of the polymerization products.

On the other hand, in order to evaluate the relative merits of different syntheses of prebiotic polymers it may be very important to know whether the polymers obtained are indeed comparable to contemporary biopolymers. In fact, polycondensation of amino acids, though a rather simple process, may easily lead to random sequences of peptide residues with different configuration or even to peptide linkages different from the $\alpha$-peptide linkage characteristic of proteins. This is so because racemization is favoured by the conditions of some of the known prebiotic polymerizations and because some amino acids have reactive groups in the side chain, *e.g.* the $\beta$, $\gamma$ and $\varepsilon$ groups of aspartic and glutamic acids and of lysine.

The presence of constitutional errors may be particularly relevant in the case of the so-called proteinoids described by Fox (1971) since aspartic acid, glutamic acid and lysine are essential in thermal polycondensations. Kovacs and Konyves (1954) studied the structure of polyaspartic acid by means of chemical degradation even before interest arose in thermal polypeptides in relation to chemical evolution. These authors demonstrated that the product obtained by pyrocondensation of aspartic acid, followed by mild alkaline hydrolysis, contains both $\alpha$- and $\beta$-peptide linkages in comparable amounts.

We thought it interesting to check these data by means of physical methods, free of the ambiguities of chemical degradation and to extend such an investigation to thermal co-polymers containing Asp and Glu residues. Among the possible physical techniques nmr spectroscopy seemed best suited because it can give detailed information on groups differing only slightly in chemical environment. On the other hand, the more traditional optical spectroscopic methods (UV, ORD, CD) are of little use if the polypeptide is not composed of residues of the same configuration. Besides, there existed in the literature a specific example of the use of nmr to distinguish between $\alpha$- and $\beta$-linkages of aspartic residues and $\alpha$- and $\gamma$-linkages of glutamic residues (Sheinblatt, 1966, 1967).

## 2. Experimental

All polymers were prepared according to procedures described by Fox and Harada (1966), with only minor modifications of the reaction vessels designed to ensure a more uniform heating through the entire mass of solid amino acid. Substantial increases in yields resulted from these modifications. Polyaspartic acid was also prepared by heating an intimate mixture of aspartic acid and kaolinite.

Purification and isolation were based, according to Rohlfing (1967), on extensive dialysis of slightly alkaline solutions but were improved by preliminary extraction of the unreacted amino acids with water in a Soxhlet apparatus and by final lyophilization of the dialyzed solution. Several copolymers containing an apolar residue (*e.g.* copolyglutamic acid-leucine) were not soluble enough to be dialyzed as true solutions; consequently we can not be sure of their purity and no structural data will be discussed in this communication. Only polyaspartic acid, copoly-(aspartic acid, glutamic acid) and copoly-(glutamic acid, glycine) could be adequately purified to give reliable structural data.

Amino acids were all reagent grade from various sources. Authentic samples of poly- $\alpha$-L-aspartic acid and poly- $\alpha$-L-glutamic acid were purchased from Sigma Chemical Company, St. Louis.

The pmr spectra were recorded on a Varian HA-100-15 spectrometer at 29 °C, using external HMS for $D_2O$ solutions. The $^{13}C$ spectra were recorded on a Bruker WH-90 spectrometer at room temperature, using external TMS as reference.

All pH measurements were performed on a Radiometer PHM 26 pH meter immediately before spectral runs.

## 3. Results and Discussion

As described by Kovacs and Konyves (1954), anhydropolyaspartic acid, the product of condensation of aspartic acid, which is insoluble in water or in acidic aqueous solutions, should have the structure of a poly-imide.

The pmr spectrum of a trifluoroacetic acid (TFA) solution of this product, shown in Figure 1, is consistent with such a structure. Only two resonances are present, a singlet for the CH group (with a half-height line-width typical of a polymer) at $\delta = 5.57$ ppm, and a rather blurred quartet at $\delta = 3.40$ ppm assignable to the $CH_2$ group. The two protons of the $CH_2$ group, adjacent to an asymmetric center, can only give rise to an AB type quartet if part of a fairly rigid structure. This is precisely the case for the rings of the polyimide shown above.

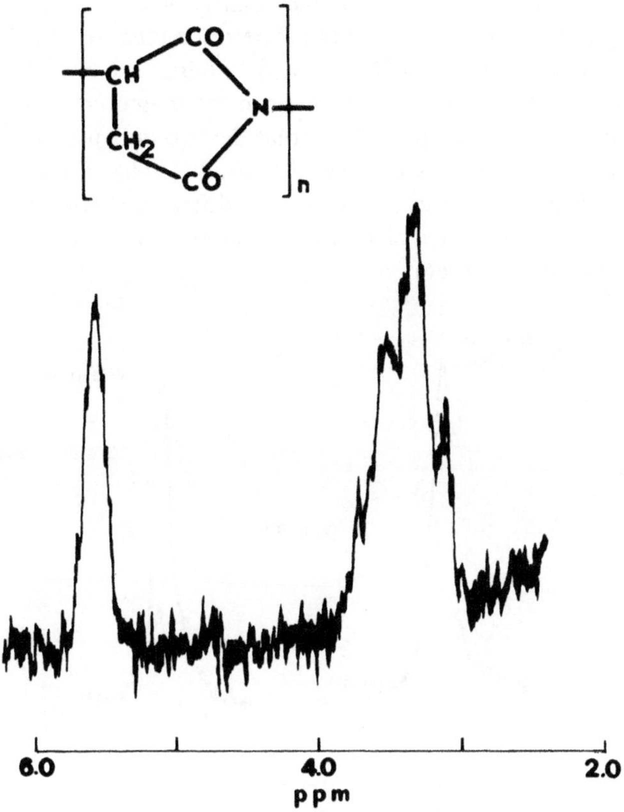


Fig. 1. 100 MHz pmr spectrum of a TFA solution of anydropolyaspartic acid.

The product of hydrolysis of the polyimide can be compared with the sodium salt of an authentic sample of poly-α-aspartic acid. The spectra of TFA solutions of both salts are apparently identical in all respects ($\delta CH_2 = 3.24$ ppm, $\delta CH = 5.15$ ppm). Owing to the marked similarity of the chemical environments of the CH and $CH_2$ groups in α- and β-linkages, however, this can not be taken as a conclusive proof that our polymer is poly-α-aspartic acid because it is not possible to exclude accidental coincidence of the resonances in a single solvent system. As mentioned in the introduction, Sheinblatt (1967) has shown that it is possible to distinguish α- and β-linkages in aspartic acid peptides by changing the degree of ionization of the carboxyl group. The variation of electron density connected with the protonation of the carboxylate group induces a downfield shift of 20 to 30 Hz in protons of groups one bond apart and half this shift in protons of groups two bonds apart. Besides, the carboxyl groups of α- and β-linked residues (adjacent, respectively, to a $CH_2$ and to a CH group) should have different pK's and, accordingly, the shifts should occur around different pH's. The pmr spectra in $D_2O$ at high pH (Figure 2) show that the $CH_2$ groups of the two polymers have different chemical shifts: that of authentic poly-α-aspartic acid is at 2.99 ppm, that of the thermal polyaspartic acid is at 3.10 ppm. The resonances of the CH groups are not observable because of the large HDO peak. Figure 3 shows the variation of the $CH_2$'s chemical shifts versus pH values. Both the magnitudes of the shifts and the positions of the inflection points indicate that the two polymers are different and that the thermal polymer is entirely composed of β-units. The correctness of this assignment is also backed by the observation that the difference between the pK's of Figure 3 is indeed of the same order of magnitude of that observed in small peptides of aspartic acid (Sheinblatt, 1966).

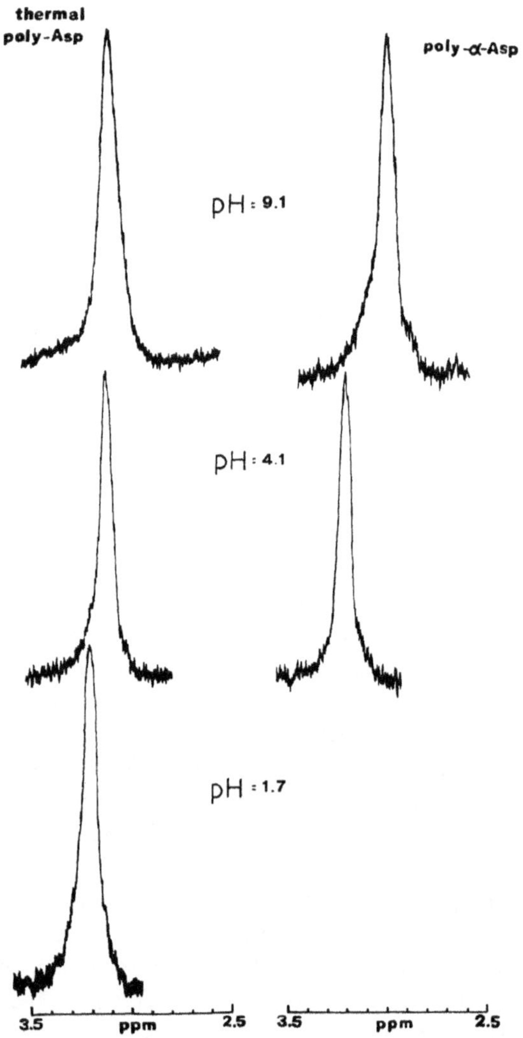


Fig. 2. 100 MHz pmr spectra of $CH_2$ groups of thermal polyaspartic and poly-α-aspartic acids of different pH values in $D_2O$ solutions.

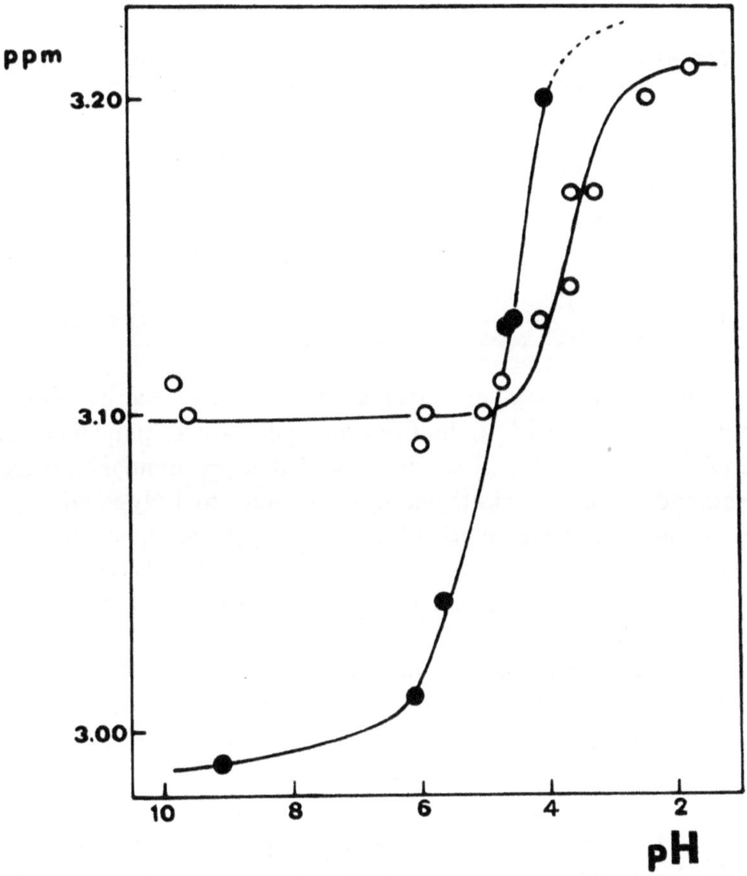


Fig. 3. pH dependence of $CH_2$ chemical shift of thermal polyaspartic acid (○) and of poly-α-aspartic acid (●)

Polymers prepared from L-aspartic, (D,L)-aspartic and a mixture of L-aspartic and kaolinite gave all the same result, *i.e.* they are all formed by β-aspartic units.

This structure was further confirmed by $^{13}C$ spectra of $D_2O$ solutions of thermal polyaspartic and poly-α-aspartic acids at several pH's. As reported in Table I the $^{13}C$ chemical shifts of α-C and β-C are consistent with the trends of the titration curves for the protons attached to the same carbons (see Figure 3).

Our result is in sharp contrast with the chemical degradation study of Kovacs and Konyves (1954) but the discrepancy is probably only apparent. The quoted degradation study involves a series of reactions before the actual final degradative step; one of these reactions, perhaps the esterification in dry methanol, may well cause a partial transpeptidation, thus transforming a sequence of β-peptide linkages into a random sequence of α- and β-units.

Rather, if one thinks of the way the thermal polyaspartic acid is synthesized, it

TABLE I

22.6 MHz $^{13}C$ chemical shifts of $D_2O$ solutions of poly-α-aspartic acid and thermal poly-aspartic acid at different pH's. All figures are ppm from external TMS.

| pH | α-C | | β-C | |
|---|---|---|---|---|
| | Thermal polyaspartic | poly-α-aspartic | Thermal polyaspartic | poly-α-aspartic |
| 1.0 | 51.1 | | 38.1 | |
| 4.5 | 53.3 | 52.6 | 39.3 | 39.2 |
| 10.0 | 53.3 | 52.9 | 39.3 | 40.2 |

is very surprising to find a regular sequence of one type of peptide bond only. In fact, thermal polyaspartic acid results from the hydrolysis of polyimide containing two very similar N—CO bonds; it seems unlikely that preferential cleavage of only one of these bonds be due to electronic reasons. Since hydrolysis takes place in a suspension of the solid material it appears that conformational preferences are probably responsible for the cleavage of one of the imide bonds only.

The two copolymers Glu-Asp and Glu-Gly were studied along the same lines followed for polyaspartic acid.

The only difficulty in both cases was due to the rather low proportions of glutamic acid residues incorporated in the copolymers (*ca.* 10% in Glu-Asp and *ca.* 15% in Glu-Gly) that made the resonances of the Glu $CH_2$ groups very weak with respect to other resonances. The Asp $CH_2$ of the copolymer behaves exactly like that of the homopolymer as a function of pH, *i.e.* it goes from $\delta=3.11$ ppm at high pH to $\delta=3.20$ at low pH so that it can be considered as typical of a $\beta$ structure. The peaks of the Glu $CH_2$'s instead, shift as those of authentic poly-α-L-glutamic acid.

At high pH the $\beta CH_2$ has a $\delta=2.35$ ppm and the $\gamma CH_2$ has a $\delta=2.55$ ppm; at low pH they shift respectively to $\delta=2.46$ ppm and $\delta=2.82$ ppm consistently with a structure of α-peptide linkages.

On the other hand the Glu residues of copoly-(Glu, Gly) are apparently linked as $\gamma$ units. In this case, in fact, it was not possible to record pmr spectra at many different pH values because the polymer precipitates at neutral or acid pH's. However, even at very high pH's the chemical shifts of both $CH_2$ peaks are downfield with respect to those of poly-α-L-glutamic acid; at pH 12.5 the $\beta CH_2$ is at $\delta=2.40$ ppm and the $\gamma CH_2$ is at $\delta=2.75$ ppm. Since a similar chemical shift difference was observed, at high pH, for the $CH_2$ peaks of α and β polyaspartic acid, it can be inferred that the Glu residues of copoly-(Glu, Gly) are linked as γ-peptide units.

## 4. Concluding Remarks

In conclusion, our nmr study, although restricted to only a few polymers seems to indicate that costitutional errors are rather likely in thermal polycondensation of amino acids. In fact, it must be noted that since the products of condensation of

Asp and Glu monomers are imides and not amides it is possible, in principle, to find different conditions for hydrolysis leading to α-peptides instead of β- or γ-peptides.

However, until such conditions are found it is our opinion that thermal condensation should be regarded as less probable than other polymerizations as source of prebiotic polypeptides.

## 5. Acknowledgements

We gratefully acknowledge Bruker-Physik AG of West Germany for the use of their WH-90 spectrometer in recording $^{13}C$ spectra.

This work has been carried out with the support of Research grant (71/01561) from the Italian CNR. A partial support from LCFMIB of CNR (Arco Felice) is also gratefully acknowledged.

## References

Fox, S. W.: 1971, in A. P. Kimball and J. Oro (eds.), *Prebiotic and Biochemical Evolution*, North Holland, Amsterdam-London, p. 8.

Fox, S. W. and Harada, K.: 1966, in P. Alexander and H. P. Lundgren (eds.), *Analytical Methods of Protein Chemistry*, Vol. 4, Pergamon Press, Oxford, p. 129.

Kovacs, J. and Konyves, I.: 1954, *Naturwiss.* **14**, 333.

Rohlfing, D. L.: 1967, *Nature* **216**, 657.

Scheinblatt, M.: 1966, *J. Amer. Chem. Soc.* **88**, 2845.

Scheinblatt, M.: 1967, in A. Ehrenberg, B. G. Malmstrom and T. Vanngard (eds.), *Magnetic Resonance in Biological Systems*, Pergamon Press, Oxford, p. 35.

# THE ORIGIN OF PROTEINS: HETEROPOLYPEPTIDES FROM HYDROGEN CYANIDE AND WATER

CLIFFORD N. MATTHEWS

*Dept. of Chemistry, University of Illinois at Chicago Circle, Chicago, Ill. 60680, U.S.A.*

**Abstract.** Evidence from laboratory and extraterrestrial chemistry is presented consistent with the hypothesis that the original heteropolypeptides on Earth were synthesized spontaneously from hydrogen cyanide and water without the intervening formation of α-amino acids, a key step being the direct polymerization of atmospheric hydrogen cyanide to polyaminomalononitrile (IV) via dimeric HCN. Molecular orbital calculations (INDO) show that the most probable structure for $(HCN)_2$ is azacyclopropenylidenimine. Successive reactions of hydrogen cyanide with the reactive nitrile side chains of IV then yield heteropolyamidines which are converted by water to heteropolypeptides. To study this postulated modification of a homopolymer to a heteropolymer, poly-α-cyanoglycine (IX) was prepared from the N-carboxyanhydride of α-cyanoglycine. Hydrolysis of IX, a polyamide analog of the polyamidine IV, yielded glycine. However, when IX was hydrolysed after being treated with hydrogen cyanide, other α-amino acids were also obtained including alanine, serine, aspartic acid and glutamic acid, suggesting that the nitrile groups of IX (and therefore of IV) are indeed readily attacked by hydrogen cyanide as predicted. Further theoretical and experimental studies support the view that hydrogen cyanide polymerization along these lines is a universal process that accounts not only for the past formation of primitive proteins on Earth, but also for the yellow-brown-orange colors of Jupiter today and for the presence of water-soluble compounds hydrolyzable to α-amino acids in materials obtained from environments as diverse as the moon, carbonaceous chondrites and the reaction chambers used to simulate organic synthesis in planetary atmospheres.

During the past two decades considerable research on chemical evolution has led to the widespread belief that the prebiotic formation of primitive proteins on Earth occurred in two stages, α-amino acid synthesis initiated by the action of natural high-energy sources on the components of a reducing atmosphere, followed by polycondensation of the accumulated monomers in the oceans or on land (Fox and Dose, 1972; Gabel and Ponnamperuma, 1972). A more critical examination of the evidence for the second step, however, shows that the inherent thermodynamic barrier to spontaneous polymerization of α-amino acids has been experimentally overcome only by invoking specific conditions – anhydrous, high-temperature or acidic – not necessarily characteristic of a young, developing planet. An alternative route for the origin of proteins obviates this problem by assuming the direct synthesis of heteropolypeptides from hydrogen cyanide and water without the intervening formation of α-amino acids (Kliss and Matthews, 1962; Matthews and Moser, 1966, 1967; Matthews, 1971), thus showing how primeval Earth could have been covered with a proteinaceous matrix able to assist the synthesis and further evolution of nucleic acids and other macromolecules essential for life today.

A highly compressed argument follows in support of this view, with particular reference to recent reports of the existence of abiotic water-soluble compounds hydrolyzable to α-amino acids in materials obtained from the moon (Harada *et al.*, 1971; Fox *et al.*, 1972), carbonaceous chondrites (Cronin and Moore, 1971), photochemical experiments simulating organic synthesis in planetary atmospheres (Sagan and Khare, 1971), and various hydrogen cyanide reactions in our laboratory (Minard and Matthews, 1972). The solid extracts yield glycine, alanine, aspartic acid, glutamic acid and other common α-amino acids (Table I) only after complete hydrolysis with 6N HCl. It is proposed here that the water-soluble products from each of these investigations are essentially the same and that they are the outcome of a universal process for the spontaneous generation of heteropolypeptides *via* hydrogen cyanide polymerization that may have been predominant on Earth over four billion years ago.

According to this hypothesis (Matthews and Moser, 1966, 1967) a low-energy pathway exists that allows hydrogen cyanide I to polymerize readily to polyaminomalononitrile IV *via* dimeric hydrogen cyanide II and its polymer III (Figure 1). Semi-empirical quantum-mechanical calculations by the INDO method suggest that II is probably azacyclopropenylidenimine (Yang *et al.*, 1973) rather than iminoacetonitrile or aminocyanocarbene, other possible structures that could also lead to III (Moser *et al.*, 1967).

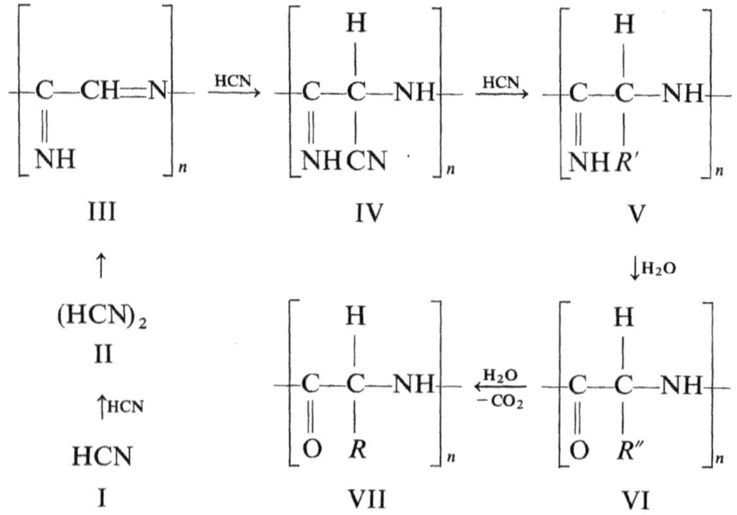

Fig. 1. Proposed route for heteropolypeptide synthesis from hydrogen cyanide and water (I → VII). *R″* and *R′* are precursors of protein side chains *R*.

Successive reactions of hydrogen cyanide with the reactive nitrile groups of IV then yield heteropolyamidines V which on contact with water are converted to heteropolypeptides VI and VII following a series of hydrolysis and decarboxylation steps (Figure 1). To study this postulated modification of a homopolymer to a heteropolymer, poly-α-cyanoglycine (IX) was prepared from the N-carboxyanhydride of

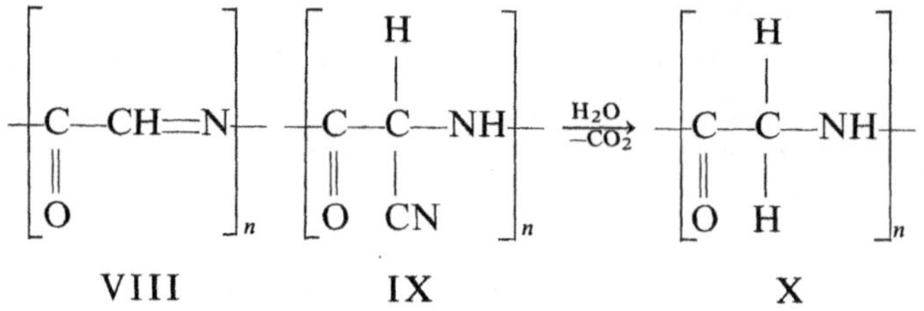

TABLE I

Protein α-amino acids found in acid hydrolyzates of aqueous extracts from the Moon, meteorites, photochemical vapor phase products, and hydrogen cyanide polymers

| General source | Specific source | Amino acids | | | | | | | | | | References |
|---|---|---|---|---|---|---|---|---|---|---|---|---|
| | | gly | ala | asp | glu | ser | thr | pro | val/ival | leu/ileu | cys | |
| Moon | Apollo 11 | × | × | × | × | × | × | | | | | Harada *et al.* (1971) |
| | Apollo 12 | × | × | × | × | × | | | | × | | |
| | Apollo 14 | × | × | × | × | × | | | | | | Fox *et al.* (1972) |
| Carbonaceous chondrites | Murchison | × | × | × | × | | | × | × | | | Cronin and Moore (1971) |
| | Murray | × | × | × | × | | | × | × | | | |
| Photochemical vapor phase products | <50°, 2537 Å $CH_4$, $C_2H_6$, $NH_3$, $H_2S$, $H_2O$ | × | × | × | × | × | | | | | × | Sagan and Khare (1971) |
| Hydrogen cyanide polymers | <50°, 2537 Å HCN | × | × | × | × | × | | | | | | Minard and Matthews (1972) |

α-cyanoglycine (Warren and Matthews, 1971). Hydrolysis of IX, a yellow polyamide analog of the polyamidine IV, yielded glycine, presumably following nitrile hydrolysis and decarboxylation to polyglycine X. However, when IX was hydrolyzed after first being treated with hydrogen cyanide, other α-amino acids were also obtained such as alanine, serine, aspartic acid and glutamic acid (GC-MS analysis) suggesting that the nitrile groups of IX, and therefore of IV, are indeed readily attacked by hydrogen cyanide as predicted (Minard *et al.*, 1973). Polymers possessing up to fifteen of the twenty amino acid side chains commonly found in proteins would be expected to be synthesized by these reactions involving only hydrogen cyanide and water. Formation of the remaining five residues, aromatic or sulfur-containing, would require further interaction of hydrogen cyanide and polyaminomalononitrile with acetylene or hydrogen sulfide.

Other evidence consistent with this proposed sequence has been provided by several types of experiments involving base-catalyzed reactions of hydrogen cyanide introduced directly as a starting material (Matthews and Moser, 1967) or formed indirectly *in situ* (either by elimination from compounds such as aminoacetonitrile (Moser and Matthews, 1968), aminomalononitrile (Moser *et al.*, 1968a), and diaminomaleonitrile (Moser *et al.*, 1968b) or by synthesis after applying electric discharges to anhydrous mixtures of methane and ammonia (Matthews and Moser, 1966)). In either case polymers could be isolated, or their presence inferred, following extraction by cold water of the reaction products. Acid hydrolysis of the polymers yield many of the common protein α-amino acids (IEC or GLC analysis), particularly glycine, alanine, aspartic acid, glutamic acid and serine, as well as some non-protein amino acids. The presence of these five residues in a representative polymer sample has been confirmed by two different gas chromatographic procedures (Minard and Matthews, 1972). It seems likely that the amino acids arising from alkaline cyanide studies of other workers including Oró and Kamat (1961), Lowe *et al.* (1963), Abelson (1966), and Ferris *et al.* (1973) are, similarly, secondary products derived from such polymers. The yellow-brown aqueous extracts usually give a positive biuret test, although instead of the expected purple color, green is observed (purple plus yellow), together with an increase in the absorption at 550 nm (Widing *et al.*, 1972). It should be noted, however, that the absorption maxima of these and other synthetic polypeptides such as polyglycine and polyglutamic acid depend on molecular weight and the nature of the side chains as well as on experimental conditions. After treatment with water, the cyanide polymers probably consist of linked segments of several kinds: heteropolypeptides VI and VII, polyglycine X formed from IV by hydrolysis and decarboxylation, and unsaturated polyamides VIII arising from incomplete conversion of III to IV by hydrogen cyanide addition. Ready susceptibility to hydrolysis predicted for VIII may account for the low molecular weights (300–400) of some of these solids after extraction with water. While extensive enzymatic cleavage of such irregular polymers would not be expected, ninhydrin-positive compounds are formed in some cases after treatment with subtilisin, carboxypeptidase A and leucine aminopeptidase (Glogovac and Matthews, 1969).

These observations suggest the need for most careful interpretation of research directed towards the question of the origin of proteins. For example, in atmospheric simulation experiments where methane, ammonia and water (or comparable reducing mixtures) are subjected to high energy radiation of various kinds, it is possible that the α-amino acids obtained are not primary products, as is generally assumed, but rather are formed by the hydrolytic breakdown of peptides generated directly from hydrogen cyanide and water, particularly since hydrogen cyanide is known to be a major intermediate in such reactions. When reaction conditions are severe, as in the pioneering experiments of Miller (1955) with electric discharges, it is not surprising that only free amino acids were detected since the boiling water in their apparatus would soon have brought about total hydrolysis of peptides. Under milder conditions, however, as in Grossenbacher and Knight's spark experiments (1965) where water at about 40° was used, or the series of experiments conducted by Ponnamperuma (1971) peptides as well as free amino acids were found. Indeed, when the water in such reactions was kept at room temperature we were able to detect α-amino acids only after acid hydrolysis of the solid residues obtained by evaporation of the products. (Stiffel *et al.*, 1973). A similar temperature effect was noticed in the photochemical studies already mentioned (Sagan and Khare, 1971) where solids hydrolyzable to amino acids were obtained below 50° but only free amino acids within the range 215–400°. Taken as a whole, these results could well indicate not that "the amino acids appear to have been linked together as soon as they were synthesized" (Ponnamperuma, 1971) but rather that the process most relevant to the question of the origin of proteins is the direct synthesis of heteropolypeptides from hydrogen cyanide and water.

Perhaps the optimum conditions for such chemistry existed when Earth was a young planet with a developing atmosphere of methane, ammonia and water. Photochemical reactions in the stratosphere leading to macromolecule formation would have been of prime importance rather than fragmentation reactions in the ionosphere such as those simulated by Huntress *et al.* (1969) in their ion cyclotron experiments. The prebiotic synthesis of molecules essential for life would then have proceeded in three overlapping stages (Kliss and Matthews, 1962, 1973) defined by the relative volatility of each component. First, hydrocarbons were formed following the photolysis of methane to acetylene and the tautomeric $C_3H_4$ molecules methylacetylene and allene. Isoprene oligomers and polymers – the foundation for today's Isoprene Rule – were important products, as were unreactive aliphatic and aromatic compounds that eventually accumulated to become the main constituents of ancient petroleum deposits. During the second stage, with ammonia becoming more involved in atmospheric photochemistry, hydrogen cyanide and cyanoacetylene (Sanchez *et al.* 1966) were major reactants. Polymeric peptide precursors IV and V were formed in the stratosphere together with nitrogen heterocycles possessing the basic skeletons of purines, pyrimidines, pyridines and porphyrins. Hydrogen sulfide also present took part in the modification of side chains of the polymers. When most of the ammonia had been used up – its action as an ultraviolet screen (McGovern, 1969) thereby being considerably diminished – photolysis of water vapor that had been confined to lower levels became possible leading to the third stage when formaldehyde and then sugars were synthesized as well as phosphates from phosphine*. Meanwhile, as the macromolecules and heterocycles of the second stage gradually settled, contact with water introduced oxygen functionality mainly by hydrolysis of azomethines to carbonyl groups and of nitriles to amides and carboxylic acids. An important result was the conversion of heteropolyamidines to heteropolypeptides possessing essentially the same side chains found in proteins today. As Earth's surface became covered with this proteinaceous material and the other products of atmospheric photochemistry, catalyzed polymerization of nucleotides to nucleic acids took place while protein-fatty acid aggregation prepared the way for cell formation. Increasingly specific interaction of organic and inorganic molecules in the oceans then led to the evolution of enzymatic reaction chains (Horowitz, 1945) and the emergence of life.

With this model for terrestrial molecular evolution as a guide, the ubiquitous role of hydrogen cyanide in extraterrestrial chemistry can perhaps be better understood. Some of the characteristic colors of Jupiter, for example, a planet with an atmosphere rich in methane and ammonia, could be caused by the presence of yellow-brown-orange polymers of hydrogen cyanide (Matthews and Moser, 1966; Woeller and Ponnamperuma, 1969). Structures II–V would be expected in the upper Jovian atmosphere and structures VI–X at lower levels after contact with water, whether present as vapor, liquid or ice. On the Moon, also, polymers of either class could, in principle, have been synthesized from HCN or deposited by meteoritic impact, although the absence of a lunar atmosphere suggests that any indigenous cyanide polymers would consist of heterogenous mixtures rather than the regular hetero-polyamidines of high molecular weight postulated for atmospheric chemistry on primeval Earth. The lunar extracts hydrolyzable to $\alpha$-amino acids (Harada *et al.*, 1971) are, indeed, reported to resemble the crude, water-soluble polymers of HCN previously known from laboratory studies. It remains to be shown whether the oxygen-containing functional groups of the extracted lunar material were originally present on the moon, indicating that water was involved in early lunar chemistry to give structures VI–X or whether the groups were introduced by hydrolysis of the non-oxygenic structures III–V during the extraction and analysis steps undertaken on Earth.**

* Biochemical structures are based essentially on the six elements – H, C, O, N, S, P – whose hydrides do not react with water. These volatile hydrides could therefore take part in atmospheric photochemistry as outlined here. The original synthesis of the major molecular structures of geochemistry can then be viewed as the outcome of reactions of water with the remaining hydrides (particularly those of the abundant elements silicon, aluminum and magnesium) that were formed during the aggregation of cosmic matter leading to the genesis of planets.

** Following further rigorous analysis (IEC, GLC) of lunar extracts hydrolyzable to $\alpha$-amino acids, Gehrke *et al.* (1973) conclude that terrestrial contamination cannot be ruled out. Whether indigenous or foreign, it is argued here that cyanide polymers are precursors of any abiotic amino acids – free or bound – obtained by extraction of Moon dust.

Similar clarification of the nature of the aqueous extracts from the Murchison and Murray meteorites (Cronin and Moore, 1971) may give clues as to the early history of these bodies.* In this connection, following the observation (Hare *et al.*, 1970) that "amino acids can be recovered from some geological samples by extracting with water first and then by hydrolyzing, whereas they are not found when the sample is hydrolyzed directly, presumably due to decomposition catalyzed by the high proportion of minerals in the presence of hot HCl" it has become accepted (Fox *et al.*, 1972) that "the use of hot water to extract samples, followed by the hydrolysis of the aqueous extract, appears to be the method of choice for identification and quantification of amino acid precursors in extraterrestrial sources." A more important reason for extraction – to isolate peptides – is, of course, clearly indicated by the cyanide model. It therefore seems reasonable to suggest that in all future analyses of extraterrestrial material an essential early step should be repeated extraction (ultrasonic?) with *cold* water, instead of the customary reflux treatment, to effect maximum removal of polymers while minimizing hydrolysis at this stage. Parallel experiments involving extraction with suitable non-aqueous solvents and with $H_2O^{18}$ might then show unambiguously whether the original materials had ever been in contact with extraterrestrial water.

The unifying hypothesis outlined here for the spontaneous generation of informational macromolecules from among the simplest and most prevalent molecules in the universe clearly has great heuristic value both for chemical evolution studies and for research on new approaches to heteropolymer synthesis. Based on no particular set of physical conditions, the model as a whole succeeds in bringing order to a situation of increasingly confusing eclecticism. Controversial aspects should certainly be challenged by experiment and theory for if the fundamental conjectures underlying the model cannot be refuted, a new paradigm of innate protein structure will have emerged that surely must lead to further insights into the origin of the genetic code, hierarchies of molecular order, and the universal phenomenon of life.

**Note added in proof.** P. R. Griffiths, P. J. Schumann, and E. R. Lippincott (*Space Sci.* **4** (1973) 279) have restated the hypothesis presented in this paper while reporting further evidence for the synthesis of heteropolypeptides from hydrogen cyanide and water.

## Acknowledgements

I thank Dr Asher Hyatt, Dr Vernon Ingram and my co-workers at the University of Illinois at Chicago Circle for critically reviewing earlier versions of this manuscript.

* In earlier investigations where free α-amino acids were detected in these meteorites (Kvenvolden *et al.*, 1970; Lawless *et al.*, 1971), no attempts to isolate water-soluble materials hydrolyzable to amino acids were reported.

## References


Abelson, P. H.: 1966, *Proc. Natl. Acad. Sci. U.S.* **55**, 1365.

Cronin, J. R. and Moore, C. B.: 1971, *Science* **172**, 1327.

Ferris, J. P., Donner, D. B., and Lobo, A. P.: 1973, *J. Mol. Biol.* **74**, 499.

Fox, S. W., Harada, K., and Hare, P. E.: 1972, *Space Life Sci.* **3**, 424.

Fox, S. W. and Dose, K.: 1972, *Molecular Evolution and the Origin of Life*, W. H. Freeman, San Francisco.

Gabel, N. W. and Ponnamperuma, C.: 1972, in C. Ponnamperuma (ed.), *Exobiology*, North-Holland Publ. Co., Amsterdam, p. 94.

Gehrke, C. W., Zumwalt, R. W., Kuo, K. C., Ponnamperuma, C., Chen, C-N., and Shimoyama, A.: 1973, *Proc. Fourth Lunar Science Conference.*

Glogovac, S. V. and Matthews, C. N.: 1969, unpublished work.

Grossenbacher, K. A. and Knight, C. A.: 1965, in S. W. Fox (ed.), *The Origins of Prebiological Systems*, Academic Press, New York, p. 173.

Harada, K., Hare, P. E., Windsor, C. R., and Fox, S. W.: 1971, *Science* **173**, 433.

Hare, P. E., Harada, K., and Fox, S. W.: 1970, *Proc. Apollo 11 Lunar Sci. Conf., Geochim. Cosmochim. Acta Suppl. 1* **2**, 1799.

Horowitz, N. H.: 1945, *Proc. Natl. Acad. Sci. U.S.* **31**, 153.

Huntress, W. T., Jr., Baldeschwieler, J. D., and Ponnamperuma, C.: 1969, *Nature* **223**, 468.

Kliss, R. M. and Matthews, C. N.: 1962, *Proc. Natl. Acad. Sci. U.S.* **48**, 1300.

Kliss, R. M. and Matthews, C. N.: 1973, Ms., in preparation.

Kvenvolden, K., Lawless, J., Pering, K., Peterson, E., Flores, J., Ponnamperuma, C., Kaplan, I. R., and Moore, C.: 1970, *Nature* **228**, 923.

Lawless, J., Kvenvolden, K., Peterson, E., Ponnamperuma, C., and Moore, C.: 1971, *Science* **173**, 626.

Lowe, C. U., Rees, M. W., and Markham, R.: 1963, *Nature* **199**, 219.

Matthews, C. N.: 1971, in R. Buvet and C. Ponnamperuma (eds.), *Molecular Evolution I: Chemical Evolution and the Origin of Life*, North-Holland Publ. Co., Amsterdam, p. 231.

Matthews, C. N. and Moser, R. E.: 1966, *Proc. Natl. Acad. Sci. U.S.* **56**, 1087.

Matthews, C. N. and Moser, R. E.: 1967, *Nature* **215**, 1230.

McGovern, W. E.: 1969, *J. Atmospheric Sci.* **26**, 623.

Miller, S. L.: 1955, *J. Am. Chem. Soc.* **77**, 2351.

Minard, R. D. and Matthews, C. N.: 1972, unpublished work.

Minard, R. D., Yang, W., and Matthews, C. N.: 1973, *Abstr. Am. Chem. Soc.*, 166th Natl. Meeting, Chicago, Illinois, Orgn. 117.

Moser, R. E., Fritsch, J. M., Westman, T. L., Kliss, R. M., and Matthews, C. N.: 1967, *J. Am. Chem. Soc.* **89**, 5673.

Moser, R. E., Claggett, A. R., and Matthews, C. N.: 1968a, *Tetrahedron Letters* 1599.

Moser, R. E., Claggett, A. R., and Matthews, C. N.: 1968b, *Tetrahedron Letters* 1605.

Moser, R. E. and Matthews, C. N.: 1968, *Experientia* **24**, 658.

Oró, J. and Kamat, S.: 1961, *Nature* **190**, 442.

Ponnamperuma, C.: 1971, *Quart. Rev. Biophys.* **4**, 77.

Sagan, C. and Khare, B. N.: 1971, *Science* **173**, 417.

Sanchez, R. A., Ferris, J. P., and Orgel, L. E.: 1966, *Science* **154**, 784.

Stiffel, P., Minard, R. D., and Matthews, C. N.: 1973, unpublished work.

Warren, C. B. and Matthews, C. N.: 1971, *Abstr. Am. Chem. Soc.*, 161st Natl. Meeting, Los Angeles, California, Orgn. 120.

Widing, R. W., Stiffel, P., Minard, R. D., and Matthews, C. N.: 1972, unpublished work.

Woeller, F. and Ponnamperuma, C.: 1969, *Icarus* **10**, 386.

Yang, W., Minard, R. D., and Matthews, C. N.: 1973, *J. Chem. Soc., Chem. Comm.*, 435.

# PREBIOTIC NUCLEOTIDE SYNTHESIS-DEMONSTRATION OF A GEOLOGICALLY PLAUSIBLE PATHWAY

ALAN W. SCHWARTZ, M. VAN DER VEEN, T. BISSELING and G. J. F. CHITTENDEN

*Dept. of Exobiology, University of Nijmegen, Nijmegen, The Netherlands*

**Abstract.** Mineral phosphate (apatite) is activated for the synthesis of nucleotides when dilute solutions containing nucleoside and ammonium oxalate are evaporated in its presence. A natural, igneous fluorapatite was found to be even more effective in nucleotide synthesis than the more soluble hydroxylapatite. The phosphorylation is considerably more efficient if urea or cyanamide is also present. Hydrolysis of solutions of cyanogen to form oxalate and urea among other products is a spontaneous process that provides a geologically plausible model for nucleotide synthesis on the primitive earth.

## 1. The Apatite-Oxalate Model

The theoretical background for this model is to be found in Schwartz (1971) and Schwartz and Deuss (1971). It was predicted that the problem of the insolubility of apatite (Gulick, 1955; Miller and Urey, 1959) might be overcome by means of the complexing of calcium by oxalate, particularly at slightly reduced pH. The applicability of the pathway to the phosphorylation of uridine was demonstrated in model 'evaporating pond' experiments (Schwartz, 1972). As part of a more systematic study of the phosphorylation process, we have examined the phosphorylation of thymidine-2-$^{14}$C(5mCi mmole$^{-1}$). Solutions which were $10^{-4}$ M in nucleoside and $10^{-3}$ M in ammonium oxalate were evaporated and heated in the presence of 10 mg of hydroxylapatite or fluorapatite in 15 ml centrifuge tubes at 90 °C for one week. In those experiments in which an organic reagent was also added, its concentration was $10^{-3}$ M.

Quantitative yields were determined by electrophoresis of water extracts of the residues (0.02 M sodium acetate containing 0.008 M EDTA, pH 6.3, 18 V cm$^{-1}$ for 2 hr). Liquid scintillation counting was carried out on 1.0 cm sections of each strip. Table I summarizes some of the results. Peaks representing components more highly anionic than TMP have been summed and are referred to as 'higher products'. The nature of these products is described in the caption to Figure 1, which also illustrates another method of separation.

In contrast to the results obtained for the slow evaporation of large volumes of dilute uridine solutions (Schwartz, 1972), Table I shows no positive effect of the addition of imidazole. This observation is in agreement with the results of Stillwell *et al.* (1972) in a study of the phosphorylation of glucose. It should be noted that the rate of evaporation of 1 ml volumes is such that the tubes appear to be dry by 16 hr. The previous experiments on the phosphorylation of uridine were carried out in volumes of 2.5 l, requiring from 4 to 7 days for evaporation. Besides possible differences in results due to evaporation rates, we have also noted differences due to the nature of the nucleoside and this effect is now under further investigation.

TABLE I

Phosphorylation of thymidine with apatite[a]

| Addition | Apatite[b] | TMP (%)[c] | Higher products (%) | Total yield (%) |
|---|---|---|---|---|
| Dicyandiamide | FA | 30.3 (±2.3) | 50.6 (±3.4) | 81 |
| Dicyandiamide | HA | 45.7 (±0.9) | 19.8 (±5.8) | 66 |
| Urea | FA | 24.4 (±1.8) | 10.8 (±1.8) | 35 |
| Urea | HA | 18.5 (±0.4) | 2.2 (±0.3) | 21 |
| Imidazole | FA | 18.2 (±0.5) | 2.0 (±0.1) | 20 |
| Imidazole | HA | 9.5 (±2.1) | – | 10 |
| None | FA | 16.8 (±2.4) | 5.4 (±0.8) | 22 |
| None | HA | 15.5 (±2.3) | 1.5 (±0.2) | 17 |
| Cyanogen hydrolysates: | | | | |
| Solution A | HA | 28.0 (±1.2) | 54.5 (±2.2) | 83 |
| Solution B | HA | 27.4 (±0.8) | 51.9 (±0.9) | 79 |

[a] For conditions see text. All yields are based on conversion of thymidine and are averages of four replicate experiments.

[b] HA is synthetic hydroxylapatite (Riedel-de Haën). FA is an igneous, macrocrystalline fluorapatite (Durango), ground in a Tema mill to approximately the same particle size range as the hydroxylapatite.

[c] Yields of TMP represent mixtures of 3′ and 5′ nucleotides.

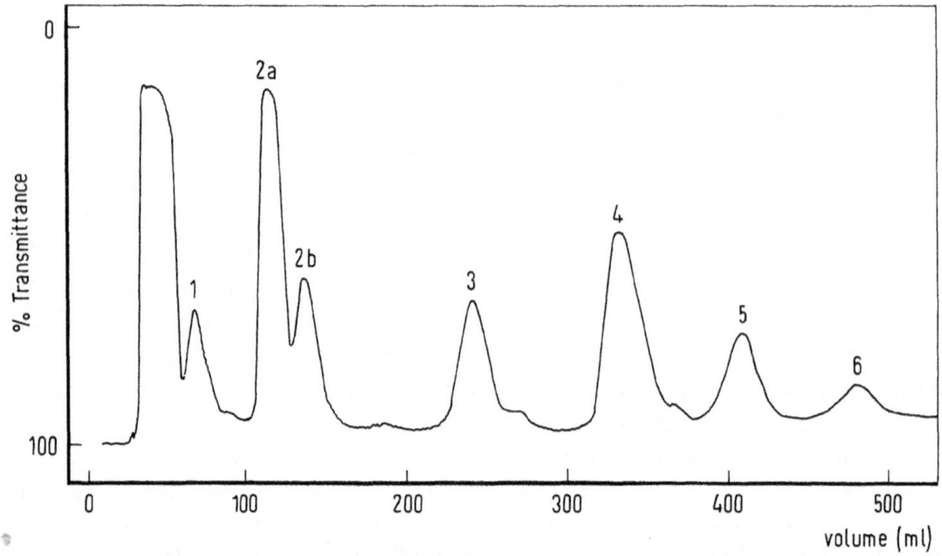


Fig. 1. Initial fractionation of the products of the phosphorylation of thymidine with hydroxylapatite in the presence of dicyandiamide. DEAE-Sephadex (1.27 × 18 cm); linear gradient of 0.05 to 0.6 M triethylammonium bicarbonate at pH 7.6 (800 ml). Peaks 2a and 2b (TMP) are thymidine 5′- and 3′-monophosphate, respectively (molar ratio approximately 5:1). All higher peaks appear to be mixtures of products. One of the components in peak 3 appears to be thymidine-5′-pyrophosphate (ppT). The predominant product in peak 4 is thymidine-3′, 5′-dimonophosphate (pTp). Work is still underway on the full characterization of these mixtures.

Of particular interest in Table I is the effect of utilizing a natural, igneous fluorapatite in place of synthetic hydroxylapatite. Not only were yields generally increased, but a higher proportion of multiply phosphorylated products was obtained. Since the particle size distributions of the two preparations were similar, it is possible that differences in surface structure may be responsible. It is especially noteworthy that increasing the proportion of hydroxylapatite causes a decrease in yield, particularly of multiply phosphorylated products (Figure 2). This result is suggestive of the transphosphorylation of intermediate polyphosphates (Krane and Glimcher, 1962). The most significant observation, however, is the failure of the natural mineral to show the same inhibitory behavior (Figure 2). Whether this property is a general one for fluorapatites or a peculiarity of this particular mineral is yet to be established.

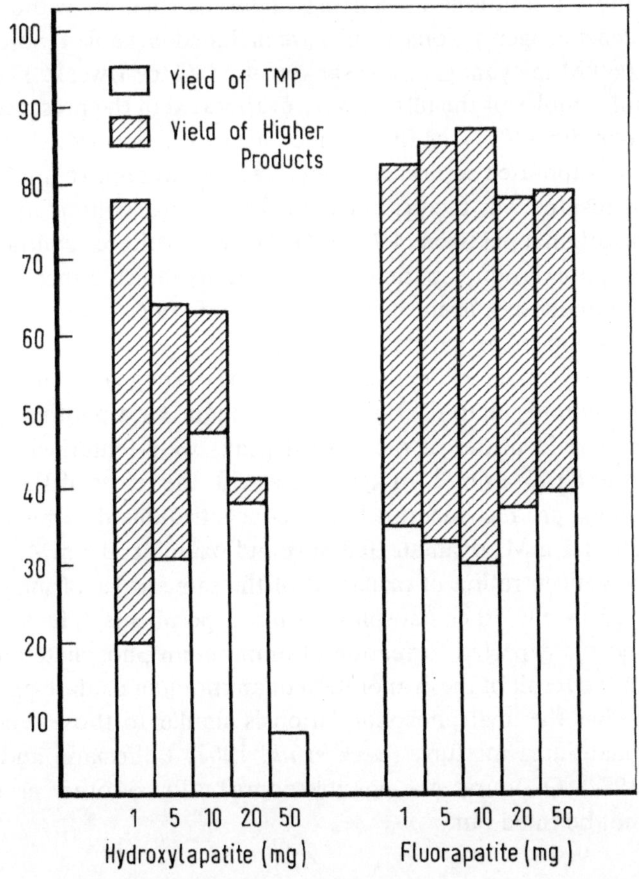


Fig. 2. The effect of quantity and source of apatite on the products of a phosphorylation reaction. Phosphorylation was carried out with the indicated quantity of apatite added to 1 ml of $10^{-4}$ M thymidine-2-$^{14}$C ($10^{-4}$ M) containing $10^{-3}$ M dicyandiamide and $10^{-3}$ M ammonium oxalate. Reaction conditions and quantitative analyses were as described in the text. Each bar of the graph represents the average of four replicate experiments.

## 2. Phosphorylation by Cyanogen Hydrolysates

Hydrogen cyanide and cyanogen have been repeatedly implicated as key intermediates in prebiotic organic synthesis (see, for example: Miller, 1957; Sanchez *et al.*, 1967; Lohrman and Orgel, 1968). The presence of these substances in interstellar dust clouds and in comets (Oró, 1972) increases the probability of their having played a central role in the formation of primitive organic matter. Ammonium oxalate is a product of the hydrolysis of cyanogen (Sidgwick, 1942). Recently, oxalate has been reported to be formed, together with large amounts of urea, in the oligomerization of hydrogen cyanide and also (after hydrolysis) from cyanogen solutions (Ferris, *et al.*, 1973a, b).

We have tested the possibility of the phosphorylation of nucleosides with apatite using only cyanogen solutions as starting materials (that is, without addition of oxalate or condensing agent). Some results are included in Table I. Solution A, which was initially 0.060 M in cyanogen, was shaken at 55 °C for 1 week. The solution was filtered and 1 ml samples of the filtrate were evaporated in the presence of hydroxylapatite and thymidine-2-$^{14}$C in the manner already described. A portion of the filtrate was also evaporated to dryness under vacuum to remove any remaining free cyanogen. This residue was reconstituted to the original volume and titrated with ammonia to the original pH (from 4.7 to 6.6). The reconstituted solution (B) was also utilized for phosphorylation experiments. The results indicate that free cyanogen is not acting as a condensing agent in this system, although both solutions A and B serve as efficient condensing agents.

Trimethylsilyl derivatives of dried samples of the cyanogen filtrates were prepared for gas chromatographic analysis (Butts, 1972). Figure 3 is a photograph of a typical chromatogram. The identities of the labeled peaks were tentatively established by coinjection of standard samples. In agreement with the result of Ferris *et al.* (1973) urea was the major product identified. The concentrations determined were: urea, 8.4 mM; oxalate, 1.1 mM; oxamate, 0.8 mM and oxamide, 0.5 mM.

Note that the concentration of oxalate is of the same order of magnitude as that usually employed in our other phosphorylation experiments. The large amount of urea present and the expected formation of ammonium phosphate and subsequent pH reduction as the result of the evaporation of ammonium oxalate-apatite mixtures, makes it likely that the final phosphorylation is similar to those reported for ammonium phosphate-urea mixtures (Beck *et al.*, 1967; Lohrmann and Orgel, 1971; Bishop *et al.*, 1972). Of course, the possible contribution of other products, not yet identified, cannot be ruled out.

## 3. Prebiotic Abundance of Oxalate

In view of the facile formation of oxalate from cyanogen or hydrogen cyanide, it may be superfluous to mention other chemical evolution experiments which have also led to its synthesis. However, in this connection the experiment of Hasselstrom and

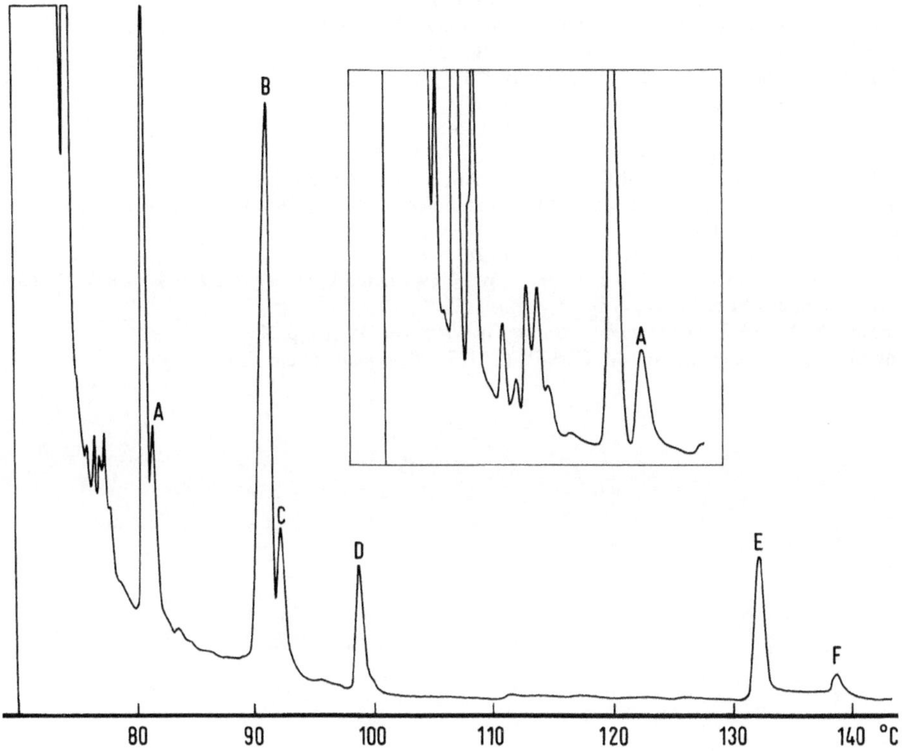


Fig. 3. Gas chromatogram of trimethylsilyl derivatives of products of the hydrolysis of cyanogen. The column (1.5 m × 3 mm I.D.) was packed with 3% OV-1 on Chromosorb G; nitrogen carrier gas at 26 ml min$^{-1}$; linear temperature program from 70° to 250 °C at 2° min$^{-1}$. Peak A is oxalic acid, B is urea, C is oxamic acid and D is oxamide. Peaks E and F have not yet been identified. The insert shows a better separation of oxalic acid from a reagent peak, performed isothermally at 70 °C.

Henry (1956) is of interest since it demonstrated the production of oxalate even under non-reducing conditions, by ultraviolet irradiation of carbonate solutions. However, the production of oxalate from the cosmically abundant CN moiety certainly seems more pertinent. In this context, it may not be amiss to point to the little-quoted fact that oxalic acid may be the singly most abundant organic compound to be identified in water extracts and hydrolysates of lunar samples (Gehrke *et al.*, 1972).

## References

Beck, A., Lohrmann, R., and Orgel, L. E.: 1967, *Science* **157**, 952.
Bishop, M. J., Lohrmann, R., and Orgel, L. E.: 1972, *Nature* **237**, 162.
Butts, W. C.: 1972, *Anal. Biochem.* **46**, 187.
Ferris, J. P., Donner, D. B., and Lobo, A. P.: 1973a, *J. Mol. Biol.* **74**, 499.
Ferris, J. P., Donner, D. B., and Lobo, A. P.: 1973b, *ibid.*, 511.
Gehrke, C. W., Zumwalt, R. W., Kuo, K., Rash, J. J., Aue, W. A., Stalling, D. L., Kvenvolden, K. A., and Ponnamperuma, C.: 1972, *Space Life Sci.* **3**, 439.
Gulick, A.: 1955, *Amer. Scientist* **43**, 479.

Hasselstrom, T. and Henry, M. C.: 1956, *Science* **123**, 1038.
Krane, S. M. and Glimcher, M. J.: 1962, *J. Biol. Chem.* **237**, 2991.
Lohrmann, R. and Orgel, L. E.: 1968, *Science* **161**, 64.
Lohrmann, R. and Orgel, L. E.: 1971, *Science* **171**, 490.
Miller, S. L.: 1957, *Biochim. Biophys. Acta* **23**, 480.
Miller, S. L. and Urey, H. C.: 1959, *Science* **130**, 245.
Oró, J.: 1972, *Space Life Sci.* **4**, 507.
Sanchez, R. A., Ferris, J. P., and Orgel, L. E.: 1967, *J. Mol. Biol.* **30**, 223.
Schwartz, A. W.: 1971, in R. Buvet and C. Ponnamperuma (eds.), *Chemical Evolution and the Origin of Life*, North-Holland, p. 207.
Schwartz, A. W.: 1972, *Biochim. Biophys. Acta* **281**, 477.
Schwartz, A. W. and Deuss, H.: 1971, in A. W. Schwartz (ed.), *Theory and Experiment in Exobiology*, Vol. 1, Wolters-Noordhoff, p. 73.
Sidgwick, N. V.: 1942, *The Organic Chemistry of Nitrogen*, Oxford, p. 300.
Stillwell, W., Steinman, G., and McCarl, R. L.: 1972, *Bioorganic Chem.* **2**, 1.

# MODELS OF PREBIOLOGICAL PHOSPHORYLATION

M. HALMANN

*Dept. of Isotope Research, The Weizmann Institute of Science, Rehovot, Israel*

**Abstract.** The hypothesis that contemporary metabolic pathways evolved from analogous chemical reaction sequences on the primitive Earth leads to a reexamination of models of prebiological phosphorylation. Present-day phosphate uptake by algae and bacteria seems to involve two transport systems: (a) An active transport process occurring at low external phosphate concentrations (as in unpolluted natural waters), with a transport constant $K_s$ of $10^{-7}$ to $10^{-6}$ $M$ $P_i$. (b) Another (probably diffusive) process at higher phosphate concentrations ($>10^{-6}$ $M$) (as in the interstitial water of reducing sediments). Laboratory model experiments are described for the reaction of reducing sugars with orthophosphate in the presence of cyanogen, producing glycosyl phosphates. These reactions proceed with appreciable yields only at high phosphate concentrations ($>10^{-3}$ $M$), and may thus possibly serve as simulations of prebiological phosphorylation with diffusive transport, as it may have occurred in the interstitial water of reducing sediments.

A plausible and useful hypothesis of chemical evolution may be that present-day metabolic pathways, which now require enzymes as catalysts, evolved gradually from similar chemical reaction sequences, which occurred already on the primitive Earth, even before the appearance of protein enzymes (Oparin, 1968; Degani and Halmann, 1967). Among the primary steps in such reaction sequences, one of the most crucial must have been the conversion of inorganic phosphate to esters, amides and acid anhydrides of phosphoric acid (Miller and Parris, 1964; Gabel, 1965; Rabinowitz, 1972).

Previous work on models of prebiological phosphorylation has either involved solid-phase condensation of acid-salts of phosphoric acid at relatively high temperatures (Ponnamperuma and Mack, 1965; Waehnelt and Fox, 1967; Rabinowitz *et al.*, 1968; Bishop *et al.*, 1972; Osterberg and Orgel, 1972; Handschuh *et al.*, 1973) or formation in aqueous solutions of reactive derivatives of phosphoric acid, such as the condensation products of phosphate with nitrilic reagents, in particular cyanamide or cyanogen (Lohrman and Orgel, 1968; Halmann *et al.*, 1969; Degani and Halmann, 1971; Schwartz, 1972).

The drawback of this type of model, requiring high-temperature surfaces such as hot volcanic rock or hot springs, is that it could have appeared only in isolated geographic regions. Such a process would have to appear separately and independently in each location. It would seem hard to reconcile such separate processes of chemical evolution with the generally accepted concept of a common origin of life – as demonstrated in the uniformity of metabolic pathways – until one considers the universality of physical laws.

Inorganic polyphosphates have been proposed as favorable and even necessary for chemical evolution (Gabel and Ponnamperuma, 1972). Some support for this hypothesis may be found in the universal occurrence of polyphosphates in the biosphere (Gabel, 1972) and in the observation that various contemporary microorganisms use polyphosphates as a reserve store of phosphorus, which they accumulate under conditions of 'luxury consumption', and which then suffice for several cell doublings under conditions of phosphate starvation (Harold, 1965).

One model of phosphorylation in aqueous solutions on which we have worked in some detail, is the cyanogen-induced activation of orthophosphate. In this reaction, preferential phosphorylation of reducing sugars occurs, to produce the corresponding glycosyl phosphates, as shown in the following scheme,

$$C_2N_2 + HPO_4^{--} \rightarrow [NC{-}C(=NH)\,OPO_3^{--}]$$

$$ROH + [NC{-}C\,(=NH)\,OPO_3^{--}] \rightarrow ROPO_3^{--} + (NC{-}CO{-}NH_2)$$

Thus, with D-glucose, the main products have been shown to be α- and β-D-glucose 1-phosphate (Degani and Halmann, 1971). Also in a small yield, α-glucose 1,6-diphosphate was formed, which in contemporary metabolism serves as coenzyme in the phosphoglucomutase catalyzed isomerization of glucose 6-phosphate and α-glucose 1-phosphate. This formation of glucose 1,6-diphosphate is probably a secondary reaction, since we had found that α-glucose 1-phosphate itself, in aqueous solution in the presence of cyanogen produces α-glucose 1,6-diphosphate, in about 2.7% yield (Degani and Halmann, 1972). As previously reported, the cyanogen-induced phosphorylation occurred only with reducing sugars, the non-reducing sugars, sucrose, trehalose and glycerol being inert. With aldoses, the main products were the aldose 1-phosphates. In the case of the keto-sugar fructose, the sugar phosphate produced was not fructose 1-phosphate nor fructose 6-phosphate, but only a fructose 2-phosphate, (Kawatsuji, 1973). On the basis of its rate of acid-catalyzed hydrolysis, ($k_1 = 0.050 \pm 0.005$ min$^{-1}$ for hydrolysis in 0.24 $M$ acetate buffer of pH 4.0 at 37°), the results of oxidation by periodic acid (formation of formic acid), as well as the optical rotation, it seems that the product is α-fructose 2-phosphate, probably in the pyranose ring form, as shown in Figure 1 (Kawatsuji, 1973).

This formation of α-fructose 2-phosphate is of considerable interest, because of recent reports (Taniguchi and Nakamura, 1972) of the isolation from the Jerusalem artichoke of an uridine diphosphate 2-fructose, which serves as the fructosyl transfer agent in the biosynthesis of inulin, an important plant sugar consisting mainly of fructose residues. In this fructosyl transfer reagent, an α-fructose 2-phosphate is connected to uridine monophosphate.

While the above results on the cyanogen-induced activation of orthophosphate seem useful as a hint on possible prebiological pathways of phosphorylation, one major obstacle is that of concentrations. These reactions succeed with a reasonable yield (at least a few percent) only in reagent concentrations higher than $10^{-3}$ $M$. Thus, in the phosphorylation of fructose, even with very high concentrations of reactants (e.g. 0.3 $M$ fructose, 0.4 $M$ $C_2N_2$ and 1 $M$ $Na_2HPO_4$) the yield of fructose phosphate was only 6% (Kawatsuji, 1973). Such concentrations exceed by many orders of magnitude the solubility of apatite minerals, and thus cannot be geochemically realistic. A very intriguing proposal to overcome this difficulty is that of an evaporating solar pond, in which a high concentration of some calcium com-

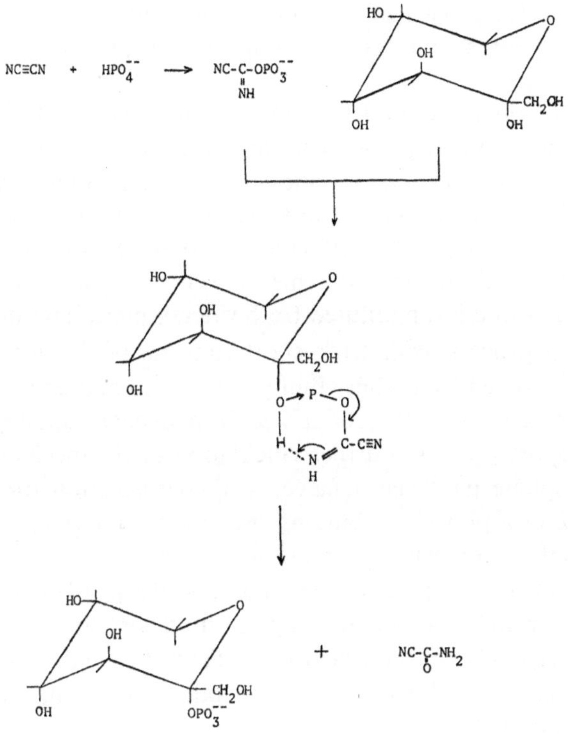


Fig. 1. Mechanism of cyanogen-induced phosphorylation of D-fructose.

plexing agent, such as ammonium oxalate, enables the solubilization of apatite – thus achieving phosphate concentrations of up to $10^{-3}$ *M* (Schwartz, 1972). With such concentrations, and using various condensing agents (such as cyanamide, urea or imidazole), excellent yields of phosphorylation of nucleosides to nucleotides could be achieved. One possible objection to an 'evaporating solar pond' model may be its reliance on geographic isolation, as with the local hot surfaces.

It is of considerable interest to consider the concentrations of phosphate that are actively taken up in contemporary life by bacteria and algae from natural waters. Many of these organisms thrive on phosphate in as low as $10^{-8}$ *M* concentrations, while $10^{-6}$ *M* phosphate is already sufficient to cause the well-known algal blooms or 'red tides'. Recent experiments on the rate $v$ of uptake of $^{32}$P-labelled phosphate in natural waters as a function of the concentration $S$ of phosphate have shown that it can be described by a Michaelis-Menten type equation (Monod, 1942)

$$v = v_{\max} S/(K_s + S)$$

in which $v_{\max}$ is the maximal rate, while $K_s$, the transport-constant, has values of 0.2–0.8 $\mu M$ (Halmann and Stiller, 1974). Not only uptake of phosphate, but also the growth of microorganisms as a function of phosphate concentration can be described by a Michaelis-Menten type equation. A summary of some data from the literature is presented in Table I, which includes values both of $K_m$, the half-saturation constant for *growth* of microorganisms under conditions of phosphate limitation, as well as of $K_s$, the transport constant for phosphate uptake by microorganisms.

In attempting to create a model of the prebiological evolution of phosphate metabolism, it is an essential prerequisite to understand the molecular mechanism of phosphate uptake as it occurs in contemporary cells. Some clarification of this mechanism has recently become available. Phosphate transport into bacteria (Medveczky and Rosenberg, 1970, 1971) and into algae (Muller, 1972) was shown to involve two kinetically distinct systems: at very low concentrations of external phosphate (such as those in unpolluted freshwaters), phosphate uptake is probably an active transport process, with a transport constant of the order of $K_s = 0.7\ \mu M$, and presumably mediated by a 'high-affinity' protein component. At higher external phosphate concentrations (above about 1 $\mu M$), phosphate uptake is probably due to a passive transport process. Further elucidation of the mechanism of the active transport of phosphate has been achieved with the isolation from the membrane of *Escherichia coli* of a phosphate-binding protein, of molecular weight 42000, able to bind one phosphate molecule per molecule of protein, and having a phosphate-binding constant of 0.8 $\mu M$ (Medveczky and Rosenberg, 1970, 1971). The binding constant is thus seen to be within the range of transport constants which had been found for microorganisms in natural waters (Table I). Neither the active site, nor other details on the molecular mechanism of phosphate transport by this protein have yet been reported.

It is of interest to speculate which of the two transport systems may have evolved earlier, the one able to extract phosphate from very low concentrations (below $10^{-6}\ M$), thus requiring some high-affinity transport mechanism (possibly already involving some lipid membrane), or the one operating at higher phosphate concentrations, and depending only on diffusive transport.

A further clue to the mechanism of active transport of phosphate came with the

TABLE I

Comparison of maximum rates and saturation constants for growth and phosphorus uptake for phosphorus limited microorganisms

| | Growth | | Uptake | | Reference |
|---|---|---|---|---|---|
| | $\mu$ $hr^{-1}$ | $K_m$ $\mu$molar | $v_m$ $\mu mol\ l^{-1}\ hr^{-1}$ | $K_s$ $\mu$molar | |
| Pseudomonas sp. (bacterium) | 0.16 | 0.085 | | | Kunicka-Goldfinger (1972) |
| Marine diatom | | 0.12 | | | Thomas-Dodson (1968) |
| Phaeodactylum tricornutum | | 0.34 | | | Harvey (1963) |
| Nitzschia actinastroides | 0.013 | 0.03 | | | Müller (1972) |
| L. Kinneret plankton | 0.02 | | 0.04–1.1 | 0.2–0.8 | Halmann and Stiller (1974) |

recent observation, that the above 'phosphate-binding-protein' of E. coli is identical with the R2a protein (Gerdes, 1973), which had previously been shown to be a regulator gene product of E. coli, regulating the synthesis of alkaline phosphatase (Garen and Otsuji, 1964). Both this R2a protein and alkaline phosphatase have the property of strongly binding orthophosphate. The substantial release of alkaline phosphatase by microorganisms under conditions of phosphate starvation had been observed both in the marine (Strickland and Solorzano, 1966) and in the freshwater environment (Berman, 1970). It will be very important to establish what role these two proteins, alkaline phosphatase and R2a, play in the uptake of phosphate under natural conditions of phosphate-limited growth, and to identify the sequence of amino acids in the binding-site of the proteins. Once this sequence is known, it will be possible to synthesize model peptides, in order to simulate the phosphate-binding process under laboratory conditions.

A protein component of the cellular membrane was implicated in the phosphate transport into chlorella (Gaudin *et al.*, 1973). In earlier work, it was considered that phosphate turnover, both in red blood cells (Prankerd and Altman, 1954; Bartlett, 1958) and in yeast (Goodman and Rothstein, 1957), involves as primary step the esterification of orthophosphate by the glyceraldehyde dehydrogenase promoted reaction with glyceraldehyde-3-phosphate.

$$\text{Glyceraldehyde-3-phosphate} + \text{Pi} + \text{NAD}^{+} \rightleftharpoons$$
$$\text{1,3-Diphosphoglycerate} + \text{NADH} + \text{H}^{+}.$$

The dehydrogenase enzyme located in the plasma membrane may mediate the direct incorporation of extracellular phosphate into the glycolytic pathway (Harold *et al.*, 1965). Phosphate transport into human erythrocytes was shown to be due to a 'membrane-phosphate-acceptor' (Schauer and Hillman, 1961). In yeast cells, phosphatidylglycerolphosphate was suggested to serve as intermediate in the transfer of a phosphate moiety from polyphosphate to a carrier-phosphate complex (Deierkauf and Booij, 1968). Polyphosphate-protein complexes were proposed to play an important role in the transfer of phosphate into neural cells (Gabel, 1972).

In case we assume that prebiological phosphorylation operated by diffuse phosphate transport, one possible water body providing the necessary high concentrations of phosphate could be the 'evaporating solar pond' mentioned above (Schwartz, 1972, 1973). A more obvious site, existing in many locations even now, is the interstitial water found in the bottom mud of anaerobic water bodies, in bays, estuaries, lakes and on the continental shelf. In these, the solubilization of iron as ferrous ions enables the solubilization of phosphate. Thus, in recent measurements, the concentrations of dissolved iron and phosphate in such waters were reported to reach 300 and 600 $\mu M$, respectively (Bray *et al.*, 1973). At these high concentrations, diffusive phosphate transport must be sufficient, and no active transport system is required.

It is therefore proposed that the site of prebiological phophorylation may have been in the interstitial water of sediments – under reducing conditions. It is also proposed that the site to search for forms of life on other planets is in the interstitial water, which may possibly be found, e.g. in the canyons of Mars.

The author wishes to acknowledge partial support of this study by the U.S. National Aeronautics and Space Administration.

## References

Bartlett, G. R.: 1958, *Ann. N.Y. Acad. Sci.* **75**, 110.
Berman, T.: 1970, *Limnol. Oceanogr.* **15**, 663.
Bishop, M. J., Lohrmann, R., and Orgel, L. E.: 1972, *Nature* **237**, 162.
Bray, J. T., Bricker, O. P., and Troup, B. N.: 1973, *Science* **180**, 1362.
Degani, Ch. and Halmann, M.: 1967, *Nature* **216**, 1207.
Degani, Ch. and Halmann, M.: 1971a, *J. Chem. Soc.* 1459.
Degani, Ch. and Halmann, M.: 1971b, 'Mechanism of the Cyanogen-Induced Phosphorylation of Sugars in Aqueous Solutions', in *Chemical Evolution and the Origin of Life* (ed. by R. Buvet and C. Ponnamperuma), North Holland Publ., Amsterdam, p. 224.
Degani, Ch. and Halmann, M.: 1972, *Nature New Biol.* **235**, 171.
Deierkauf, F. A. and Booij, H. L.: 1968, *Biochim. Biophys. Acta* **150**, 214.
Gabel, N. W.: 1965, *Life Sci.* **4**, 2085.
Gabel, N. W.: 1972, *Perspectives Biol. Medicine* **15**, 640.
Gabel, N. W. and Ponnamperuma, C.: 1972, in *Exobiology* (ed. by C. Ponnamperuma) North Holland Publ., Amsterdam, p. 125.
Garen, A. and Otsuji, N.: 1964, *J. Mol. Biol.* **8**, 841.
Gaudin, C., Jeanjean, R., and Blasco, F.: 1973, *Compt. Rend. Acad. Sci. N.Y.* **277**, 301.
Gerdes, H.: 1973, *Proc. Austr. Biochem. Soc.* **6**, 50.
Goodman, J. and Rothstein, A.: 1957, *J. Gen. Physiol.* **40**, 915.
Halmann, M., Sanchez, R. A., and Orgel, L. E.: 1969, *J. Org. Chem.* **34**, 3702.
Halmann, M. and Stiller, M.: 1974, *Limnol. Oceanogr.*, in press.
Handschuh, G. J., Lohrmann, R., and Orgel, L. E.: 1973, *J. Mol. Evolution* **2**, 251.
Harold, F. M.: 1966, *Bacteriol. Rev.* **30**, 772.
Harvey, H. W.: 1957, *Chemistry and Fertility of Seawaters*, Cambridge Univ., p. 96.
Kawatsuji, M.: 1973, *Cyanogen-Induced Phosphorylation of D-Fructose*, M.Sc. Thesis, Feinberg Graduate School, The Weizmann Institute of Science, Rehovot.
Kunicka-Goldfinger, W. and Kunicki-Goldfinger, W. J. H.: 1972, *Acta Microbiol. Polon.* **4**, 49.
Lohrmann, R. and Orgel, L. E.: 1968, *Science* **161**, 64.
Medveczky, N. and Rosenberg, H.: 1970, *Bioch. Biophys. Acta* **211**, 158; 1971, *ibid.* **241**, 494.
Miller, S. L. and Parris, M.: 1964, *Nature* **204**,
Monod, J.: 1942, *Recherches sur la croissance des cultures bacteriennes*, Paris, Hermann.
Müller, H.: 1972, *Arch. Hydrobiol. Suppl.* **38**, 4, 399.
Oparin, A. I.: 1968, *Genesis and Evolutionary Development of Life*, Academic Press, New York, p. 95.
Osterberg, R. and Orgel, L. E.: 1972, *J. Mol. Evol.* **1**, 241.
Ponnamperuma, C. and Mack, R.: 1965, *Science* **148**, 1221; Abstr. 150th Meeting Amer. Chem. Soc., Atlantic City, C.44.
Prankerd, T. A. J. and Altman, K. I.: 1954, *Biochem. J.* **58**, 622.
Rabinowitz, J., Chang, S., and Ponnamperuma, C.: 1968, *Nature* **218**, 442.
Rabinowitz, J.: 1972, *Chimia* **26**, 350.
Schauer, R. and Hillman, G.: 1961, *Hoppe-Seyler's Z. Physiol. Chem.* **325**, 9.
Schwartz, A. W.: 1972, *Biochim. Biophys. Acta* **281**, 477.
Schwartz, A. W., van der Veen, M., Bisseling, T., and Chittenden, G. J. F.: 1973, *Biosystems* **5**, 119.
Strickland, J. and Solorzano, L.: 1966, in *Some Contemporary Studies in Marine Science* (ed. by H. Barnes), Allen and Unwin, London, p. 665.
Taniguchi, H. and Nakamura, M.: 1972, *Agr. Biol. Chem.* **36**, 2185, 2373.
Thomas, W. H. and Dodson, A. N.: 1968, *Biol. Bull.* **134**, 199.
Waehneldt, T. V. and Fox, S. W.: 1967, *Biochim. Biophys. Acta* **134**, 1.

# CONDITIONS OF OCCURRENCE FOR PRIMEVAL PROCESSES OF TRANSPHOSPHORYLATIONS

ELISABETH ETAIX and RENÉ BUVET

*Laboratoire d'Energétique Biochimique, Université Paris Val de Marne, 94000 Creteil, France*

**Abstract.** Model reactions of the phosphate group transfer through hydrolysis-condensation processes without enzymic catalysts are studied in dilute aqueous solutions. Tripolyphosphate and acetylphosphate are used as energy and phosphate donors to perform syntheses of pyrophosphate. Results are discussed in terms of the variation versus pH of the overall standard free enthalpy change.

## 1. Introduction

Different kinds of biochemical reactions proceed through the formation and the use of condensed phosphorylated compounds. Among the most important of these reactions, the three following types will be examined more or less extensively here:

- phosphorylations coupled with redox reactions,
- transphosphorylations from polyphosphates to orthophosphate,
- transphosphorylations from acylphosphates to orthophosphate.

With respect to the problem of the primary occurrence and achievement of such processes, we have investigated model reactions, in aqueous solutions, and under the simplest possible experimental conditions, employing the most elementary reagents and none of the biological sophistications such as ultrastructures or enzymic catalysts.

## 2. Results

### 2.1. Template phosphorylation coupled with an oxido-reduction reaction

Calvin (1962) reported one model of oxidative phosphorylation to have been successfully achieved by Barltrop. We shall just mention here that we have evidenced no pyrophosphate synthesis in aqueous solutions at different buffered pH at 25 °C, from the simple oxidation of $10^{-1}$ M ferrous sulfate by $2.5 \times 10^{-1}$ M hydrogen peroxide or other inorganic oxidants, in the presence of $10^{-2}$ M and $5\,\mu$ Ci ml$^{-1}$ labelled orthophosphate. However, other attempts are being made in our laboratory to obtain polyphosphates through the transfer of redox energy using inorganic as well as organic redox reactions such as those employing the couples $NAD^+/NADH$ (Barltrop *et al.*, 1963) and organic disulfide/thiol (Wieland and Bäuerlein, 1967).

### 2.2. Model transphosphorylations

We shall more extensively consider the following phosphate transfer model reactions:

$$ATP + P \rightleftharpoons ADP + P_2 \tag{1}$$

or

$$P_3 + P \rightleftharpoons 2P_2 \tag{2}$$

and

$$AcOP + P \rightleftharpoons AcOH + P_2 \tag{3}$$

which all lead to the synthesis of inorganic diphosphate (pyrophosphate) $P_2$ through the coupling of the exergonic hydrolysis of the phosphate donor (ATP, or inorganic tripolyphosphate $P_3$, and acetylphosphate AcOP) and the endergonic condensation of orthophosphate P on itself.

The three transphosphorylation reactions have a negative overall energy-balance, $\Delta G^{\circ\prime}$, at any pH, as is shown in Figure 1 where we have assumed that from the energetic viewpoint $P_3$ is very similar to ATP at pH 7. The variations with pH of the standard free enthalpy changes, $\Delta G^{\circ\prime}$, for the coupled reactions are deduced from the differences in the positive $\Delta G^{\circ\prime}$ region between the formation curves of pyrophosphate and

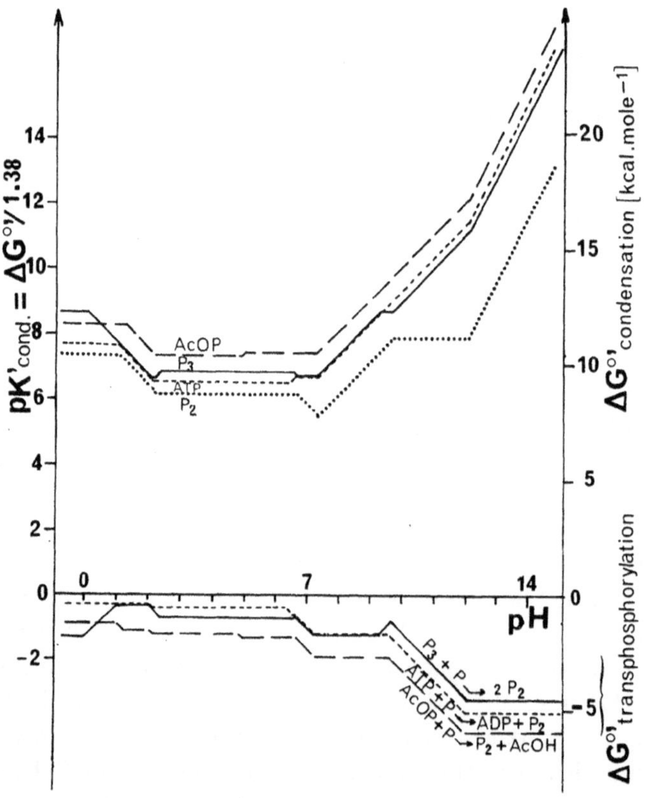


Fig. 1. $\Delta G^{\circ\prime}$ vs pH diagrams of some phosphorylated compounds, in the absence of alkaline earth cations. (a) In the positive $\Delta G^{\circ\prime}$ region: condensation reactions at 25 °C (according to data at pH 7 by Jencks (1968)):

- - - - - $AcOH + P \rightleftharpoons AcOP + H_2O$ (acetylphosphate formation)
——— $P_2 + P \rightleftharpoons P_3 + H_2O$ (tripolyphosphate formation)
-------- $ADP + P \rightleftharpoons ATP + H_2O$ (ATP formation)
. . . . . . . $P + P \rightleftharpoons P_2 + H_2O$ (pyrophosphate formation)

(b) In the negative $\Delta G^{\circ\prime}$ region: coupled transphosphorylation reactions at 25 °C:

-------- $ATP + P \rightleftharpoons ADP + P_2$
——— $P_3 + P \rightleftharpoons 2P_2$
- - - - - $AcOP + P \rightleftharpoons AcOH + P_2$

the relevant phosphate donor. The diagrams show that pyrophosphate formation from tripolyphosphate or acetyl-phosphate would be most easily obtained towards strongly basic media, above pH 10 where the transfer standard enthalpy changes reach the most negative values.

In preliminary transphosphorylation studies in aqueous solution with divalent metal ions as the only catalysts, the efficiency of tripolyphosphate for performing pyrophosphate synthesis at pH 9 and 39 °C was found of the same order of magnitude to that of ATP (Lowenstein, 1958; Le Port *et al.*, 1971).

In both cases, the best yields were obtained with calcium ions, whereas they were the lowest with magnesium ions. However, the order of catalytic activity of the cations was slightly modified, whether ATP or $P_3$ was concerned. Also, with calcium ions, the pyrophosphate yields increased with pH in the range 7–10, and with temperature between 0 °C and 39 °C.

Figure 2 reports the yields of labelled soluble pyrophosphate obtained as a function of time, as percent of initial labelled orthophosphate at two different concentrations, in the presence of inorganic tripolyphosphate and the four cations found to be the

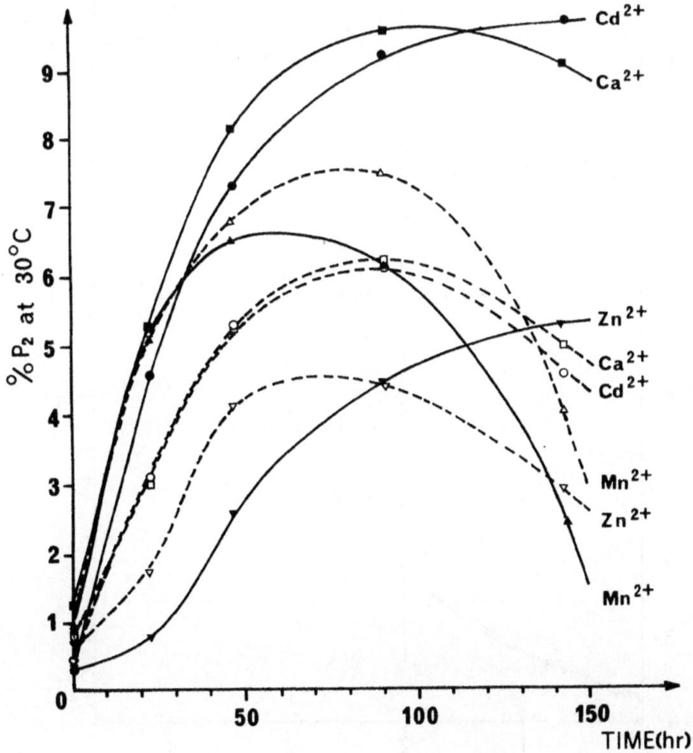


Fig. 2. Percentage of radioactive orthophosphate incorporated into pyrophosphate $P_2$ during transphosphorylation: $P_3 + P \rightleftharpoons 2P_2$ at 30 °C and buffered pH 9.5; $5 \times 10^{-2}$ M tripolyphosphate, $5 \times 10^{-2}$ M metal ion $M^{2+}$, $5 \times 10^{-3}$ M (solid lines) or $1.6 \times 10^{-5}$ M (dashed lines) orthophosphate, 5 $\mu$Ci ml$^{-1}$ $P^{32}$ radioactivity.

most efficient in $P_2$ synthesis. The yields of condensed $P_2$ are higher when orthophosphate is initially more concentrated. It should be noted that transphosphorylation without any metal ion as catalyst occurs with pyrophosphate yields around only 0.1%. In most curves, the yield-decrease after three days (72 hr) is concomitant with the appearance with time of a labelled phosphate precipitate in the solution, since only the radioactivity of the liquid phase was measured. For calcium and cadmium ions, we also noticed the formation of labelled tripolyphosphate and of two other unidentified labelled polyphosphates in a few percent concentration.

We shall now examine the exergonic transphosphorylation with acetylphosphate in buffered aqueous solution at 30 °C, according to:

$$\mathrm{AcOP} + \mathrm{P} \rightleftharpoons \mathrm{AcOH} + \mathrm{P}_2, \quad \Delta G'^{\circ} < 0.$$

The initial rate of AcOP hydrolysis was measured at every pH by pH titration of the

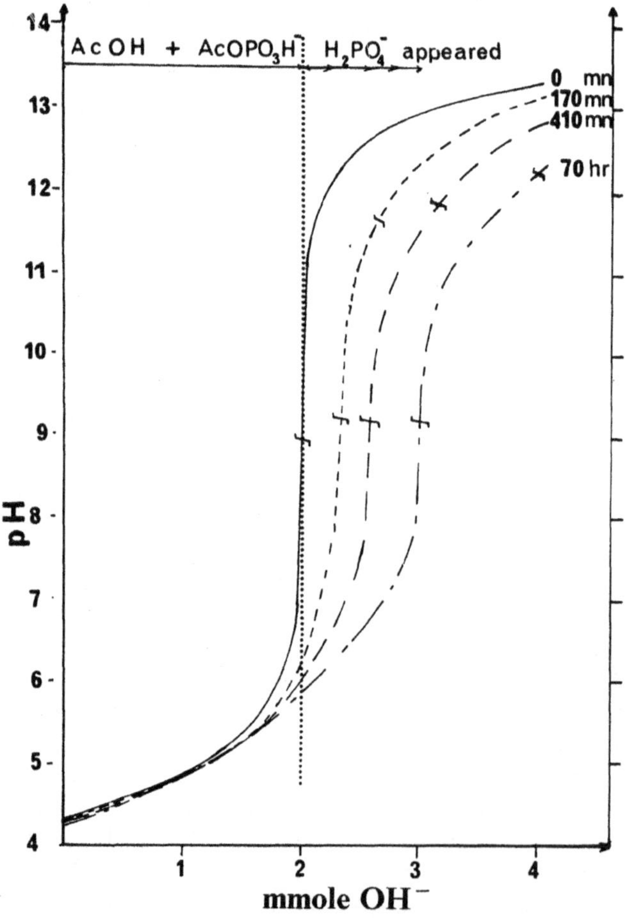


Fig. 3. Analytical measurement of the rate of hydrolysis of $2 \times 10^{-1}$ M acetylphosphate by pH titration: example of a solution buffered at pH 4.4 by $2 \times 10^{-1}$ M AcOH, $2 \times 10^{-1}$ M AcO$^-$; 1 millimole $CH_3COOPO_3H^-$ titrated.

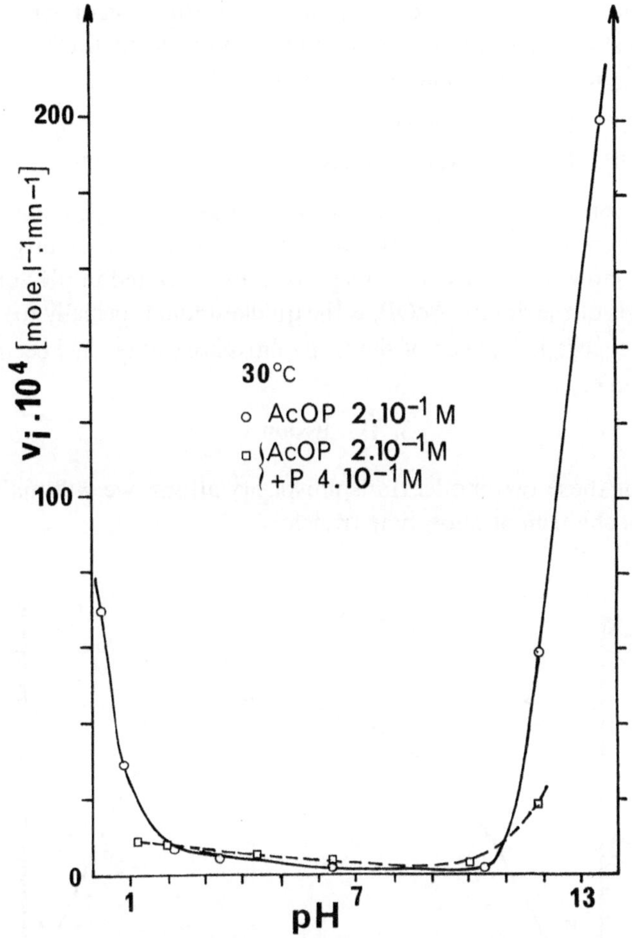


Fig. 4. Effect of pH on the initial rate of hydrolysis of $2 \times 10^{-1}$ M acetylphosphate in the absence (circles) or the presence (squares) of $4 \times 10^{-1}$ M initially added orthophosphate at 30 °C.

orthophosphate released (Figure 3) (the transphosphorylation yields remaining low), and is slightly enhanced when orthophosphate is initially present (Figure 4), except in extreme pH conditions, proving that the reaction is catalyzed by the addition of orthophosphate. In both cases (with or without initial orthophosphate), we have evidenced the formation of pyrophosphate by thin-layer chromatography, at a concentration level of at the most 0.5% of the total orthophosphate released, which explains why the presence of pyrophosphate is not detected by the pH metric measurements of the rate of hydrolysis and is actually often overlooked in acylphosphate hydrolysis experiments (Koshland Jr, 1952; Di Sabato and Jencks, 1961).

Pyrophosphate appears as soon as the hydrolysis of AcOP starts and, at given pH, its concentration remains fairly constant until completion of hydrolysis, whereas inorganic tripolyphosphate and unidentified polyphosphates appear on the chromatograms. The steady state of $P_2$ concentration over the duration of hydrolysis suggests that, once the first transphosphorylation into $P_2$ has occurred, further transphosphorylations into higher polyphosphates are easier:

$$AcOP + P_2 \rightleftharpoons P_3 + AcOH$$
$$AcOP + P_3 \rightleftharpoons P_4 + AcOH \text{ etc.} \ldots$$

$P_2$ concentration varies with pH, according to the bell-shaped curve in Figure 5, with a maximum in the pH range 6–10. In the two extreme pH regions, only traces of $P_2$ are detected showing that $P_2$ condensation is not favoured in pH zones where the rate of hydrolysis of the donor, AcOP, is the quickest and especially towards alkaline pHs where the energetic balance of the transphosphorylation has been shown to be the most exergonic.

## 3. Discussion

In conclusion of these two model transphosphorylations, we will make three comments on the mechanism of phosphate transfer:

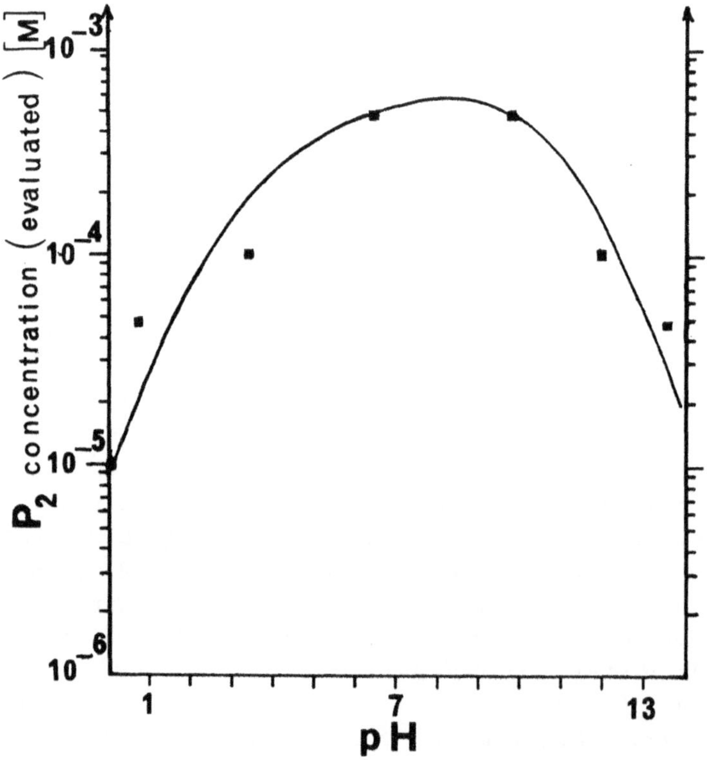


Fig. 5. Qualitative evaluation of pyrophosphate concentration vs pH, during hydrolysis of $2 \times 10^{-1}$ M acetylphosphate in aqueous buffered solutions. $P_2$ concentration remains constant over the hydrolysis at a given pH.

(1) From direct experimental evidence, the adenosine residue of ATP does not seem to have any influence on the phosphate transfer with tripolyphosphate. We shall also remark that the presence of adenosine does not much alter the pKa values of the three phosphate groups either (George and Rutman, 1960).

(2) The presence of cations greatly improves the non enzymic phosphorylation of orthophosphate from $P_3$ ($P_2$ yields go from 0.1% to 10% in the absence and the presence of metal ions, respectively). However, the pyrophosphate syntheses from acetylphosphate have been performed without cations with maximum yields of 0.5%.

(3) In the case of acetylphosphate as energy donor, the best pyrophosphate yields occur near neutrality. Our experimental results seem in discrepancy with the kinetics and energetics of AcOP hydrolysis which afford the best conditions in the alkaline pH range.

This fact cannot be ascribed to the hydrolysis of $P_2$ which is very slow in extreme pH ranges (respective $P_2$ half-life times at pH 1 and 13:1 month and 10 years (Griffith, 1959)).

The transphosphorylation could however be interpreted on the basis of the data obtained by Le Port, from this laboratory, on transacylation reactions. Le Port observed that any coupled hydrolysis-condensation reactions can only occur in pH ranges where either the donor or the acceptor bears at least one acidic hydrogen (i.e. an easily ionizable hydrogen), so that in a condensation where water is a product, a minimum number of extra protons or hydroxyls are utilized from the buffer.

In our case, the condensation of two orthophosphates

$$P + P \rightleftharpoons P_2 + H_2O$$

will be more or less facilitated according to the protonation state of orthophosphate, depending on the pH.

For example, at pHs > 12, where both orthophosphate and pyrophosphate are completely ionized (with respective pKas of $H_3PO_4$ and $H_4P_2O_7$ being 2.1 – 7.2 – 12 and 1 – 2 – 6.6 – 9.6 (George and Rutman, 1960)), for the condensation to take place, two protons from buffer will be needed as shown below;

$$PO_4^{3-} + PO_4^{3-} + 2H^+ \rightleftharpoons P_2O_7^{4-} + H_2O.$$

Although the energy balance of this condensation is strongly negative ($\Delta G_{12}^{\circ\prime} = -4.5$ k cal mole$^{-1}$, Figure 1), this reaction is kinetically unfavourable, since not only does its stoichiometry imply a quatermolecular reaction, but an elevation of pH will displace the reaction towards the left. Thus, in alkaline media, pyrophosphate condensation is disfavoured altogether.

On the other hand, at 6.6 < pHs < 7.2, condensation will be performed between dihydrogenated species according to:

$$H_2PO_4^- + H_2PO_4^- \rightleftharpoons H_2P_2O_7^{2-} + H_2O$$

so that pyrophosphate formation (a bimolecular reaction, in this case) needs no extra protons from the medium, and the water release will be kinetically good.

## 4. Experimental

### 4.1. Reagents

Hexahydrate of sodium tripolyphosphate was obtained from Rhone Progil research laboratories (38 – Decines, France) and purified by repeated precipitations in alcohol according to Van Wazer (1961). Chromatographic analysis on DOWEX 1X8 anion-exchange resin gave the following percentages: ortho- and pyrophosphates 0%, $Na_5P_3O_{10}$ 98.1%, $(NaPO_3)_3$ 1.9%. Acetylphosphate, monoacid monopotassium salt, 100% pure, was prepared in our laboratory by a rapid procedure (Le Port and Etaix, 1973). Acetylphosphate monoanion purity was checked by thin-layer chromatography of a $10^{-2}$ M aqueous solution, and pH titration of a molar solution. Its second pKa = 4.9 was determined on the titration curve.

### 4.2. Separation method for phosphates

Ortho-, di-, and polyphosphates as well as acetylphosphate were all separated on cellulose by thin-layer chromatography (Aurenge *et al.*, 1964). Location of spots was detected by autoradiography for radioactive $P^{32}$, or by spraying the Hanes-Isherwood ammonium molybdate reagent. Spots turned spontaneously to deep blue without chemical reducer, simply by exposing the chromatograms to day-light.

### 4.3. Determination of hydrolysis rates of acetylphosphate monoanion (Figure 3)

The hydrolysis reaction is, for example at $2 < \text{pHs} < 4.9$

$$\underset{1-x}{CH_3COOPO_3H^-} + H_2O \rightarrow \underset{x}{AcOH} + \underset{x}{H_2PO_4^-}$$

$x < 1$ represents the millimoles of AcOP hydrolyzed. Since $AcOPO_3H^-$ and AcOH have very close $pK_as$, they are titrated together and the first equivalent point, for 2 mmoles $OH^-$, corresponds to the mixture of 1 mmole (from buffer) $+x$ mmole (from AcOP hydrolysis) of AcOH and $(1-x)$ mmole of AcOP, the total sum of which makes 2 mmoles. So, whatever the extent of hydrolysis, 2 mmoles of $OH^-$ are consumed by the two acyl compounds. Then, as the base titration is carried on, $x$ mmole $OH^-$ is necessary to neutralize $x$ mmole $H_2PO_4^-$ into $HPO_4^{2-}$. The end of $H_2PO_4^-$ neutralization has a great pH variation and is accurately determined on the titration curve. After 70 hr hydrolysis is completed and 1 mmole of $H_2PO_4^-$ is titrated.

## References

Aurenge, J., De Georges, M., and Normand, J.: 1964, *Bull. Soc. Chim.* **87**, 508.
Barltrop, J. A., Grubb, P. W., and Hesp, B.: 1963, *Nature* **199**, 759.
Calvin, M.: 1962, in: *Horizons in Biochemistry* (eds. M. Kasha and B. Pullman), Academic Press, New York, p. 51.
Di Sabato, G. and Jencks, W. P.: 1961, *J. Amer. Chem. Soc.* **83**, 4400.
George, P. and Rutman, R. J.: 1960, *Progress in Biophysics* **10**, 1.
Griffith, E. J.: 1959, *Ind. Eng. Chem.* **51**, 240.

Jencks, W. P.: 1968, in: *Handbook of biochemistry* (ed. H. A. Sober) The Chemical Rubber Co, Cleveland, Ohio, p. 148.
Koshland, D. E. Jr.: 1952, *J. Amer. Chem. Soc.* **74**, 2286.
Le Port, L., Etaix, E., Godin, F., and Buvet, R.: 1971, in: *Molecular Evolution* **1** (eds. R. Buvet and C. Ponnamperuma), North Holland, Amsterdam, p. 197.
Le Port, L. and Etaix, E.: 1973 unpublished results.
Lowenstein, J. M.: 1958, *Biochem. J.* **70**, 222.
Van Wazer, J. R.: 1961, *Phosphorus and Its Compounds* **1**, Interscience, New York, p. 648.
Wieland, Th. and Bäuerlein, E.: 1967, *Monatsch chem.* **98**, 1381.

# PART IV

# STRUCTURAL AND THERMODYNAMIC CONSIDERATIONS ON THE ORIGIN OF LIFE

# EXPERIMENTS ON THE ORIGINS OF OPTICAL ACTIVITY

WILLIAM A. BONNER and JOSE J. FLORES

*Chemistry Dept., Stanford University, Stanford, Calif. 94305, and Ames Research Center, NASA, Moffett Field, Calif. 94035, U.S.A.*

**Abstract.** Two recent reports claim that (1) aqueous L-aspartic acid polymerizes faster than D-Asp in the presence of kaolin at 90°, and (2) L-phenylalanine is adsorbed by kaolin more extensively than D-Phe at pH 6 (the reverse being true at pH 2). The novelty of these observations and their potential significance for the origin of optical activity has prompted us to duplicate these experiments using more sensitive methods. L- and D, L-Asp in 0.01 *M* solution were incubated with kaolin at 90° for 8 days. Careful examination of the aqueous residues from such experiments failed to demonstrate any preferential polymerization of L-Asp over D-Asp, or indeed any significant gross polymerization of Asp at all.

In other experiments 0.001 *M* solutions of D, L-Phe at pH 6 and pH 2 were stirred with large excesses of kaolin for 24 hr, and the aqueous extracts from these mixtures were examined for gross adsorption using the amino acid analyzer. No significant gross adsorption was noted. We then looked for asymmetric adsorption in the aqueous residues using optical rotatory dispersion, gas chromatography and thin layer chromatography. By none of these analytical criteria could we find any evidence whatsoever for the preferential adsorption of D- versus L-Phe from either pH 6 or pH 2 solutions.

Finally, in experiments bearing on the origin of optical activity by parity violation during $\beta$-decay, we have irradiated solid samples of D-, L- and D, L-leucine in a 61700 Ci Sr-90 source at Oak Ridge National Lab. for 1.34 yr (total dose: $4.2 \times 10^8$ rad). Gas chromatographic examination of the (appropriately derivitized) recovered samples showed that the L-Leu was 16.7% decomposed, the D-Leu 11.4% and the D,L-Leu 13.8% decomposed. The recovered D,L-Leu sample had a gas-chromatographically determined enantiomeric composition of 50.8% D-leu and 49.2% L-Leu. These data, though very close to experimental error, may indicate a slight preferential radiolysis of L-Leu compared to D-Leu by the Bremsstrahlung from Sr-90 $\beta$-decay. These high intensity irradiation experiments are being continued on a prolonged basis in order to reach more definitive conclusions.

Two recent reports in the literature have struck us as novel and potentially important enough to require further experimental substantiation. These involve the claims that kaolin (a) adsorbs L-phenylalanine from $10^{-3}$ *M* aqueous solution more extensively than D-phenylalanine at pH 6 (the reverse being true at pH 2) (Jackson, 1971a, b) and (b) induces the polymerization of L-aspartic acid faster than D-aspartic acid in $10^{-2}$ *M* solution at 90° (Degens *et al.*, 1970; Jackson, 1971a). The potential significance of these observations for the origin of optically active compounds in nature, as well as the absence of any chiral crystal lattice in kaolin which might theoretically induce such results, have prompted us to duplicate these experiments using more sensitive and varied experimental techniques.

## 1. Attempted Asymmetric Adsorption of Phenylalanine by Kaolin

We first attempted to observe the asymmetric adsorption of phenylalanine enantiomers by kaolin under the literature conditions (Bonner and Flores, 1973). D,L-Phenylalanine in $10^{-3}$ *M* pH 6 solution was stirred with kaolin, then centrifuged. The supernatant (Fraction 6-1; Table I) was quantitatively analyzed for phenylalanine using an amino acid analyzer, the kaolin residue was extracted with 4 *M* $NH_4OH$, then was centrifuged, and the second supernatant (Fraction 6-2) was similarly analyzed. An analogous adsorption experiment was conducted using pH 2 HCl solvent, again providing two fractions (2-1; 2-2; Table I). The 'adsorption' by kaolin and the total recovery of phenylalanine between the two fractions of each experiment is shown in Table I. The pH 6 adsorption values are considerably lower than those previously reported (Jackson, 1971a, b).

TABLE I

Retention of D,L-Phenylalanine by kaolin in aqueous solution

| Fraction | pH | % Phenylalanine retained with | | Total recovery, % |
|---|---|---|---|---|
| | | Supernatant[a] | Kaolin[b] | |
| 6-1 | 6 | 93.2 | 6.8 | 102 |
| 6-2 | 6 | 8.8 | 0 | |
| 2-1 | 2 | 68.6 | 31.4 | 81.6 |
| 2-2 | 2 | 13.0 | 18.4 | |

[a] By amino acid analyzer analysis.
[b] By difference.

The solute from each fraction in Table I was recovered and examined in a spectropolarimeter to determine its optical rotatory dispersion curve in the 280–625 nm region. If asymmetric adsorption of D,L-phenylalanine had indeed occurred, it was hoped that this could be detected by measurable optical activity at those wavelengths corresponding to maximum (or minimum) specific rotations of authentic samples of L-phenylalanine at pH 6 and its hydrochloride at pH 2. In none of the recovered samples, however, was an observed rotation in excess of $+0.005°$ (from that of the solvent blank) noted at any wavelength used, suggesting little, if any, optical activity. Unfortunately, however, such ORD data could not be considered definitive, since the 'predicted observed rotations' of our samples (based on Jackson's (1971a, b) reported adsorption values and the known solute concentrations in our ORD solutions) were within or close to the observed rotation values which we found. We therefore sought alternative criteria for asymmetric adsorption in the four fractions in Table I.

The phenylalanine in each of the above ORD fractions was recovered by freeze drying and each sample was converted to a diastereomeric mixture of its N-trifluoroacetyl-(+)-2-butyl ester derivative (Pollock *et al.*, 1965). These mixtures, along with similar D,L-phenylalanine derivative controls, were analyzed quantitatively for their D- and L-enantiomeric compositions by gas chromatography on a $150' \times 0.02''$ capillary column (OV-225 phase). These analyses are summarized in Table II, where we see that each fraction has an enantiomeric composition identical within experimental error to that of the D,L-phenylalanine control. By contrast, Jackson's (1971a, b) adsorption data would predict enantiomeric compositions for the recovered phenylalanine of 51.2% D- and 48.8% L- for the pH 6 experiment, and 48.4% D- and 51.6% L- for the pH 2 experiment.

TABLE II

GC analyses for enantiomeric composition of phenylalanine fractions isolated after kaolin treatment

| Fraction | % Phenylalanine enantiomer | |
|---|---|---|
| | D | L |
| 6-1 | 49.9 | 50.1 |
| 6-2 | 49.7 | 50.3 |
| D,L-Control | 49.8 | 50.2 |
| 2-1 | 50.2 | 49.8 |
| 2-2 | 50.0 | 50.0 |
| D,L-Control | 50.8 | 49.2 |

To confirm the absence of asymmetric adsorption of phenylalanine by kaolin we lastly resorted to thin layer chromatography. Two TLC plates were coated with kaolin as adsorbant, and samples of D-, L- and D,L-phenylalanine were spotted in quadruplicate on each. One plate was developed with water and one with pH 2 HCl, whereupon the plates were dried and the spots visualized with ninhydrin. Differences in adsorption of the two enantiomers by the kaolin adsorbant should, of course, show up as different $R_f$ values. All twelve spots on each plate, however, had $R_f$ values on that plate identical within experimental error, indicating again no preferential adsorption of either phenylalanine enantiomer and no resolution of the racemic phenylalanine by kaolin.

### 2. Attempted Asymmetric Polymerization of Aspartic Acid by Kaolin

We next investigated (Flores and Bonner, 1973) the reported (Degens *et al.*, 1970; Jackson, 1971a, b) asymmetric polymerization of aspartic acid by kaolin in aqueous suspension at 90°. 0.01 *M* Solutions of D,L- (0.05 mmole) and L-aspartic acid (0.10 mmole) were treated with kaolin in sealed tubes for 8 days at 90°, duplicating literature conditions. The kaolin was filtered and washed with water (Fractions 1, Table III), then was reextracted thrice with 2 *M* $NH_4OH$ (Fractions 2) to remove any remaining adsorbed aspartic acid. The fractions were each treated with known quantities of D,L-threonine as an internal standard and were then analyzed each in triplicate for their Asp/Thre ratio on an amino acid analyzer, thus providing accurate assays of the aspartic acid content of each fraction. As summarized in Table III, we see that both for D,L- and L-aspartic acid the total recovery of *unchanged* material was essentially quantitative (>95%), suggesting that only trivial – if any – polymerization had actually occurred. These data are in sharp contrast to Jackson's reports (1971a, b), which maintain that L-aspartic acid is 40% and D-aspartic acid 15% polymerized under these conditions. To see if the slightly below quantitative recovery of L-aspartic acid in the L-1 fraction of Table III might actually be due

TABLE III

Recovery of D,L- and L-aspartic acid after treatment with kaolin

| Fraction | mMoles | % of total | |
|---|---|---|---|
| D L-1 | 0.0485 | 97.0 ± 1.0 | ~98.2 |
| D,L-2 | 0.0006 | 1.2 | |
| L-1 | 0.0942 | 94.2 ± 3.2 | ~95.4 |
| L-2 | 0.0012 | 1.2 | |
| L-1 (Hydrolyzed) | 0.0959 | 95.9 ± 2.8 | |

to the presence of poly-aspartic acid, we subjected this fraction to 20-hr hydrolysis with refluxing 6 *N* HCl, then re-analyzed for aspartic acid using a threonine internal standard as before. As seen in Table III the aspartic acid content of Fraction L-1 was unchanged after hydrolysis, thus indicating again that little if any polymer was present.

For confirmation, we next looked for the asymmetric polymerization of aspartic acid by a completely independent technique, namely, examination of the D/L ratio in the residual aspartic acid recovered after treatment of D,L-aspartic acid with kaolin at 90° for 8 days. If, as reported, L-aspartic acid polymerizes faster than D-aspartic acid, the D/L ratio in the recovered sample should be $>1$, (D/L $=1.417$ on the basis of Jackson's data). Our above results, of course, would predict that the D/L ratio should be 1.000.

We determined the desired D/L ratios by quantitatively converting the recovered kaolin-treated aspartic acid into diastereomeric mixtures of L-Leu-D-Asp and L-Leu-L-Asp after reaction with excess L-leucine N-carboxy anhydride (Manning and Moore, 1968). The diastereomers were readily and quantitatively separated on the amino acid analyzer, whereupon peak area measurements on the triplicately-run chromatograms allowed accurate determination of the desired D/L ratios. As seen in Table IV (No. 1) the recovered D,L-aspartic acid had a D/L ratio within experimental error of 1.000. This was unchanged after acid hydrolysis (No. 2), indicating no increase in either enantiomer due to acid hydrolysis of any polymer which might have been present.

Since the action of kaolin on D- and L-aspartic acid solutions at 90° has been reported to cause racemization (Degens *et al.*, 1970; Jackson, 1971a, b), it was of course possible that the 1:1 D/L ratios for Nos. 1 and 2 in Table IV were due to racemization of both unpolymerized aspartic acid enantiomers during the kaolin treatment, rather than to the absence of asymmetric polymerization. To test this possibility we repeated the experiment with L-aspartic acid, analyzing the recovered product again for its D/L ratio. The product proved 31.3% converted into D-aspartic acid (No. 3, Table IV), indicating extensive (62.6%) racemization by the kaolin treat-

TABLE IV

Enantiomeric composition of kaolin-treated aspartic acid samples before and after acid hydrolysis

| No. | Sample | D-Asp/L-Asp | Enantiomeric composition | |
|---|---|---|---|---|
| | | | %-D | %-L |
| 1 | D,L | $1.008 \pm 0.009$ | 50.2 | 48.8 |
| 2 | D,L (Hydrolyzed) | $1.000 \pm 0.003$ | 50.0 | 50.0 |
| 3 | L | $0.455 \pm 0.014$ | 31.3 | 68.7 |
| 4 | L (Hydrolyzed) | $0.461 \pm 0.006$ | 31.5 | 68.5 |
| 5 | (D,L-Standard)[a] | $(1.000 \pm 0.006)$ | (50.0) | (50.0) |

[a] Control, untreated with kaolin.

ment. Finally, the latter sample was hydrolyzed with HCl to see if this changed the enantiomeric ratio. The D/L ratio of No. 4 (Table IV), however, was identical to that of the unhydrolyzed sample No. 3. This clearly confirms that the L-aspartic acid had not polymerized appreciably, since if it had, acid hydrolysis of the polymer would have liberated additional L-aspartic acid, thus *decreasing* the D/L ratio in No. 4. All of our experimental data thus accord with the conclusions that (a) there is no significant gross polymerization of aspartic acid in 0.01 *M* solution by kaolin during 8-day treatment at 90°, and (b) there is likewise no selective polymerization of L- over D-aspartic acid under these same conditions. Since completing our studies we have learned that McCullough and Lemmon (1973) have attempted similar aspartic acid polymerization experiments, and have come to identical conclusions using analytical techniques completely different from ours.

### 3. Strontium-90 Bremsstrahlung Radiolysis of Leucine

In 1959 Vester and coworkers proposed a novel hypothesis for the natural origin of optically active compounds, based on the known violation of the parity principle during $\beta$-decay of certain radioactive isotopes (Wu *et al.*, 1957). In this mechanism the predominantly 'left-handed' longitudinally polarized electrons emitted during $\beta$-decay interact with matter to form circularly polarized $\gamma$- and X-ray Bremsstrahlung photons, which in turn may induce absolute asymmetric syntheses or degradations in appropriate organic molecules, thus yielding optically active products. Early attempts to demonstrate this mechanism experimentally (Vester *et al.*, 1959; Ulbricht and Vester, 1962; Garay, 1968), however, have led to negative or ambiguous results, presumably due to the relatively weak $\beta$-sources employed and to the short exposure times. We have attempted to examine this mechanism experimentally using a considerably more powerful $\beta$-ray Bremsstrahlung source, namely, 61 700 Ci of strontium-90 housed at Oak Ridge National Laboratory. This source was calibrated on February 17, 1971 and found to provide a Bremsstrahlung dose rate of $3.56 \times 10^4$ rad $hr^{-1}$. After demonstrating that $\gamma$-radiation was in fact capable of causing degradation of amino acids in the solid state (Bonner, 1973), we placed samples of D-, L- and D,L-leucine, both as solids and as salts in solution, in the above $^{90}Sr$ source, and irradiated them for a total of 11 736 hr (1.34 yr; total dose $4.11 \times 10^8$ rad). The recovered amino acid samples were converted to diastereomeric mixtures of their volatile N-trifluoroacetyl (+)-2-butyl esters (Pollock *et al.*, 1965), then were assayed for percent degradation gas chromatographically, using the technique of enantiomeric markers (Bonner, 1973). The recovered D,L-leucine samples were also examined for enantiomeric inequality by optical rotatory dispersion measurements and again by gas chromatography of their N-trifluoroacetyl (+)-2-butyl esters. The results of these experiments are summarized in Tables V and VI.

Examination of Table V indicates that there is no preferential radiolysis of either enantiomer of the leucine salts in aqueous solution, even after total degradation has proceeded to the extent of almost 50%. Such an observation, while contrary to the report of Garay (1968), however, is what one would probably anticipate for radiolysis conducted in aqueous solution. Here the initial mechanism for degradation of the amino acids would presumably be attacked by symmetrical free radicals generated by the Bremsstrahlung irradiation of the solvent. It is difficult to see how such aqueous solution photochemistry could support any asymmetric bias, and the data of Table V are accordingly perhaps to be expected. In the *solid* state, however, such a mechanism could not be operative, and one might therefore reasonably anticipate direct asymmetric interaction between the circularly polarized Bremsstrahlung photons and the amino acid substrate. This *appears* to be borne out in the %-Decomposition column of Table VI, where we see that the L-leucine has undergone the greatest amount of radiolysis, the D-leucine least, and the D,L-leucine an intermediate amount. From the % Decomposition of D- and L-leucine in Table VI one can calculate that the irradiated D,L-leucine should be about 14.0% decomposed and should have an enantiomeric composition of about 51.5% D- and 48.5% L-leucine. Actual gas chromatographic measurements of the %-Decomposition (13.8%) and enantiomeric composition (50.7% D-, 49.3% L-; average values in Table VI) of the irradiated D,L-leucine gave values close to those calculated. These data thus

TABLE V

Radiolysis of solutions of D,L-leucine salts by strontium-90 bremsstrahlung[a]

| Sample | Solvent | %-Decomposition | Enantiomeric composition, % | | | | | |
|---|---|---|---|---|---|---|---|---|
| | | | Sample | | | D,L-Standard | | |
| | | | D | L | (±) | D | L | (±) |
| D,L-Leu Na Salt | $H_2O$[b] | 48.6 | 49.4 | 50.6 | 0.2 | 49.3 | 50.7 | 0.2 |
| D,L-Leu HCl Salt | $H_2O$[c] | 34.6 | 50.2 | 49.8 | 0.3 | 50.3 | 49.7 | 0.1 |

[a] $\beta$-Source: 61.7 kCi $^{90}Sr$; Dose: $4.11 \times 10^8$ rad.
[b] With 5% excess NaOH; conc. ~1 M.
[c] With 5% excess HCl; conc. ~1 M.

TABLE VI

Radiolysis of solid D-, L- and D, L-leucine samples by strontium-90 bremsstrahlung[a]

| No. | Sample | % Decomposition | Enantiomeric composition, % | | | | | | |
|---|---|---|---|---|---|---|---|---|---|
| | | | Sample | | | D,L-Standard | | | Difference |
| | | | D | L | (±) | D | L | (±) | $D\text{-}D_{st}$ |
| 1 | D | 11.4 | | | | | | | |
| 2 | L | 16.7 | | | | | | | |
| 3 | D,L | 13.8 | 50.8[b] | 49.2[b] | 0.05 | 50.1[c] | 49.9[c] | 0.21 | 0.7 |
| 4 | D,L | 13.8 | 50.6[d] | 49.4[d] | 0.25 | 49.8[d] | 50.2[d] | 0.29 | 0.8 |
| | | (14.0)[e] | (51.5)[e] | (48.5)[e] | | | | | |

[a] $\beta$-Source: 61.7 kCi $^{90}$Sr; Dose: $4.11 \times 10^8$ rad.
[b] Average of 3 G.C. Analyses.
[c] Average of 4 G.C. Analyses.
[d] Average of 5 G.C. Analyses.
[e] Calculated from % Decomposition in Nos. 1 and 2.

appear to substantiate the %-Decomposition data, and again to suggest that L-leucine in the solid state undergoes $^{90}$Sr radiolysis slightly faster than does D-leucine. Unfortunately, however, the differences in the Enantiomeric Composition (Table VI) between the irradiated sample and the D,L-leucine standard are so close (in our experience) to the experimental error of our gas chromatographic quantitative analytical technique (Bonner, 1972), that we doubt whether a definitive conclusion as to the validity of the Enantiomeric Composition data is possible at the present time. While the %-Decomposition data in Table VI are presumably reasonably accurate (Bonner, 1973), it should be noted that the relatively small differences observed between the D-, L- and D,L-samples could arise alternatively as a result of radiation field inhomogeneities in the $^{90}$Sr source employed during the 1.34 yr irradiation period. Thus, in themselves, the %-Decomposition data in Table VI can unfortunately not be considered definitive either. Such ambiguities might hopefully be resolved in the future by longer irradiation times. At present we have a similar set of samples which have now been irradiated in the same $^{90}$Sr source for a period of some 2.4 yr. We propose to examine these samples by similar techniques after a total 3.5–4 yr exposure time, whereafter the inconclusive results above can perhaps be clarified, and the question of the origin of molecular chirality by parity violation in $\beta$-decay can perhaps be answered definitively.

## References

Bonner, W. A.: 1972, *J. Chromat. Sci.* **10**, 159.
Bonner, W. A.: 1973, *J. Chromat. Sci.* **22**, 101.
Bonner, W. A. and Flores, J. J.: 1973, *Currents Mod. Biol.* **5**, 103.
Degens, E. T., Matheja, J., and Jackson, T. A.: 1970, *Nature* **227**, 492.
Flores, J. J. and Bonner, W. A.: 1974, *J. Mol. Evolution* **3**, 49.
Garay, A. S.: 1968, *Nature* **219**, 338.
Jackson, T. A.: 1971a, *Experientia* **27**, 242.

Jackson, T. A.: 1971b, *Chem. Geol.* **7**, 295.
Manning, J. M. and Moore, S.: 1968, *J. Biol. Chem.* **243**, 5591.
McCullough, J. J. and Lemmon, R. M.: 1974, *J. Mol. Evolution* **3**, 57.
Pollock, G. E., Oyama, V. I., and Johnson, R. D.: 1965, *J. Gas Chromatog.* **3**, 174.
Ulbricht, T. L. V. and Vester, F.: 1962, *Tetrahedron* **18**, 629.
Vester, F., Ulbricht, T. L. V., and Krauch, H.: 1959, *Naturwiss.* **46**, 68.
Wu, C. S., Ambler, E., Hayward, R. W., Hoppes, D. D., and Hudson, R. P.: 1957, *Phys. Rev.* **105**, 1413.

# A TECHNIQUE FOR THE DETERMINATION OF ENANTIOMERIC AMINO ACIDS IN BIOLOGICAL SAMPLES

K. D. HAEGELE, P. Y. HOWARD and W. PARR

*Dept. of Chemistry, University of Houston, Houston, Tex. 77004, U.S.A.*

**Abstract.** Racemic amino acids can be separated into their enantiomers by means of gas-liquid chromatography. The most applied technique, today, is the conversion of chiral compounds into diastereoisomers with optically active reagents and subsequent chromatography on conventional optically inactive stationary phases. In previous studies it has been realized that this technique is associated with various problems. We studied the use of optically active stationary phases for separating enantiomers directly via a diastereoisomeric association complex. The optically active stationary phases employed are N- and C-terminal substituted dipeptides of the type N-trifluoroacetyl-dipeptide-cyclohexyl esters and have been synthesised by the I-hydroxibenztriazole dicyclohexylcarbodiimide method.

The quality of these phases with respect to separation factors, resolution factors, and thermodynamical properties have been evaluated. All synthetic phases show excellent properties; however, when attempting separation of mixtures of naturally occurring amino acids extensive overlap in the elution diagram was detected. Only one phase – N-TFA-L-α-amino-n-butyryl-L-α-amino butyric acid cyclohexyl ester – gave complete resolution of the naturally occurring amino acids alanine, valine, glycine, threonine, leucine, isoleucine, serine and proline on a 400 ft × 0.02 in capillary column. Less volatile amino acids such as aspartic acid, phenylalanine, methionine, glutamic acid, tyrosine, arginine, and tryptophan can be resolved at a 100 ft × 0.02 in column.

## 1. Introduction

Whenever a biologically active peptide has to be synthesized, one of the main criteria for the amino acids used for this purpose is their optical purity. For example, only 1% D isomer in the amino acids would limit the yield of a synthesized peptide containing 100 amino acid residues, to merely 37% (as judged by the optical purity). A problem of even higher importance is racemization during synthesis. Since Corrigan (1969) showed that D-amino acids also are widely distributed in nature, especially in lower organisms, it is even more important to have reliable methods in determining the optical isomers of amino acids in biological material.

Up to now several methods are available: (1) Polarimetric methods lack sensitivity and cannot distinguish between optical or chemical impurity (Greenstein and Winitz, 1961). (2) Enzymatic methods are characterized by the high stereospecificity of the enzymes but lack chemical specificity. (3). Chromatography on Sephadex columns has been successfully applied to separate diastereomeric dipeptides (Wieland and Bende, 1965; Krebs and Schumacher, 1966). (4) Thin-layer chromatography is the most common approach for the separation of optical isomers and is widely used as a qualitative measurement (Feltkamp and Pfrommer, 1965). (5) In paper chromatography the optical isomers are separated through the interaction with the asymmetric centers of cellulose (Rhuland *et al.*, 1955; Dalgiesh, 1952). (6) By far the best method, by matters of sensitivity, reproducibility and efficiency, is gas liquid chromatography. Since about 1960 this method has been investigated by several laboratories and developed into a highly versatile analytical tool. Principally there are two possibilities of resolving amino acid and peptide enantiomers by gas chromatography: Diastereoisomerization, followed by chromatography on an optically inactive phase and, direct injection of the derivatized enantiomers onto optically active phases. Both techniques, though clearly different, have their own merits, depending on the type of analysis, the available material, analysis time and accuracy of quantitation. Scheme A rationalizes the two principal ways.

SCHEME A

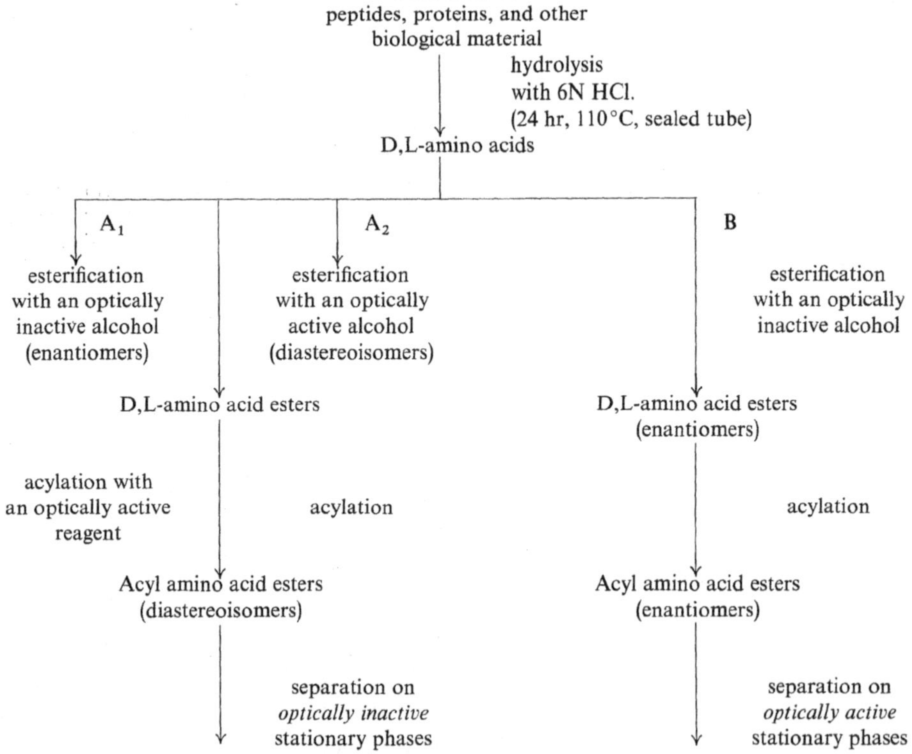


## 1.1. Resolution of Diastereoisomers

Weygand *et al.* (1960) first resolved a pair of diastereoisomeric dipeptides as their methyl esters and N-trifluoroacetyl derivatives on a 2 m column. Using a 50 m capillary column the authors were able to resolve 34 peptides. In addition to resolvability, the diastereosiomers employed satisfied the other conditions for the separation of diastereoisomers, which are: (a) easy and quantitative formation with conservation of configuration of the originally employed enantiomers, (b) thermal stability, which also means that no racemization occurs during chromatography, and (c) possible reconversion of the diastereoisomers in good yield into the constituting enantiomers. The separation of amino acid enantiomers *via* diastereoisomers can be achieved through either reacting the amino-group with an asymmetric acylating reagent or by reacting the carboxyl-group with an asymmetric alcohol (Scheme A).

In general, the first approach has found wider application. For the derivatization of the amino function, in most cases the trifluoroacetyl-derivative, as originally introduced by Weygand *et al.* (1960) is used. This is followed by esterification with the optically active alcohol, as for example 2-butanol. Pollock and Oyama (1966) could resolve 21 amino acids as their N-TFA and 2-butyl esters on a 50 m column coated with Carbowax 1540. Other optically active alcohols used include 2-octanol (Gil-Av *et al.*, 1965), and menthol (Vitt *et al.*, 1965; Halpern and Westley, 1965). The latter two alcohols have the advantage of their higher optical purity in comparison to 2-butanol, but have the disadvantage of being less volatile than the 2-butyl esters, which so far limits their applications.

As optically active reagents for the amino-group 2-chloroalkanoic acids have been introduced by Halpern and Westley (1965), where 2-chloro-3,3-dimethylbutanoic acid was found to be the best resolving agent. Other reagents include N-trifluoro-acetyl-prolyl-chloride ((Westley and Halpern, 1969) and (−)-menthylchloroformate (Westley and Halpern, 1968). The main disadvantage of all these procedures is the lack of optical purity of the reagent to be used for forming the diastereoisomers.

### 1.2. Resolution of enantiomers on optically active stationary phases

Gil-Av *et al.* (1966, 1967) reported the resolution of 18 pairs of enantiomeric N-TFA-amino acid esters by using N-TFA-L- or D-isoleucine lauryl ester and N-TFA-L-phenylalanine cyclohexyl esters as stationary phases. For better performance and higher temperature stability N-TFA-L-valyl-valine cyclohexyl ester (Gil-Av and B. Feibush, 1967; Nakaparksin *et al.*, 1970) and N-TFA-L-phenylalanyl-L-leucine cyclohexyl ester (Koenig *et al.*, 1970) have been introduced as stationary phases. In all cases, the L-enantiomer emerged first on the D-phases, but last on the L-phases. The mechanism is postulated to involve formation of a hydrogen bonded diastereo-isomeric association complex, which has just a slightly higher stability towards the enantiomer of the same configuration to allow its retardation and subsequently to give it the longer retention time than the enantiomer of the opposite configuration (Parr *et al.*, 1970; Parr and Howard, 1972). This formation of a diastereoisomeric association complex was confirmed by Corbin *et al.* (1971) and by Parr and Grohmann (1972). By using N-TFA amino acid cyclohexyl esters of various amino acids, which only differed in the length of the α-carbon chain, Parr and Howard (1973) found only a slight inclination toward decreasing solute-solvent interaction, while the length of the α-carbon side chain was increasing. This is explained by free rotation between the α and β carbons of the solute, which is not affected by the length of the side chain of the solute.

Changing α-carbon side chains of the solvent, by holding the α-carbon side chains of the solute constant resulted in a general increase of solute-solvent interaction. So, identical modifications in the side chains of solvent and solute cause different effects on their ability to form diastereoisomeric association complexes.

The greatest advantage using stationary phases as a way of separating amino acid enantiomers, is the simple derivatization of the amino acids by using non-optically

TABLE I

Analytical data and yields in the preparation of new N-TFA-L-L-Dipeptide Cyclohyxyl Esters as optically active stationary phases

| Amino Acid | Amino Acid Cyclohexyl Ester × HCl | | L-L Dipeptide Cyclohexyl Ester × TFA | | N-TFA-L-L-Dipeptide Cyclohexyl Esters | | Elemental Composition of N-TFA-L-L Dipeptide Cyclohexyl Esters | | | | |
|---|---|---|---|---|---|---|---|---|---|---|---|
| | % Yield | °C, Mp | % Yield | °C, Mp | % Yield | °C, Mp | | % C | % H | % N | % F |
| Ala | 90 | 50– 51 | 63 | a | 81 | 121–123 | found: | 49.75 | 6.27 | 8.25 | 16.80 |
| | | | | | | | calc.: | 49.70 | 6.26 | 8.29 | 16.85 |
| Aba | 93 | 147–150 | 63 | 172–173 | 79 | 110–112 | found: | 52.44 | 6.88 | 7.65 | 15.66 |
| | | | | | | | calc.: | 52.45 | 6.88 | 7.59 | 15.56 |
| Nval | 90 | 138–140 | 71 | 196–197 | 92 | 88– 90 | found: | 54.83 | 7.33 | 7.08 | 14.52 |
| | | | | | | | calc.: | 54.81 | 7.41 | 7.10 | 14.45 |
| Leu | 86 | 203–205 | 55 | 190–193 | 84 | 77– 80 | found: | 56.95 | 7.83 | 6.54 | 13.52 |
| | | | | | | | calc.: | 56.86 | 7.87 | 6.63 | 13.49 |
| Jle | 90 | 134–136 | 60 | 187–189 | 73 | 93– 94 | found: | 56.73 | 7.90 | 6.74 | 13.43 |
| | | | | | | | calc.: | 56.86 | 7.87 | 6.63 | 13.49 |
| NLeu | 77 | 94– 97 | 74 | 182–183 | 62 | 86– 88 | found: | 56.70 | 7.74 | 6.70 | 13.60 |
| | | | | | | | calc.: | 56.80 | 7.87 | 6.63 | 13.49 |

a – obtained as an oil.

active reagents. Thus no D-D artifacts, which may arise from optically impure reagents, will be found. In general as amino protecting group, the trifluoroacetyl function is used. The suitable alcohols for the esterification are either methanol, isopropanol or *tert.* butanol. As shown by Nakaparksin *et al.* (1970), the resolution increases in the order primary, secondary, tertiary alcohols. Because of the difficulty in preparing *tert.* butyl esters of amino acids, most commonly TFA-amino acid isopropyl esters are used. These isopropyl esters have a rather low volatibility. In particular the N-TFA derivatives of cysteine, tryptophan, histidine, cystine and lysine could not be resolved by GC on an optically active stationary phase. To overcome these difficulties we used N-TFA-phenylalanyl-phenylalanine cyclohexyl ester as a stationary phase, which enabled us to operate the column up to 170 °C, but gave relatively poor resolution (Parr *et al.*, 1969). The volatility of the derivatives to be examined was greatly enhanced by using pentafluoropropionyl amino acid isopropyl esters (Parr *et al.*, 1971). This made it possible to resolve seventeen pairs of natural amino acids, including lysine, cysteine, tryptophane and arginine with retention times of up to 47% less, than with N-TFA-amino acid isopropyl esters.

The preparation of the most commonly used TFA-L-L-dipeptidecyclohexyl esters is illustrated by the synthesis of the dipeptide phase N-TFA-L-α-amino-butyryl-L-α-aminobuturyl acid cyclohexyl ester: N-Boc-L-α-amino-n-butyric acid and L-α-n-butyric acid cyclohexyl ester are coupled by way of the carbodiimide method, as described by Koenig and Geiger (1970). After removal of the Boc-group with HCl/Dioxan the free amino function was acylated with trifluoroacetic anhydride.

Table I summarizes the analytical data and the yields obtained for the synthesized

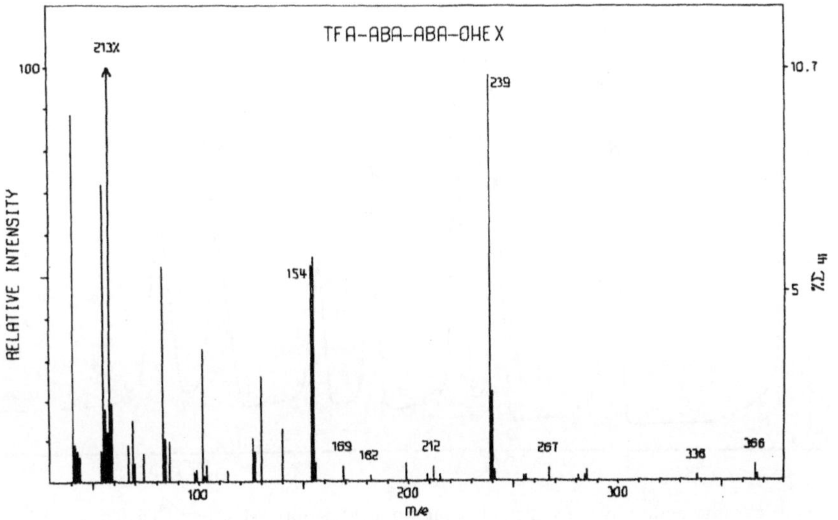


Fig. 1. Mass Spectrum of N-TFA-L-α-amino-n-butyryl-L-α-amino-n-butyric acid cyclohexyl ester. LKB-9000, ionizing voltage 70 eV, accelerating voltage −3.5 kV; ion-source temperature 270°C, trap current 60 μA.

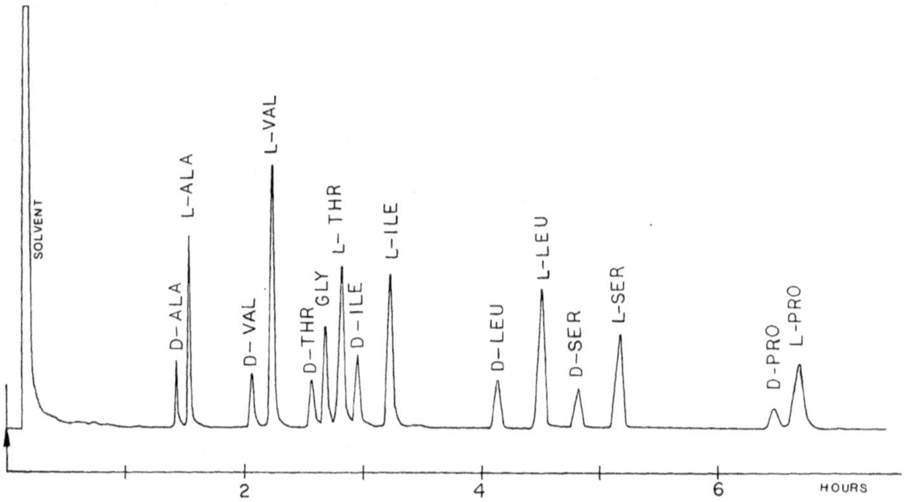


Fig. 2. Gas chromatograms of TFA-amino acid isopropyl esters (D: L ratio 1:3) Varian 1200-l, column: 122 m × 0.508 mm, i.d., stainless steel capillary coated with N-TFA-L-α-amino-n-butyryl-L-α-amino-n-butyric acid cyclohexyl ester, temperature 110 °C, isothermal; injectors 180 °C, detector 275 °C, carrier gas, Helium at 0.84 Kp $cm^{-2}$. Integration of the peak areas with an automatic integrator (Varian 480) gave a precision of detection of either isomer of ±1%.

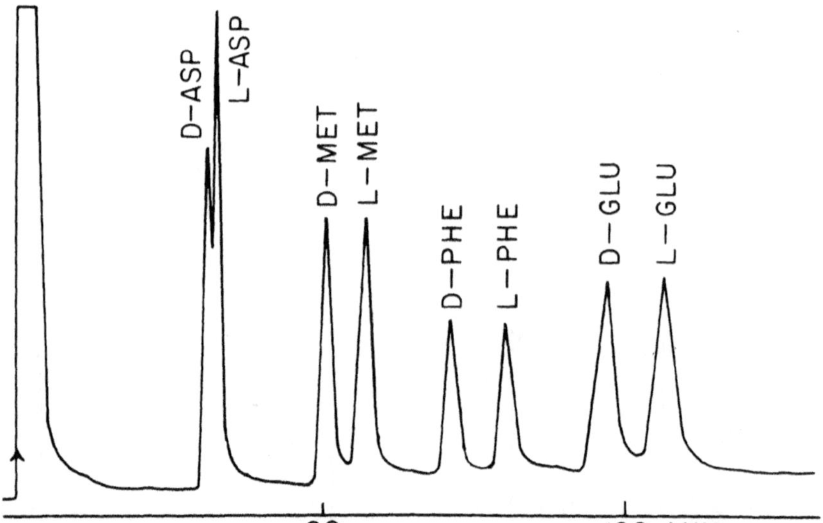


Fig. 3. Gas chromatogram of racemic TFA-amino acid isopropyl esters, Varian 1200-1 column: 30.5 m × 0.508 mm, i.d., stainless steel capillary coated with N-TFA-L-phenylalanyl-L-leucine cyclohexyl ester, temperature 110 °C isothermal, injector 180 °C, detector FID 250 °C, carries gas He, pressure at 0.35 Kp $cm^{-2}$. Integration of the peak areas with an automatic integrator (Varian 480) gave a precision of detection of either isomer of ±1%.

phases. The purity of the phases was controlled by thin-layer chromatography and by GC/MS on 1.8 m, 1% SE-30 column. Figure 1 shows the mass spectrum for N-TFA-L-α-amino-n-butyryl-L-α-amino-n-butyric acid cyclohexyl ester. The molecular ion of the compound is found at m/e 366. The first fragment occurs via MacLafferty rearrangement to eliminate $C_2H_4$ from the Cα-side chain to form the ion at m/e 338. Expulsion of cyclohexen gives raise to the ion at m/e 285 and 284. Loss of the O-hexyl moiety yields the ion at m/e 267 and loss of —COOHexyl gives raise to the abundant ion at m/e 239. The sequence ions for the N-terminus are observed at m/e 182 and m/e 154, for the C-terminus at m/e 212 and m/e 169. The base peak of the spectrum corresponds to $H_2N^{\oplus}{=}CH{-}CH_2{-}CH_3$, m/e 58. This phase, stable up to 120 °C, showed the best properties in resolving enantiomer-mixtures of the more volatile amino acids. It was no problem to separate the optical isomers of a mixture of leucine, isoleucine or tert. leucine and within a total analysis time of about 6½ hr the enantiomeric mixture of 16 naturally occurring more volatile amino acids could be resolved completely (Figure 2). For the less volatile amino acids, especially, Tyr, Arg, Lys the more temperature stable phase N-TFA-phenylalanyl-leucine cyclohexyl ester is more successful. This phase can be operated up to 140 °C without substantial

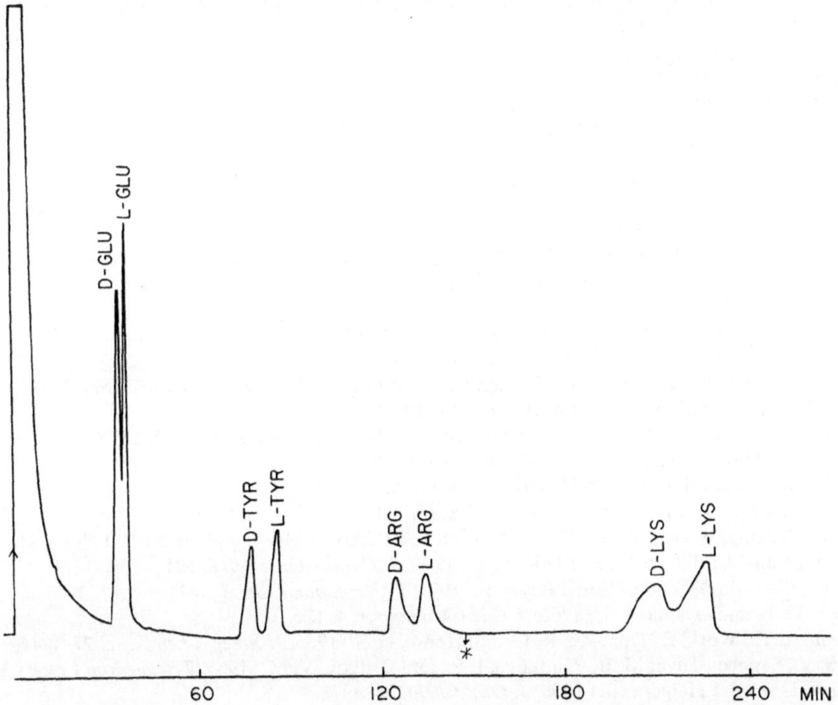


Fig. 4. Gas chromatogram of racemic pentafluoropropionyl-amino acid isopropyl esters Varian 1200-1 column: 30.5 m × 0.508 mm, i.d., stainless steel capillary coated with N-TFA-L-phenylalanyl-L-leucine cyclohexyl ester, temperature 125 °C isothermal, injector 180 °C, detector FID 250 °C, carrier gas He, pressure at 1.12 Kp cm$^{-2}$. Manual pressure increase to 1.93 Kp cm$^{-2}$. Integration of the peak areas with an automatic integrator (Varian 480) gave a precision of detection of either isomer of ±1%.

bleeding (Koenig *et al.*, 1970; Parr *et al.*, 1971). Figures 3 and 4 show chromatograms of the less volatile TFA amino acid isopropylesters on a 30.5 m column. Aspartic acid, methionine, phenylalanine and glutamic acid emerge from this column in a reasonable time, whereas for the resolution of tyrosine, arginine and especially lysine, conditions of higher temperature and pressure are required.

An application of these procedures proved that no racemization is occurring during the solid phase peptide synthesis (Bayer *et al.*, 1970a). For the dodecapeptide (Leu-Ala)$_6$ D-isomers were found only in amounts below 0.1% (Bayer *et al.*, 1970b, Figure 2). The slightly higher amounts of D-isomers detected in synthetic desamidosecretion (up to 2.2% in the case of aspartic acid) could be explained through the rigid conditions of peptide hydrolysis, were aspartic acid racemizes to a considerable degree. So it could be shown that the solid phase synthesis of peptides proceeds with complete retention of configuration.

## References


Bayer, E., Eckstein, H., Haegele, K. D., Koenig, W. A., Brüning, W., Hagenmaier, H., and Parr, W.: 1970a, *J. Amer. Chem. Soc.* **92**, 1735.

Bayer, E., Gil-Av, E., Koenig, W. A., Nakaparksin, S., Oró, J., and Parr, W.: 1970b, *J. Amer. Chem. Soc.* **92**, 1738.

Corrigan, J. J.: 1969, *Science* **164**, 142.

Dalgiesh, C. E.: 1952, *J. Chem. Soc.* **1952**, 3940.

Feltkamp, H. and Pfrommer, H.: 1965, *J. Chromatogr.* **18**, 403.

Gil-Av, E. and Feibush, B.: 1967, *Tetrahedron Letters* **3345**.

Gil-Av, E., Charles, R., and Fischer, G.: 1965, *J. Chromatogr.* **17**, 408.

Gil-Av, E., Feibush, B., and Charles-Sigler, R.: 1966, *Tetrahedron Letters*, **1009**.

Gil-Av, E., Feibush, B., and Charles-Sigler, R.: 1967, in A. B. Littlewood (ed.), *Gas-Chromatography 1966*, Inst. of Petroleum, London, p. 227.

Greenstein, J. P. and Winitz, M.: 1961, *Chemistry of the Amino Acids*, Wiley, New York.

Grohmann, K. and Parr, W.: 1972, *Chromatographia* **5**, 18.

Halpern, B. and Westley, J. W.: 1965a, *Chem. Commun.*, 246.

Halpern, B. and Westley, J. W.: 1965b, *Chem. Commun.*, 421.

Koenig, W. and Geiger, R.: 1970, *Chem. Ber.* **103**, 788.

Koenig, W., Parr, W., Lichtenstein, H., Bayer, E., and Oró, J.: 1970, *J. Chromatogr. Sci.* **7**, 183.

Krebs, H. and Schumacher, W.: 1966, *Chem. Ber.* **99**, 1341.

Nakaparksin, S., Birrell, P., Oró, J., and Gil-Av, E.: 1970, *J. Chromatogr. Sci.* **7**, 177.

Parr, W. and Howard, P. Y.: 1972a, *J. Chromatogr.* **67**, 227.

Parr, W. and Howard, P. Y.: 1972b, *Angew. Chem. Int. Ed.* **84**, 586.

Parr, W. and Howard, P. Y.: 1973, *Anal. Chem.* **45**, 711.

Parr, W., Koenig, W., and Bayer, E.: 1969, Southwest Regional Meeting of the ACS, Tulsa, Oklahoma.

Parr, W., Yang, C., Bayer, E., and Gil-Av, E.: 1970, *J. Chromatogr. Sci.* **8**, 591.

Parr, W., Pleterski, J., Yang, C., and Bayer, E.: 1971, *J. Chromatogr. Sci.* **9**, 141.

Pollock, G. E. and Oyama, V. I.: 1966, *J. Gas Chromatogr.* **4**, 126.

Rhuland, L. E., Work, E., Denman, R. F., and Hoare, D. S.: 1955, *J. Amer. Chem. Soc.* **77**, 4844.

Vitt, S. V., Saporovskaya, M. B., Gudkova, I. P., and Belikov, V. M.: 1965, *Tetrahedron Letters* **30**, 2575.

Westley, J. W. and Halpern, B.: 1968, *J. Org. Chem.* **33**, 3978.

Westley, J. W. and Halpern, B.: 1969, in S. L. A. Harbaum (ed.), *Gas Chromatography 1968*, Institute of Petroleum, London, p. 119.

Weygand, F., Kolb, B., Prox, A., Tilak, M. A., and Tomida, I.: 1960, *Hoppe-Seyl. Z. Physiol. Chem.* **322**, 38.

Weygand, F., Prox, A., Schmidhammer, L., and Konig, W.: 1963, *Angew. Chem. Int. Ed.* **2**, 183.

Wieland, Th. and Bende, H.: 1965, *Chem. Ber.* **98**, 504.

# COACERVATE-LIKE MICROSPHERES FROM LYSINE-RICH PROTEINOID

DUANE L. ROHLFING
*Dept. of Biology, University of South Carolina, Columbia, S.C. 29208, U.S.A.*

**Abstract.** Microspheres form isothermally from lysine-rich proteinoid when the ionic strength of the solution is increased with NaCl or other salts. Studies with different monovalent anions and with polymers of different amino acid composition indicate that charge neutralization and hydrophobic bonding contribute to microsphere formation. The particles also form in sea water, especially if heated or made slightly alkaline. The microspheres differ from those made from acidic proteinoid but resemble coacervate droplets in some ways (isothermal formation, limited stability, stabilization by quinone, uptake of dyes). Because the constituent lysine-rich proteinoid is of simulated prebiotic origin, the study is interpreted to add emphasis to and suggest an evolutionary continuity for coacervation phenomena.

## 1. Introduction

Coacervate droplets (Oparin, 1957) and proteinoid microspheres (Fox and Dose, 1972) have been much studied as protocellular models. The former are usually prepared from polymers of biotic origin, while the microspheres are typically formed from thermally prepared polyamino acids, of simulated prebiotic origin, that are enriched with dicarboxylic amino acids (acidic proteinoid). The two model structures differ in several ways (Fox, 1970), including mode of formation, uniformity of size, and stability.

This paper describes the formation of microparticles from proteinoid enriched with lysine (basic proteinoid). The process used, merely increasing the ionic strength of polymer solution, is unlike that used with acidic proteinoid (see Section 2), and the resulting microparticles differ from acidic-proteinoid microspheres. The particles from lysine-rich proteinoid, however, resemble coacervate droplets in several ways,* a feature that may have evolutionary significance.

## 2. Methods

Lysine-rich proteinoids (Fox *et al.*, 1962) were prepared from A-grade amino acids by heating equal weights of lysine free base and an equimolar mixture of 17 other proteinous amino acids under nitrogen at 195 °C for 1–2 hr. Polymers of lysine plus one-to-three other amino acids were similarly prepared, as was an 'omission' proteinoid (Rohlfing and Fox, 1969) that lacked aromatic residues. A polymer prepared from the nonproteinous amino acids ε-aminohexanoic acid, sarcosine, β-alanine, norvaline, and α-amino-*n*-butyric acid was a gift from M. A. Saunders.

* Microspheres with coacervate-like properties have been described, but with basic proteinoid either absent (Smith and Bellware, 1966) or in combination with other organic material, including acidic proteinoid (Fox and Dose, 1972; Young, 1972).

Microspheres were formed from lysine-rich proteinoid by adding salt solution (usually NaCl) to an aqueous *solution* of polymer. [The procedure used with acidic proteinoid (Fox *et al.*, 1959) involves solubilizing a *suspension* of polymer by heating; microspheres form when the solution is cooled. Alternatively (Young, 1965), the particles form when the temperature is lowered from 25 to 0°C.] A second 'crop' of basic-proteinoid microspheres often appeared when the supernatant was heated (not cooled, as with acidic proteinoid) to 50–100°C. Coacervate droplets were formed by mixing 0.1% solutions of gum arabic and gelatin or protamine sulfate.

The percentage of the polymer preparation in microspherical form was estimated by comparing the absorbance at 400 nm of the polymer remaining in solution after microsphere formation to that of an aliquot of salt-free, totally soluble polymer. [The absorbance, described by Rohlfing (1967), is attributed to a nondiffusible pigment associated with proteinoid; *cf.* Weber *et al.* (1968).]

## 3. Results and Discussion

Figure 1a shows microspheres prepared by mixing equal volumes of 15% NaCl and lysine-rich proteinoid solutions. The size of the particles, less uniform than those from acidic proteinoid (Fox and Dose, 1972), increased with increasing polymer concentration, reaching 10–15 $\mu$ in diam at 4–8 mg $ml^{-1}$. The *p*H range 5–12 was effective, a broad optimum occurring from *p*H *ca.* 7–10. Acidic-proteinoid microspheres dissolve at *p*H values of 6 or higher (Fox and Dose, 1972, p. 203). Microspheres also formed when solid (amorphous) basic proteinoid was added to NaCl solution.

### 3.1. Effects of Salts

The percentage of total polymer in microspherical form increased with increasing NaCl concentration, reaching *ca.* 75% in 24% NaCl. Plots of log solubility versus ionic strength were linear (the correlations were statistically significant at the 99% confidence level), which indicates that salting-out phenomena (Mahler and Cordes, 1966) contribute to microsphere formation.

Microspheres formed in 10% KCl or $NH_4Cl$, as well as NaCl, but the chlorides of $Mg^{2+}$, $Ca^{2+}$, $Mn^{2+}$, and $Fe^{3+}$ were relatively ineffective. For example, $MgCl_2$, when tested over the concentration range 0–20%, showed its maximum effect at 5% (ionic strength = 1.6), at which concentration only *ca.* 15% of the basic proteinoid was in microspherical form. The valency of the anion was not critical qualitatively, microspheres being formed in the presence of the sodium salts of $Br^-$, $NO_3^-$, $SCN^-$, $ClO_4^-$, $H_2PO_4^-$, $CO_3^{2-}$, $SO_4^{2-}$, citrate$^{3-}$, $ADP^-$, and $RNA^{n-}$ ($NH_4^+$ salt). However, a quantitative effect was noted with monovalent anions (Figure 2). At low ionic strength, the effectiveness of $Cl^-$, $Br^-$, $NO_3^-$, $SCN^-$, and $ClO_4^-$ in causing microsphere formation closely parallels their binding affinity to quaternary ammonium resins (Robinson and Jencks, 1965). This feature suggests that charge neutralization

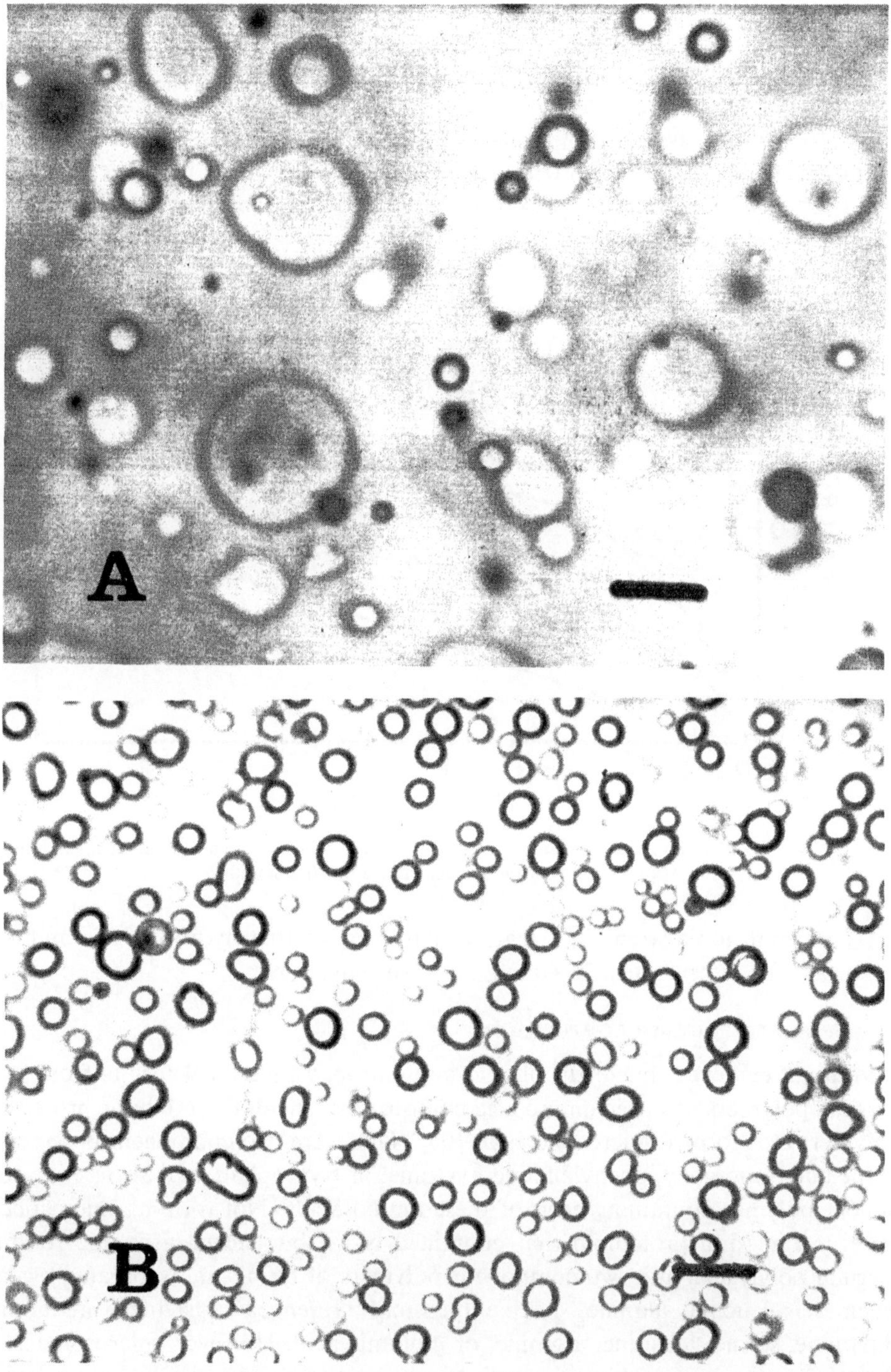


Figs. 1a–b. Microspheres prepared from lysine-rich proteinoid. (a) The smaller particles are spherical, but on contact with glass they collapse within a few minutes to form the larger, disc-shaped particles. The latter eventually coalesce to form a mosaic pattern. Polymer at 8 mg $ml^{-1}$, NaCl at 7.5%. (b) Microspheres on glass coated with lanolin, on which the particles do not collapse. In both photographs, the marker denotes ca. 10 $\mu$m.

is involved in the mechanism of microsphere formation, as with coacervate formation (Oparin, 1957, p. 307). If however, salting-out phenomena contribute to microsphere formation, then salting-in, or chaotropic, ions [ions that favor the transfer of apolar groups to water (Hatefi and Hanstein, 1969)] would be expected to limit such formation. This appears to be the case with $SCN^-$, and to a lesser extent with

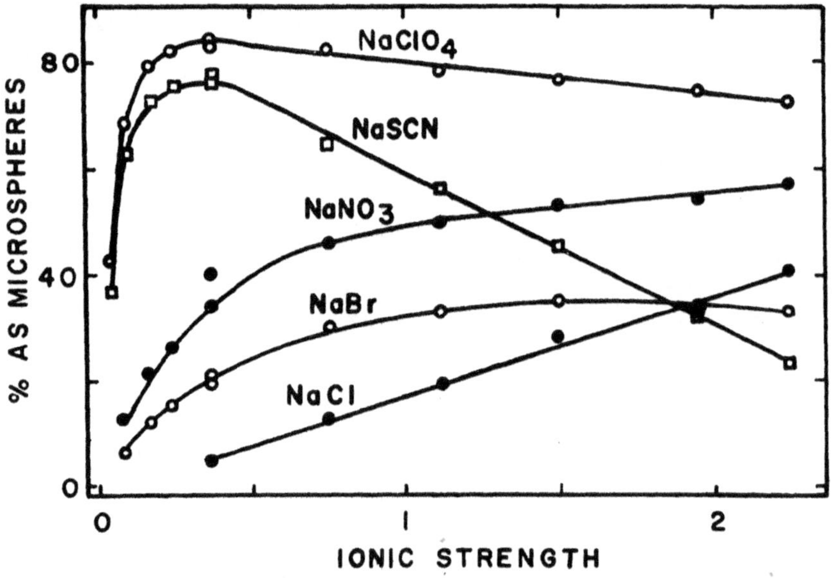


Fig. 2. Effect of different monovalent anions on the percentage of lysine-rich proteinoid in microspherical form. Total polymer concentration was 0.7 mg $ml^{-1}$.

$ClO_4^-$, at high ionic strength, a feature that suggests that hydrophobic bonding is important for morphogenesis (Hatefi and Hanstein, 1969).

### 3.2. Effect of polymer composition

Additional evidence that hydrophobic bonding may be critical results from using thermal polymers of varied amino acid composition. In addition to basic proteinoid, polymers that formed microspheres in 10% NaCl were two-component polymers of lysine and (separately) phenylalanine, tyrosine, or tryptophan, and also a four-component polymer containing each of these amino acids. Not forming microspheres were an omission proteinoid that contained lysine but lacked aromatic residues, thermal polylysine, and two-component polymers of lysine and (separately) valine, leucine, isoleucine, alanine, cystine (possibly present as cysteine), methionine, threonine, serine, histidine, arginine, or glutamic acid. However, microspheres did form from polymers lacking aromatic residues when $SCN^-$ or $ClO_4^-$ was used. A relatively nonpolar polymer, composed of nonproteinous amino acids (see Section 2) and lacking aromatic and trifunctional residues, formed microparticles in NaCl. Thus, with largely polar polymers containing much lysine, 'strong' cation binders

($SCN^-$) were required for microsphere formation if aromatic residues were absent; 'weaker' cation binders ($Cl^-$) sufficed when aromatic residues were contained or when polar residues were largely absent.

The polymer prepared from nonproteinous amino acids formed microspheres in distilled water at 40 °C or higher that dissolved on cooling, which further suggests a role for hydrophobic bonding (Kauzmann, 1959). A polymer prepared (Saunders and Rohlfing, 1972) from the proportions of amino acids found (Harada *et al.*, 1971) in a hydrolyzed extract of a lunar sample also formed microspheres in 20% NaCl (Saunders, 1972).

### 3.3. Phenomena in Sea Water

The behavior of lysine-rich proteinoids in sea water is of special interest in a prebiological context. Microspheres formed (Figure 3) to some extent at a salinity of *ca.* 30 parts per thousand and to a lesser extent at lower salinities that may have been prevalent in early oceans (but *cf.* Rubey, 1951). However, a sizable increase in the percentage of polymer in microspherical form occurred when the suspensions were heated or when the *p*H was raised (Figure 3, inset) with $NH_4OH$. Thus, even if large numbers of particles were a critical prebiological factor, an appreciable yield

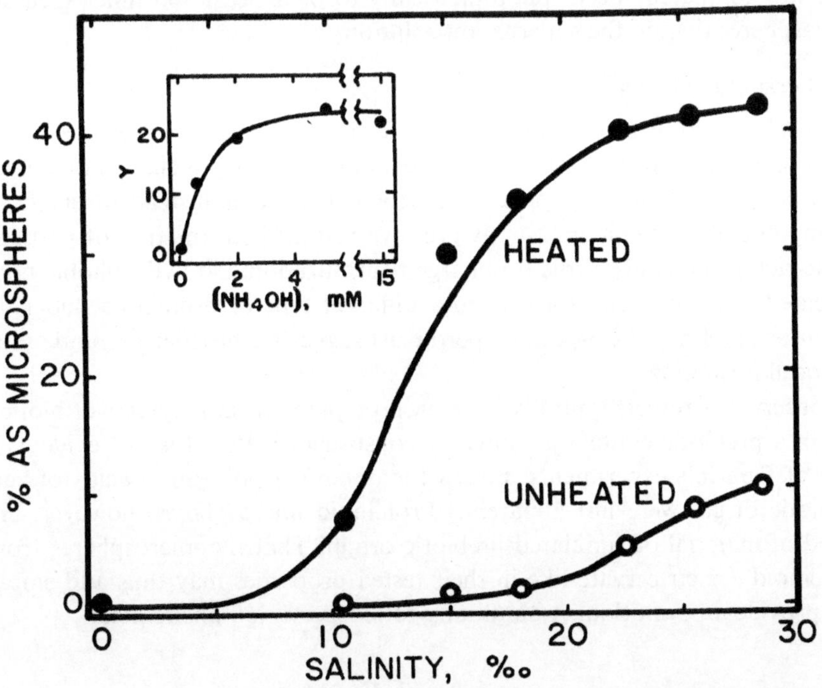


Fig. 3. Effect of salinity of sea water (parts per thousand) on the percentage of lysine-rich proteinoid in microspherical form. Total polymer concentration was 0.5 mg $ml^{-1}$. The indicated samples were heated to boiling for a few seconds. *Inset:* Effect of $NH_4OH$ concentration on the percentage of proteinoid in microspherical form (Y-axis). Total basic-proteinoid concentration was 0.5 mg $ml^{-1}$; salinity was 15 parts per thousand. Measured *p*H values (left to right) were 7.7, 8.0, 8.6, 9.0, and 9.3.

of microspheres seems plausible, since both heat and somewhat alkaline oceans are geologically relevant (Fox and Dose, 1972; Abelson, 1966). It is of interest to note that microspheres made from only acidic proteinoid apparently would dissolve at the *p*H values of Figure 3.

### 3.4. Comparison with Coacervate Droplets

Basic-proteinoid microspheres, unlike those from acidic proteinoid but like coacervate droplets, do not remain intact when centrifuged.* Other similarities between coacervates and basic-proteinoid microspheres have been noted. Both types of particles are relatively nonuniform in size. Both 'collapse' at air-liquid or glass-liquid interphases to form flattened discs (Figure 1a) that eventually coalesce into a mosaic pattern. [The collapse of microspheres on glass is prevented by coating the glass with lanolin (Figure 1b), which suggests that the collapse is mediated by charge phenomena (*cf.* Hatefi and Hanstein, 1969).] Suspensions of either type of particle settle into two separate layers on long standing in test tubes. Both types are stabilized (i.e., limited or no collapse on surfaces) when made in the presence of benzoquinone (*cf.* Evreinova *et al.*, 1972). Both coacervates and basic-proteinoid microspheres interact with and concentrate various dyes. For example, the concentration of Evans Blue was estimated spectrophotometrically to be at least 100 times greater in the microspheres than in the supernatant solution.

### 3.5. Interpretations

Microspheres form from acidic and lysine-rich proteinoids under largely different conditions. The process with acidic proteinoids, involving a change in temperature, resembles a crystallization process, except that spheres instead of crystals result. Salting-out phenomena apparently are involved in the formation of particles from lysine-rich proteinoid; a role for charge neutralization and hydrophobic bonding is indicated. That particles form from a different type of proteinoid under different conditions (*p*H may be especially pertinent) suggests a broader prebiotic setting for protocell formation.

Coacervate droplets typically have been prepared from polymers of biotic origin, and in a prebiotic context a source of constituent material is not evident. (At the time of Oparin's experiments, means for obtaining polyamino acids of simulated prebiotic origin were not apparent.) Proteinoid microspheres, however, are composed of material of simulated prebiotic origin. That the microspheres from basic proteinoid are coacervate-like in their tested properties may thus add emphasis to and provide an evolutionary continuity for coacervation phenomena.

**Acknowledgments**

Supported in part by NASA grant NGR 41-002-034. Reported in part at the 4th

* Oparin (1953) has discussed advantageous aspects of limited stability.

International Conference on the Origin of Life, Barcelona, 1973, and at a Conference on Extreme Environments, Ames Research Center, Moffett Field, Calif., 1970.

**Note added in proof:** Fractions of proteinoid prepared in the presence of seawater salts form microspheres that are stable at pH values above 8 (Snyder, W. D. and Fox, S. W.: 1973, *Federation Proc.* **32**, 640).

## References


Abelson, P. H.: 1966, *Proc. Nat. Acad. Sci. U.S.* **55**, 1365.

Evreinova, T. N., Mamontova, T. W., and Karnaukhov, V. N.: 1972, in D. L. Rohlfing and A. I. Oparin, (eds.), *Molecular Evolution: Prebiological and Biological*, Plenum, New York, p. 361.

Fox, S. W.: 1970, *J. Evol. Biochem. Physiol.* **6**, 131.

Fox, S. W. and Dose, K.: 1972, *Molecular Evolution and the Origin of Life*, W. H. Freeman and Co., San Francisco.

Fox, S. W., Harada, K., and Kendrick, J.: 1959, *Science* **129**, 1221.

Fox, S. W., Harada, K., and Rohlfing, D. L.: 1962, in M. Stahmann, (ed.), *Polyamino Acids, Polypeptides, and Proteins*, The University of Wisconsin Press, Madison, p. 47.

Harada, K., Hare, P. E., Windsor, C. R., and Fox, S. W.: 1971, *Science* **173**, 433 (Sample 12033).

Hatefi, Y. and Hanstein, W. G.: 1969, *Proc. Nat. Acad. Sci. U.S.* **62**, 1129.

Kauzmann, W.: 1959, *Adv. Protein Chem.* **14**, 1.

Mahler, H. R. and Cordes, E. H.: 1966, *Biological Chemistry*, Harper and Row, New York.

Oparin, A. I.: 1953, *The Origin of Life*, Dover, New York.

Oparin, A. I.: 1957, *The Origin of Life on the Earth*, Academic Press, Inc., New York.

Robinson, D. R. and Jencks, W. P.: 1965, *J. Am. Chem. Soc.* **87**, 2470.

Rohlfing, D. L.: 1967, *Nature* **216**, 657.

Rohlfing, D. L. and Fox, S. W.: 1969, *Adv. Catalysis* **20**, 373.

Rubey, W. W.: 1951, *Geol. Soc. Amer. Bull.* **62**, 1111.

Saunders, M. A.: 1972, Ph.D. Dissertation, University of South Carolina, Columbia, S.C.

Saunders, M. A. and Rohlfing, D. L.: 1972, *Science* **176**, 172 ('Lunar B').

Smith, A. E. and Bellware, F. T.: 1966, *Science* **152**, 362.

Weber, A. L., Wood, A., Hardebeck, H., and Fox, S. W.: 1968, *Federation Proc.* **27**, 303.

Young, R. S.: 1965, in S. W. Fox (ed.), *The Origins of Prebiological Systems*, Academic Press, Inc., New York, p. 347.

Young, R. S.: 1972, Personal Communication.

# EVOLUTION IN BIOIDS: HYPERCOMPETITIVITY AS A SOURCE OF BISTABILITY AND A POSSIBLE ROLE OF METAL COMPLEXES AS PRENUCLEOPROTIC MEDIATORS OF MOLECULAR ASYMMETRY* **

P. DECKER
*Chemical Department, Veterinary School, Hannover, F.R.G.*

**Abstract.** Spontaneous production of optically active compounds can occur through kinetic instability of an asymmetric steady state in open systems, in which two enantiomeric autocatalysts compete for a common prochiral substrate in a stereoselective reaction of order $n>2$. For the case of $n=3$, a proof of instability of a symmetric reacting state in the general case, and functions of reaction parameters ('Chemical Reynolds Numbers') governing the existence and stability of 7 different steady states are derived. The 'extinct state' (without autocatalyst) is stable; a finite amount of products is required to shift it into one of the reacting steady states. A mutation from one state into another in such systems ('bioids') involves an amplification of different 'kinds of information', as 'stochastic' (noise into dissipative structures), 'molecular' (autocatalysts), and 'stoichiometric' information. Stereospecific third order kinetics are believed to be realizable on octahedral metal complexes with two-dentated ligands and to have played a role in the prebiological evolution of optically active compounds.

## 1. Introduction

Evolution in open systems is the basic physical mechanism in the evolution of biological and supposed life-like extraterrestrial systems. Prigogine (Glansdorff and Prigogine, 1971) discussed its thermodynamical aspects, evolution of dissipative structures, and Eigen (1971) developed a special systems theory of the evolution of linear matrices. In a discussion of the role of autocatalytic reaction kinetics in the first steps of prebiological evolution it was emphasized that chemical kinetics in open systems rather than abiogenic syntheses as such played a decisive role in the origin of life (Decker and Speidel, 1968, 1972; Decker *et al.*, 1971.

Open systems which can exist in several steady states (bioids) can evolve stepwise through instabilities, 'mutations', from one state into another. The term bioid has been proposed for that class of cybernetic systems because it contains biological systems as a prominent subclass, and physical generalizations of the biological concepts of species, mutation and evolution by mutation and selection according to higher selection value were derived in terms of chemical kinetics of a bioid composed of two feedback cycles, e.g. two autocatalytic reactions, competing for a common substrate in an open system (Decker and Speidel, 1968, 1972).

Evolution of open systems into asymmetric states by bistability (bifurcation) has been discussed by Frank (1953), Seelig (1971a, b) and Decker (1973a, b). This work has shown that in contrast to current belief macromolecules, enzymes and linear matrices are not necessary prerequisites for the origin of molecular asymmetry, and

* V: Bioids; IV: Decker (1973b); III: Decker (1973a); II: Decker *et al.* (1971); I: Decker and Speidel (1972).
** For Professor A. Butenandt on his 70th birthday.

that in principle it could spontaneously arise at a much earlier state of prebiological evolution of open systems. The systems considered so far were characterized by two parts, a competitive part of two enantiomeric autocatalysts (or more generally, positive feedback cycles) competing stereospecifically for one prochiral (Seelig, 1971; Decker, 1973a) or racemic (Decker, 1973b) substrate, and an 'aggressive part' where the two autocatalysts inhibit or consume each other. Thus, a local surplus at the molecular level of one of the antipodes can multiply itself and eventually suppress its competitor. In the present communication we discuss a new type of bistable bioids.

## 2. Hypercompetitive Systems

If in a competitive bioid (Decker, 1972) the competing reactions are of more than 2nd order, instability, causing the system to switch into one of two equivalent asymmetric states, can occur even without an aggressive mechanism: Let $B_1$ and $B_2$ be two antipodes which arise from a prochiral substrate $A$ by a more than first order reaction:

$$A+(1+d)\,B_1 \xrightarrow{a_1(1+b)} (2+d)\,B_1\,; \qquad A+(1+d)\,B_2 \xrightarrow{a_1(1+b)} (2+d)\,B_2$$

where $0<d$. We also consider 'parasitic' reactions, impairing stereospecificity:

$$A+(1+d)\,B_1 \xrightarrow{a_1(1-b)} (1+d)\,B_1+B_2\,;$$
$$A+(1+d)\,B_2 \xrightarrow{a_1(1-b)} (1+d)\,B_2+B_1\,.$$

Thus $a_1$ is a mean reaction constant and $0<b<1$ a measure of stereoselectivity. Such fractional order kinetics may be assumed to occur as the overall result of a more complex mechanism, e.g. a cooperative or a chain reaction. To have an open system, let $A$ be supplied at constant rate $a_4$ (gs$^{-1}$ ml$^{-1}$), and $A$, $B_1$ and $B_2$ be removed from the system by first order kinetics giving some end products $E$ at rates $a_3$ (s$^{-1}$) assumed for simplicity to be equal. An 'aggressive' reaction between $B_1$ and $B_2$ at rate $a_2$ (Decker, 1973a) is not involved here. The change of concentrations of $A$, $B_1$ and $B_2$, $\dot{A}$, $\dot{B}_1$ and $\dot{B}_2$, is described by three differential equations (we omit brackets to denote explicitly the concentrations (g ml$^{-1}$) of the compounds $A$, $B$ as $[A]$, $[B]$, as the respective meaning is everywhere evident from the context):

$$\dot{A}=a_4-2a_1A(B_1^{(1+d)}+B_2^{(1+d)})-a_3A \tag{1.1}$$

$$\dot{B}_1=a_1A[(1+b)\,B_1^{(1+d)}+(1-b)\,B_2^{(1+d)}]-a_3B_1 \tag{1.2}$$

$$\dot{B}_2=a_1A[(1+b)\,B_2^{(1+d)}+(1-b)\,B_1^{(1+d)}]-a_3B_2 \tag{1.3}$$

where steady states occur if $\dot{A}=\dot{B}_1=\dot{B}_2=0$.

The sum of (1, $i$), $\dot{z}=a_4-a_3z$, implies for all steady states: $z=A+B_1+B_2=a_4/a_3$ and

$$A=g-x \tag{2}$$

where $x\equiv B_1+B_2$ and $g\equiv a_4/a_3$

This means that for $t\to\infty$ all trajectories approach the plane $A+B_1+B_2=g$ and for steady states, we can limit ourselves to this plane, eliminating $A$ by Equation (2).

Besides the trivial (non-reacting) steady state $B_1=B_2=0$; $A=a_4/a_3$ evidently there also can exist symmetrical steady states $B_1=B_2$ where

$$2a_1(g-x)\,x^d-a_3=0. \tag{3}$$

In order to evaluate the stability of such symmetric states we subtract Equation (1.3) from Equation (1.2) after division by $B_1$ and $B_2$ respectively:

$$\% \quad \dot{p}=\frac{\dot{B}_1}{B_1}-\frac{\dot{B}_2}{B_2}=a_1A\left[(1+b)\,(B_1^d-B_2^d)-(1-b)\,\frac{B_1^{(2+d)}-B_2^{(2+d)}}{B_1B_2}\right]. \tag{4}$$

Since $A$, $B_1$, $B_2$ and $d$ are all positive, the ratio of $B_1$ and $B_2$ will increase or decrease with time as depending on the sign of $\dot{p}=d/dt\,\ln(B_1/B_2)$. If $B_1>B_2$ entails a further increase of the ratio $B_1/B_2$ and vice versa, this would mean instability of a state $B_1=B_2$ (independently of whether it is a steady state or not). This is the case if $\dot{p}>0$. This condition can be rewritten as

$$\frac{(1+b)}{(1-b)}\geqslant\frac{B_1^{(2+d)}-B_2^{(2+d)}}{B_1B_2(B_1^d-B_2^d)}. \tag{5}$$

Considering nearly symmetrical states, $B_1-B_2=\beta$, $|\beta|\ll 1$ by expansion of Equation (5) omitting higher order terms we obtain

$$\frac{(1+b)}{(1-b)}\geqslant\frac{2+d}{(1-\beta)\,d}=\frac{2}{d}+1 \quad \text{or} \quad b\geqslant\frac{1}{1+d} \tag{6}$$

a 'chemical Reynolds number' (Decker, 1973b) delimiting the existence of asymmetric states. Graphical representation of Equation (3) as intersection of two functions of a parameter $u$, $u=g-x$ and $u=x^{-d}a_3/2a_1$, shows that there can exist up to two real solutions in the positive quadrant. The important special case, $d=1$, is resolved explicitly in the next section:

## 3. Kinetics of a Third Order Hypercompetitive Bioid

At first glance third order kinetics might appear quite improbable for a real system. However, just this case may provide a chance to find such systems: A reaction in the coordination sphere of a metal ion with three bidentate ligands may well follow overall third order kinetics.

$$A+2B+Me\rightleftharpoons MeABB\rightarrow MeBBB\rightleftharpoons Me+3B.$$

Such complexes and their asymmetric kinetics may have preceded the evolution of enzymes as catalytic mechanisms producing and reproducing molecular asymmetry.

Consider a prochiral substrate $A$ which is transformed into two possible antipodes $B_1+B_2$. Let $B_1$ and $B_2$ catalyze preferentially their own formation in a stereospecific third order reaction, e.g. in the coordination sphere of a metal ion, the concentration of which is assumed not to be limiting:

$$A+2B_1\xrightarrow{a_1(1+b)}3B_1$$
$$A+2B_2\xrightarrow{a_1(1+b)}3B_2.$$

Like in Section 2, we assume also slower, 'parasitic' reactions leading to the other antipode with a reaction constant $a_1(1-b)$, inflow of $A$ and removal of all reactants. By addition and subtraction of Equations (1.2) and (1.3) using $d=1$, $A=g-x$ and the abbreviations $x$ and $y$ for the sum and the difference of $B_1$ and $B_2$ (where $y$ represents the amount of 'chirate', the surplus of one of the chiral products as measured by the optical rotation) we obtain, because of $B_1^2+B_2^2=\frac{1}{2}(x^2+y^2)$ and $B_1^2-B_2^2=xy$, for the

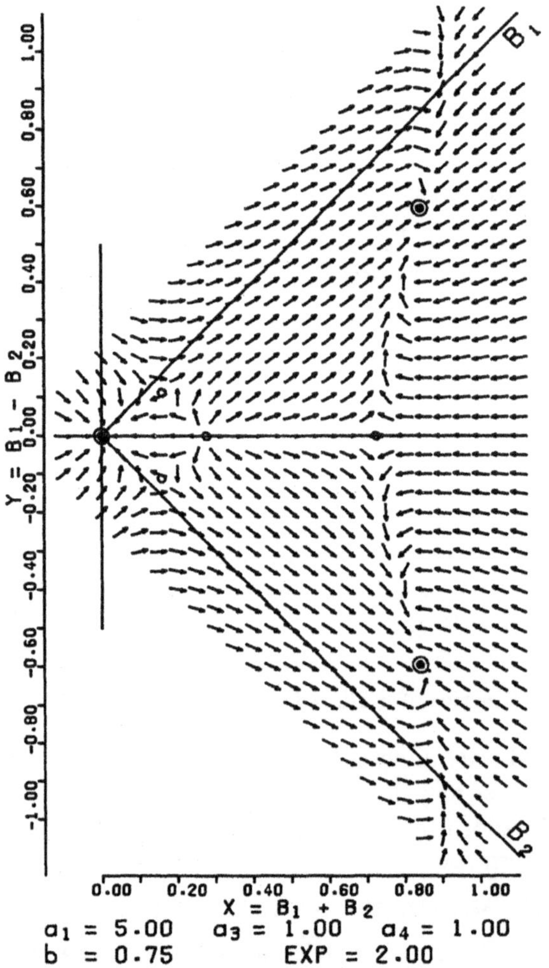


Fig. 1. Direction of trajectories in a 3rd order hypercompetitive bioid ($d+1=\exp=2$), producing inequal amounts of antipodes $B_1$ and $B_2$ from a common precursor $A$, with initial conditions on the phase plane $B_1+B_2+A=z$ projected onto the $B_1$, $B_2$-plane.

$$A+2B_1 \xrightarrow{a_1(1+b)} 3B_1; \qquad A+2B_1 \xrightarrow{a_1(1-b)} B_2+2B_1$$

$$A+2B_2 \xrightarrow{a_1(1+b)} 3B_2; \qquad A+2B_2 \xrightarrow{a_1(1-b)} B_1+2B_2$$

$$\text{Pool} \xrightarrow{a_4} A; \qquad A \xrightarrow{a_3} E; \qquad B_1 \xrightarrow{a_3} E; \qquad B_2 \xrightarrow{a_3} E$$

Circles: steady states; double circles: stable steady states.

case of steady states:

$$0=\dot{x}=a_1(g-x)(x^2+y^2)-a_3x \tag{7.1}$$
$$0=\dot{y}=2a_1b(g-x)\,xy-a_3y. \tag{7.2}$$

There are three groups of steady states (Figure 1):

(I) $x=0$; $y=0$, the *extinct state* $(x_{1,1})$ without autocatalysts;
(II) $x\neq 0$; $y=0$, symmetric states $(x_{2,i})$;
(III) $x\neq 0$; $y\neq 0$, asymmetric steady states $(x_{3,i})$.

Introducing the abbreviation $h\equiv a_3/a_1$ we obtain for the symmetric state, $y=0$, from Equation (7.1)

$$x_{2,1},\, x_{2,2}=\frac{g}{2}\pm\sqrt{\left(\frac{g^2}{2}\right)-h}. \tag{8}$$

As $g$ and $h$ are positive and real constants, both solutions are positive and real if

$$R_1^2=\left(\frac{g}{2}\right)^2-h>0. \tag{9}$$

For the asymmetric state we obtain from Equation (7.2)

$$x_{3,1},\, x_{3,2}=\frac{g}{2}\pm\sqrt{\left(\frac{g}{2}\right)^2-\frac{h}{2b}} \tag{10}$$

representing two positive solutions if

$$R_2^2=\left(\frac{g}{2}\right)^2-\frac{h}{2b}>0. \tag{11}$$

By elimination of $g-x$ from Equation (7.1) using Equation (7.2) we obtain further

$$y=\pm x\sqrt{2b-1}. \tag{12}$$

Thus, two real values of opposite sign of $y$ correspond to every value of $x_{3,i}$, if

$$Q^2=2b-1>0. \tag{13}$$

This condition means, that asymmetric states can occur only if $b>0.5$, i.e. if the reaction constant of the stereospecific reaction, $a_1(1+b)$, is at least three times larger than its parasitic counterpart, $a_1(1-b)$. Remarkably, in the case of existing asymmetric states, i.e. $2b>1$, we have $R_1<R_2$ and condition (9) is stronger than condition (11). Therefore systems where $R_1<0<R_2$ yet have asymmetric states (Equation (10)) but no symmetric states (Equation (8)) at all.

## 4. Stability of Steady States

The stability or instability of steady states depends on the signs of the roots of the characteristic equation which is obtained from the coefficient matrix of a system of differential equations after shifting the steady state into the origin by introduction of new variables and linearizing by omission of nonlinear terms (which vanish near the origin). Two negative roots imply a stable node; two positive roots an unstable node and two roots of opposite sign an unstable saddle point (Cesari, 1971).

Using the coordinates of steady states $x_s$, $y_s$, we shift the respective steady state into the origin by replacing in Equations (7.1) and (7.2) $x$ and $y$ by $\bar{x}=x-x_s$ and $\bar{y}=y-y_s$. By omission or all nonlinear terms (i.e. containing $\bar{x}^2$, $\bar{y}^2$ or $\bar{x}\bar{y}$; absolute terms cancel because of Equations (7.1), (7.2)), rearrangement and further simplification by division by $a_1$ and $2ba_1$ respectively using the abbreviation $h\equiv a_3/a_1$ (this does not alter the signs of the roots of the characteristic equation) we obtain

$$0=\frac{\dot{\bar{x}}}{a_1}=\bar{x}\left[2x_s(g-x_s)-h-x_s^2-y_s^2\right]+\bar{y}\left[2y_s(g-x_s)\right] \tag{14.1}$$

$$0=\frac{\dot{\bar{y}}}{2ba_1}=\bar{x}\left[y_s(g-x_s)-y_sx_s\right]+\bar{y}\left[x_s(g-x_s)-\frac{h}{2b}\right]. \tag{14.2}$$

For the extinct state, already in the origin ($x_s=y_s=0$), this gives the linearized coefficient matrix

$$\mathbf{A}=\begin{matrix}-h & 0\\ 0 & -\dfrac{h}{2b}\end{matrix} \tag{15}$$

and the characteristic equation

$$\mathbf{A}-\lambda E=\begin{vmatrix}-h-\lambda & 0\\ 0 & -\dfrac{h}{2b}-\lambda\end{vmatrix}=(-h-\lambda)\left(-\frac{h}{2b}-\lambda\right)=0 \tag{16}$$

with the roots $\lambda_1=-h$ and $\lambda_2=-h/2b$. Thus, the extinct state is stable and this system cannot start with an arbitrarily low amount of autocatalyst as a normal second order autocatalysis does. This holds also for the general hypercompetitive system, as Equations (1.2), (1.3) after substitution of $A$ using Equations (2) and linearization also give two negative roots. The two symmetric states (II), after substitution in Equations (14.1), (14.2)of $x_s$ by $x_{2,1}$ and $x_{2,2}$ using Equations (8), (9) and (13) and of $y_s=0$, give the characteristic equation

$$\begin{vmatrix}-2\left(R_1^2\pm\dfrac{g}{2}R_1\right)-\lambda & 0\\ 0 & Q\dfrac{h}{b}-\lambda\end{vmatrix}=0$$

with the roots $\lambda_1=-2(R_1^2\pm(g/2)\,R_1)$ and $\lambda_2=Qh/b$. In the upper steady state (positive sign in the brackets), the first root is always negative and the second root is so, if $Q<0$. Thus the upper symmetric steady state is stable if there are no asymmetric states, otherwise unstable. Because of $R_1 < (g/2)$, Equation (9), the lower steady state has always a positive first root and is an unstable node or a saddlepoint depending on whether $Q<0$ or $Q>0$ respectively.

Similarly, for the two pairs of asymmetric states (III), using Equations (10–13), after some rearrangement we obtain the characteristic equation

$$\lambda^2 + \lambda_2 [Q^2 h/2b + 2b(R_2^2 \pm R_2 g/2)] + (R_2^2 \pm R_2 g/2)\, Q^2 2h/b = 0.$$

As $0 < R_2^2 < R_2 g/2$, Equation (11), and, for existing steady states, $Q^2 > 0$, in the upper pair of steady states both coefficients are positive. This implies two negative roots and therefore, for each of both signs of $\pm Q$, a stable steady state (stable node). In the lower pair of steady states (with negative sign in the brackets) we have a negative absolute term because of $(R_2^2 - R_2 g/2) < 0$, which entails two roots of opposite sign and instability (saddle points).

## 5. Pasteur's '1st Method' (1848) of Optical Resolution and the Relation of Cooperative and Hypercompetitive Systems

The ability of hypercompetitive systems to produce spontaneously optically active compounds from inactive precursors represents the theoretical base of the sole method known till today, by which optically active compounds were produced without use of external asymmetric agents as enzymes, asymmetric reagents or asymmetric physical forces as circularly polarized light. This is the famous '1st method' devised in 1848 by Louis Pasteur, the growth of hemiedric crystals (i.e. crystals containing each only one of the antipodes) from a solution of a racemic mixture. Using an asymmetrical quaternary ammonium salt, which behaves stably in crystalline form but spontaneously racemises in solution, in 1954 Havinga even succeeded in obtaining a spontaneous nearly 100% yield of only one of the antipodes as single crystals of one optically pure species from the equilibrium mixture of the racemate.

Crystal growth is an autocatalytic process of more than second order; big crystals are less soluble than small ones. However, we can also interpret crystal growth as a special case of a cooperative process: in fact a molecule, which is held by its neighbours in the crystal lattice from more than one side will leave the surface much more rarely than a first lone molecule of a newly forming layer or the lattice; just as in a linear matrix, growth is enhanced cooperatively by a parallel complementary chain (Eigen, 1971; Babloyantz, 1972). Especially a growing helix may acquire a higher growth potential, once it surpasses the size of one turn and further growth is enhanced by the cooperative action of a neighbouring turn. Thus, kinetically, cooperative systems may be considered as an, indeed very special, subclass of hypercompetitive i.e. more than second order autocatalytic systems.

Acquisition, amplification and conservation of *coded* information in selfreproducing linear matrices (Eigen, 1971) has been shown to be a special case of *informal* information handling in the general case of bioids, where the choice of one of several possible steady states represents information (Decker and Speidel, 1968, 1972; Decker *et al.*,

1971). We note that this concept (Decker, in preparation) leads to a distinction of *different kinds of information*: Classical dissipative structures like turbulence or the mutation of a hypercompetitive system from a symmetric steady state into an asymmetric state represent an amplification of statistical noise from molecular into macroscopic dimensions (Jordan, 1944), i.e. of *stochastic information* like local arithmetic differences of great numbers of the two antipodes. In contrast, an autocatalytic reaction 'amplifies' the autocatalyst, i.e. *molecular information* (Decker and Speidel, 1968); and, in order to switch a stable asymmetric state into the opposite state, *stoichiometric information*, i.e. a finite amount of the other antipode, is required.

## Acknowledgements

This work has been supported by grants from the Deutsche Forschungsgemeinschaft, Bad Godesberg, and Verband der Chemischen Industrie, Köln. I thank Professor I. Prigogine, Professor M. Eigen and Dr A. Babloyantz for helpful discussions.

## References

Babloyantz, Agnessa: 1972, *Biopolymers* **11**, 2349.
Cesari, L.: 1971, *Asymptotic Behavior and Stability Problems in Ordinary Differential Equations*, Springer Verlag Berlin, Heidelberg, New York, p. 24.
Decker, P.: 1973a, *Nature, New Biology* **241**, 72.
Decker, P.: 1973b, *J. Molecular Evolution*, in press.
Decker, P. and Speidel, A.: 1968, Nordwestdeutsche Chemiedozententagung, Hamburg, April 1968; cf. P. Decker, *6th FEBS Meeting, Madrid, April 1969, Abstracts Vol.*, p. 146.
Decker, P. and Speidel, A.: 1972, *Zeitschr. Naturforschung* **27b**, 257.
Decker, P., Speidel, A., and Nicolai, W.: 1971, *I. Europ. Biophysics Congr., Proceedings*, Vienna, Vol. IV, p. 515.
Eigen, M.: 1971, *Naturwissenschaften* **58**, 465.
Frank, F. C.: 1953, *Biochim. Biophys. Acta* **11**, 459.
Glansdorff, P. and Prigogine, I.: 1971, *Structure, Stability and Fluctuations*, Wiley, London.
Havinga, E.: 1954, *Biochim. Biophys. Acta* **13**, 171.
Jordan, P.: 1944, *Naturwissenschaften* **32**, 309.
Pasteur, L.: 1848, *Compt. Rand. Acad. Sci.* **26**, 535.
Seelig, F.: 1971a, b, *J. Theor. Biology* **31**, 355; **32**, 93.

# ENTROPY OF THE GENETIC INFORMATION AND EVOLUTION

MASAMI HASEGAWA and TAKA-AKI YANO

*Dept. of Biophysics and Biochemistry, Faculty of Science, University of Tokyo, Hongo, Tokyo, Japan*

**Abstract.** The entropy of the amino acid sequences coded by DNA is considered as a measure of diversity or variety of proteins, and is taken as a measure of evolution. The DNA or m-RNA sequence is considered as a stationary second-order Markov chain composed of four kinds of bases. Because of the biased nature of the genetic code table, increase of entropy of amino acid sequences is possible with biased nucleotide sequence. Thus the biased DNA base composition and the extreme rarity of the base doublet $C_pG$ of higher organisms are explained. It is expected that the amino acid composition was highly biased at the days of the origin of the genetic code table, and the more frequent amino acids have tended to get rarer, and the rarer ones more frequent. This tendency is observed in the evolution of hemoglobin, cytochrome C, fibrinopeptide, immunoglobulin and lysozyme, and protein as a whole.

## 1. Entropy of Amino Acid Sequence in Protein

In a pair of papers, Zuckerkandl *et al.* (Zuckerkandl *et al.*, 1971; Vogel and Zuckerkandl, 1971) have suggested that, on the level of amino acid substitution during the evolution of proteins (cytochrome c and globins), the more frequent amino acids tend on the average to get rarer, and rarer ones more frequent, and hence the entropy of the amino acid sequence in protein tends to increase. Recently, the present authors (Yano *et al.*, 1973) have extensively studied on this subject.

The probable phylogenetic trees of several protein families are constructed, and the ancestral sequences are estimated from the contemporary protein sequences by the principle of minimum number of mutations. The protein families used are hemoglobin, cytochrome c. fibrinopeptide, immunoglobulin, lysozyme, toxin, insulin, proinsulin, growth hormone, virus coat protein, ferredoxin and calcitonin (Dayhoff, 1972). Along each branch of these trees, we count the number of the accepted point mutations, $F_{ij}$, from the $j$th amino acid to the $i$th, and the number of invariants, $F_{ii}$. The transition probability matrix of amino acid substitution, **L**, is obtained by normalizing the matrix **F**;

$$L_{ij} = F_{ij} / \sum_k F_{kj}.$$

The transition matrices are calculated for each protein family and for protein as a whole, and the latter is shown in Figure 1.

Dayhoff *et al.* (1969) have previously obtained the transition probability matrix by another way. Their method, however, does not take account of the direction of time explicitly, and, therefore, the origin of the asymmetry of the matrix is artificial and not obscure. On the other hand, our method takes the direction of time into account explicitly, and the asymmetrical matrix is automatically constructed.

Let's denote amino acid frequencies by a column vector **p**. If we multiply the initial vector $\mathbf{p}_0$, which represents amino acid frequencies of a contemporary protein family

Genetic code table

2nd→

<table>
<tr><th>1st↓</th><th>U</th><th>C</th><th>A</th><th>G</th><th>3rd↓</th></tr>
<tr><td rowspan="4">U</td><td rowspan="2">Phe</td><td rowspan="4">Ser</td><td rowspan="2">Tyr</td><td rowspan="2">Cys</td><td>U</td></tr>
<tr><td>C</td></tr>
<tr><td rowspan="2">Leu</td><td rowspan="2">term</td><td>term</td><td>A</td></tr>
<tr><td>Trp</td><td>G</td></tr>
<tr><td rowspan="4">C</td><td rowspan="4">Leu</td><td rowspan="4">Pro</td><td rowspan="2">His</td><td rowspan="4">Arg</td><td>U</td></tr>
<tr><td>C</td></tr>
<tr><td rowspan="2">Gln</td><td>A</td></tr>
<tr><td>G</td></tr>
<tr><td rowspan="4">A</td><td rowspan="3">Ile</td><td rowspan="4">Thr</td><td rowspan="2">Asn</td><td rowspan="2">Ser</td><td>U</td></tr>
<tr><td>C</td></tr>
<tr><td rowspan="2">Lys</td><td rowspan="2">Arg</td><td>A</td></tr>
<tr><td>Met</td><td>G</td></tr>
<tr><td rowspan="4">G</td><td rowspan="4">Val</td><td rowspan="4">Ala</td><td rowspan="2">Asp</td><td rowspan="4">Gly</td><td>U</td></tr>
<tr><td>C</td></tr>
<tr><td rowspan="2">Glu</td><td>A</td></tr>
<tr><td>G</td></tr>
</table>

or protein as a whole, by the corresponding transition probability matrix **M**, the resulting vector **p** will be the amino acid frequencies after an arbitrary evolutionary interval,

$$\mathbf{p} = \mathbf{M} \cdot \mathbf{p}_0 . \tag{1}$$

The entropy of amino acid sequence in protein, $H_{AA}$, is defined by

$$H_{AA} = -\sum_{i=1}^{20} p_i \log_2 p_i \quad \text{(bits)}. \tag{2}$$

The entropy $H_{AA}$ is a measure of the degree of evenness of amino acid distribution; i.e., the larger this value, the more evenly amino acids are distributed.

Figure 2 shows the behavior of entropy $H_{AA}$ as a function of times of operations of **M** on the contemporary amino acid frequencies $\mathbf{p}_0$. Increasing trend of entropy is observed for every protein family as far as we have examined and protein as a whole. This trend contradicts Ohta and Kimura (1971) suggestion that the amino acid composition of contemporary organisms represents quasi-equilibrium states of substitution among selectively neutral or nearly neutral mutations. This contradiction comes from the difference of our transition matrix from Ohta and Kimura, which is counted by Dayhoff *et al.* (1969). The latter contains some artificial assumption on the origin of the asymmetry of matrix, while ours do not.

Zuckerkandl *et al.* (Zuckerkandl *et al.*, 1971; Vogel and Zuckerkandl, 1971) have interpreted the entropy increase as a randomization process. Our interpretation, however, is different from theirs. In so far as the central dogma is correct, the random process will occur only on DNA sequence. Then, because of the biased nature of genetic

ORIGINAL AMINO ACID (columns); REPLACEMENT AMINO ACID (rows)

| | Ala | Arg | Asn | Asp | Cys | Gln | Glu | Gly | His | Ile | Leu | Lys | Met | Phe | Pro | Ser | Thr | Trp | Tyr | Val | Asx | Glx | gap |
|---|---|---|---|---|---|---|---|---|---|---|---|---|---|---|---|---|---|---|---|---|---|---|---|
| Ala | 9224 | 12 | 64 | 55 | 0 | 13 | 59 | 92 | 0 | 27 | 10 | 24 | 0 | 11 | 89 | 110 | 89 | 0 | 25 | 82 | 0 | 0 | 47 |
| Arg | 5 | 9629 | 21 | 0 | 0 | 51 | 8 | 23 | 48 | 13 | 14 | 59 | 49 | 0 | 10 | 19 | 15 | 33 | 0 | 0 | 0 | 0 | 16 |
| Asn | 21 | 0 | 9258 | 95 | 0 | 0 | 8 | 23 | 79 | 0 | 0 | 41 | 0 | 10 | 10 | 84 | 37 | 0 | 0 | 0 | 0 | 0 | 8 |
| Asp | 16 | 0 | 106 | 9422 | 0 | 0 | 269 | 41 | 16 | 0 | 0 | 12 | 0 | 0 | 30 | 45 | 7 | 0 | 25 | 0 | 0 | 0 | 55 |
| Cys | 0 | 0 | 0 | 0 | 9984 | 0 | 0 | 0 | 0 | 27 | 0 | 0 | 0 | 0 | 0 | 26 | 15 | 0 | 0 | 14 | 0 | 0 | 0 |
| Gln | 21 | 48 | 11 | 8 | 0 | 9517 | 118 | 9 | 79 | 27 | 5 | 47 | 0 | 0 | 20 | 13 | 7 | 0 | 0 | 21 | 0 | 0 | 8 |
| Glu | 53 | 12 | 32 | 206 | 0 | 89 | 9286 | 32 | 0 | 0 | 0 | 47 | 0 | 0 | 30 | 26 | 0 | 0 | 0 | 21 | 357 | 0 | 16 |
| Gly | 100 | 24 | 64 | 32 | 0 | 64 | 50 | 9592 | 0 | 0 | 10 | 24 | 0 | 0 | 20 | 58 | 22 | 0 | 25 | 27 | 0 | 0 | 55 |
| His | 5 | 24 | 42 | 24 | 0 | 64 | 0 | 5 | 9666 | 0 | 10 | 18 | 0 | 72 | 10 | 26 | 0 | 33 | 37 | 0 | 0 | 385 | 16 |
| Ile | 5 | 36 | 21 | 8 | 0 | 0 | 8 | 5 | 0 | 9295 | 87 | 6 | 0 | 51 | 20 | 26 | 37 | 33 | 0 | 178 | 0 | 0 | 16 |
| Leu | 21 | 12 | 11 | 0 | 0 | 25 | 25 | 5 | 0 | 186 | 9596 | 6 | 146 | 62 | 69 | 6 | 7 | 0 | 37 | 69 | 0 | 0 | 32 |
| Lys | 16 | 108 | 95 | 24 | 0 | 25 | 17 | 18 | 16 | 40 | 5 | 9556 | 0 | 0 | 10 | 0 | 22 | 0 | 0 | 7 | 0 | 0 | 16 |
| Met | 0 | 12 | 0 | 0 | 0 | 25 | 0 | 0 | 0 | 40 | 63 | 18 | 9563 | 10 | 0 | 0 | 7 | 33 | 0 | 55 | 0 | 0 | 0 |
| Phe | 11 | 0 | 0 | 0 | 0 | 0 | 0 | 0 | 0 | 13 | 58 | 6 | 49 | 9547 | 10 | 19 | 7 | 100 | 258 | 14 | 0 | 0 | 16 |
| Pro | 58 | 24 | 11 | 0 | 0 | 13 | 0 | 0 | 16 | 27 | 10 | 6 | 0 | 0 | 9524 | 39 | 7 | 0 | 0 | 0 | 0 | 0 | 16 |
| Ser | 211 | 24 | 170 | 32 | 16 | 25 | 0 | 73 | 16 | 0 | 14 | 36 | 49 | 21 | 69 | 9245 | 186 | 0 | 25 | 27 | 0 | 0 | 71 |
| Thr | 137 | 0 | 32 | 16 | 0 | 38 | 42 | 23 | 48 | 80 | 5 | 41 | 97 | 21 | 30 | 200 | 9441 | 0 | 0 | 34 | 0 | 0 | 16 |
| Trp | 0 | 0 | 0 | 0 | 0 | 0 | 0 | 0 | 0 | 0 | 0 | 0 | 0 | 0 | 0 | 0 | 0 | 9601 | 12 | 0 | 0 | 0 | 0 |
| Tyr | 11 | 12 | 21 | 8 | 0 | 0 | 8 | 0 | 16 | 13 | 5 | 6 | 0 | 154 | 0 | 19 | 7 | 133 | 9533 | 14 | 0 | 0 | 8 |
| Val | 53 | 0 | 32 | 16 | 0 | 13 | 50 | 14 | 0 | 213 | 91 | 24 | 49 | 31 | 30 | 19 | 52 | 33 | 12 | 9425 | 0 | 0 | 31 |
| Asx | 11 | 0 | 0 | 0 | 0 | 13 | 34 | 9 | 0 | 0 | 0 | 18 | 0 | 0 | 0 | 6 | 22 | 0 | 0 | 7 | 9643 | 0 | 24 |
| Glx | 5 | 12 | 0 | 0 | 0 | 0 | 0 | 5 | 0 | 0 | 0 | 6 | 0 | 0 | 0 | 6 | 7 | 0 | 0 | 0 | 0 | 9231 | 16 |
| gap | 16 | 12 | 11 | 55 | 0 | 25 | 17 | 32 | 0 | 0 | 19 | 0 | 0 | 11 | 20 | 6 | 0 | 0 | 12 | 7 | 0 | 385 | 9520 |

Fig. 1. Transition probability matrix for protein as a whole. The elements are shown multiplied by 10000.

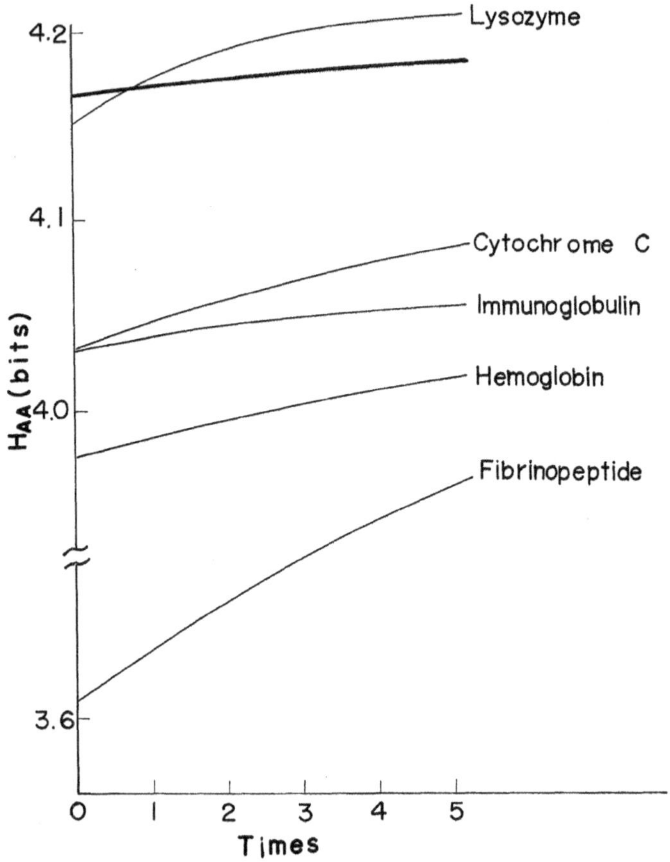


Fig. 2. Entropy $H_{AA}$ as a function of times of operation of transition matrix on the contemporary amino acid frequencies. Bold line shows protein as a whole. The time scale of the abscissa is arbitrary and it differs for different protein families.

code, the entropy $H_{AA}$ will not necessarily increase, but rather decrease. We regard the increasing trend of $H_{AA}$ as a tendency against randomization. The $H_{AA}$ value of the amino acid sequence coded by completely random DNA sequence, in which case amino acid frequency is proportional to the degree of codon degeneracy, is 4.14, while the observed $H_{AA}$ values of the contemporary total protein are higher than this value, and now increasing. This suggests the possibility that, at the days of the origin and establishment of the genetic code, the amino acid frequencies of protein were highly biased owing to the biased nature of the code (Watson, 1970), and in the course of evolution the nucleotide base compositions and base doublet frequencies have become biased so as to decrease the bias of the amino acid frequencies and to increase the entropy $H_{AA}$. We will discuss on this subject in the following section.

## 2. Genetic Code Table and Entropy

Smith (1969) has shown that the maximum in the entropy function of protein provides an insight into the question as to why all vertebrates have a base C+G composition of about 42% (Sueoka, 1965). Since then, several authors have discussed this problem (King, 1972; Gatlin, 1972a, b; Miyazawa and Miyata, personal communication), and we have further developed as the following manner (Hasegawa and Yano, 1972a, b).

Now, DNA/mRNA is assumed to be a stationary second-order Markov chain of four kinds of bases; thymine (T, uracil U in mRNA), cytosine (C), adenine (A), and guanine (G). As this base sequence is assumed to be stationary, probabilities of four kinds of bases, $D(i)$, doublet frequencies, $D(i, j)$, and triplet frequencies, $D(i, j, k)$, in which $i, j, k \in \{U, C, A, G\}$, are constant at any part of the strand. Then,

$$D(i, j) = \sum_{k \in \{U, C, A, G\}} D(k, i, j) = \sum_{k} D(i, j, k), \tag{3}$$

$$D(i) = \sum_{j} D(i, j) = \sum_{j} D(j, i), \tag{4}$$

and the normalization condition

$$\sum_{i, j, k} D(i, j, k) = 1. \tag{5}$$

There exist $64 D(i, j, k)$'s, but the conditions (3) and (5) reduce the number of independently variable triplet frequencies to 48. Here, we are concerned with mRNA and are free from Watson-Crick constraint.

The entropy of stationary second-order Markov chain of mRNA is

$$H^{2M}_{mRNA} = -\sum_{i, j, k} D(i, j, k) \log_2 \{D(i, j, k)/D(i, j)\}. \quad \text{(bits)} \tag{6}$$

Still more, we define

$$H^{M}_{mRNA} = -\sum_{i, j} D(i, j) \log_2 \{D(i, j)/D(i)\} \quad \text{(bits)}, \tag{7}$$

$$H^{R}_{mRNA} = -\sum_{i} D(i) \log_2 D(i) \quad \text{(bits)}. \tag{8}$$

Base sequence of mRNA is translated into amino acid sequence according to the genetic code. The probability of the $I$-th amino acid, $P(I)$, and probability of amino acid doublet $I$-$J$, $P(I, J)$, of the translated amino acid sequence are

$$P(I) = \sum_{(i, j, k)} D(i, j, k), \tag{9}$$

$$P(I, J) = \sum_{(i, j, k)} \sum_{(l, m, n)} D(i, j, k, l, m, n) \tag{10}$$

where summation is over the degenerate codons of each amino acid, and besides 20 kinds of amino acid, chain terminating word is included. The entropy of amino acid sequence considered as a simple Markov chain is written as

$$H_{AA}^{M} = -\sum_{I=1}^{21} P(I, J) \log_2 \{P(I, J)/P(I)\} \quad \text{(bits)} \tag{11}$$

and that considered as a random chain is

$$H_{AA}^{R} = -\sum_{I=1}^{21} P(I) \log_2 P(I) \quad \text{(bits)}. \tag{12}$$

Let's maximize the entropy function of amino acid sequence $H_{AA}^{M}$ on the phase space spanned by 48 independent variables of $\{D(i, j, k)\}$ by the Monte Carlo Method. The results are

$$D(\mathrm{U}) = 0.2630, \quad D(\mathrm{C}) = 0.1757,$$

$$D(\mathrm{A}) = 0.3171, \quad D(\mathrm{G}) = 0.2442,$$

$$H_{AA}^{R} = 4.376 \text{ bits}, \quad H_{AA}^{M} = 4.367 \text{ bits},$$

$$H_{mRNA}^{R} = 1.970 \text{ bits}, \quad H_{mRNA}^{M} = 1.865 \text{ bits}, \quad H_{mRNA}^{2M} = 1.818 \text{ bits}.$$

C+G content of $H_{AA}^{M}$ maximum DNA is 42%, which coincides with Smith's result, and around this value the base composition of higher organisms converge. In Figure 3 the deviations of dinucleotide frequencies from random expectation, dfr, defined by

$$\mathrm{dfr} = \frac{D(i, j)}{D(i)\, D(j)} - 1 \tag{13}$$

are shown for human spleen DNA, *E. coli* DNA (Josse *et al.*, 1961; Swartz *et al.*, 1962), and DNA with the maximum $H_{AA}^{M}$ under Watson-Crick constraint. It is clearly seen that the doublet pattern of human DNA (any vertebrate DNA resembles closely) is more close to that of the $H_{AA}^{M}$ maximum DNA than that of *E. coli*. An especially distinctive feature is that doublet $C_pG$ frequencies of human DNA and the $H_{AA}^{M}$ maximum DNA are both scarce. The anomalous rarity of doublet $C_pG$ in vertebrate DNA are noted by Josse *et al.* (1961), Swartz *et al.* (1962), and King and Jukes (1969). Our result gives a reasonable explanation from the standpoint of informational theory. We take the entropy $H_{AA}^{M}$ as a measure of a variety of protein per one amino acid. Evolutionary process is considered to be accompanied with diversification of proteins so as to carry out various kinds of functions. As a consequence, the base compositions and the base doublet frequencies of DNA would have tended to the direction to increase the entropy $H_{AA}^{M}$.

In Figure 4, the entropies of protein $H_{AA}$, calculated from the experimental data of 20 amino acid frequencies for various organisms, are plotted as a function of C+G content. It is clearly seen that the C+G content of DNA is related to the entropy of amino acid sequence, and higher organisms with complex structure have higher $H_{AA}$ values than those of lower organisms.

What does the increase of entropy of amino acid sequence mean? The theory of non-Darwinian evolution developed by Kimura (1968, 1969) and King and Jukes

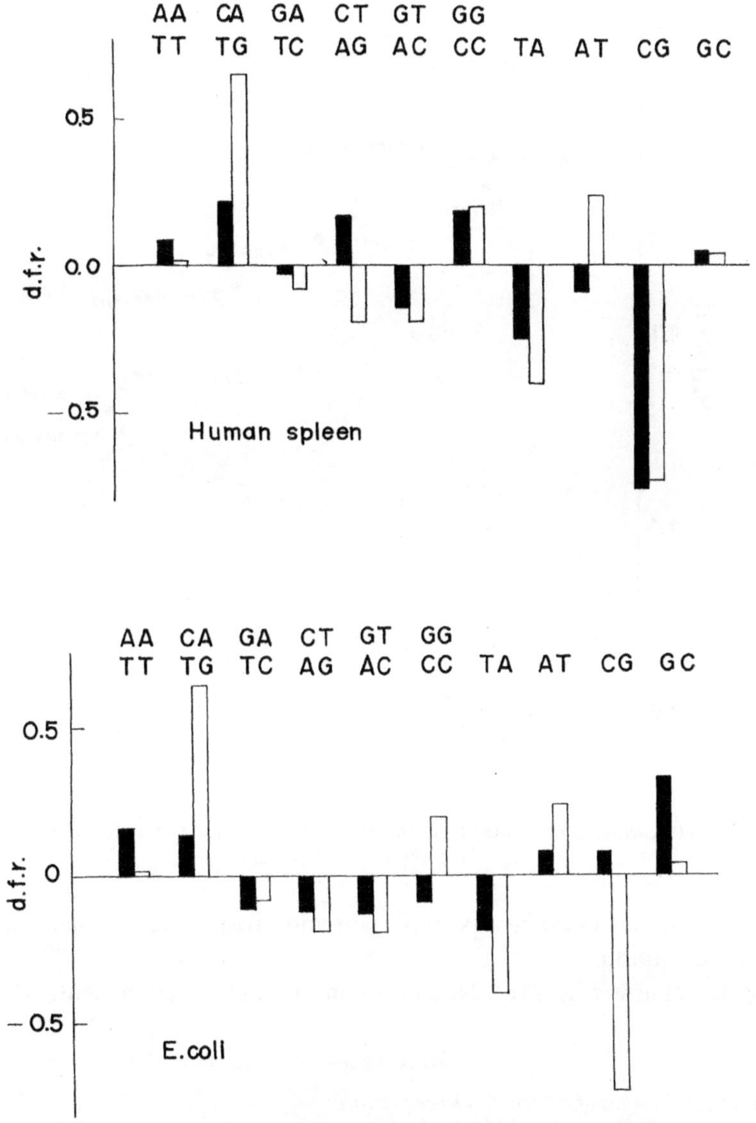


Fig. 3. Deviation of dinucleotide frequencies of DNA from random expectation. [white bar] DNA with the maximum $H^{M}_{AA}$, [black bar] (a) human spleen DNA (C+G 40.4%), (b) E. coli DNA (C+G 49.8%).

(1969) is well established in the field of molecular evolution. Yet, if neutral mutations are predominant and randomization takes place on the level of nucleotide sequence in DNA, the frequency of each amino acid will tend to be proportional to the degree of its codon degeneracy, and the entropy will decrease. However, this is not the case. Although we cannot yet give a complete and satisfactory answer to this question,

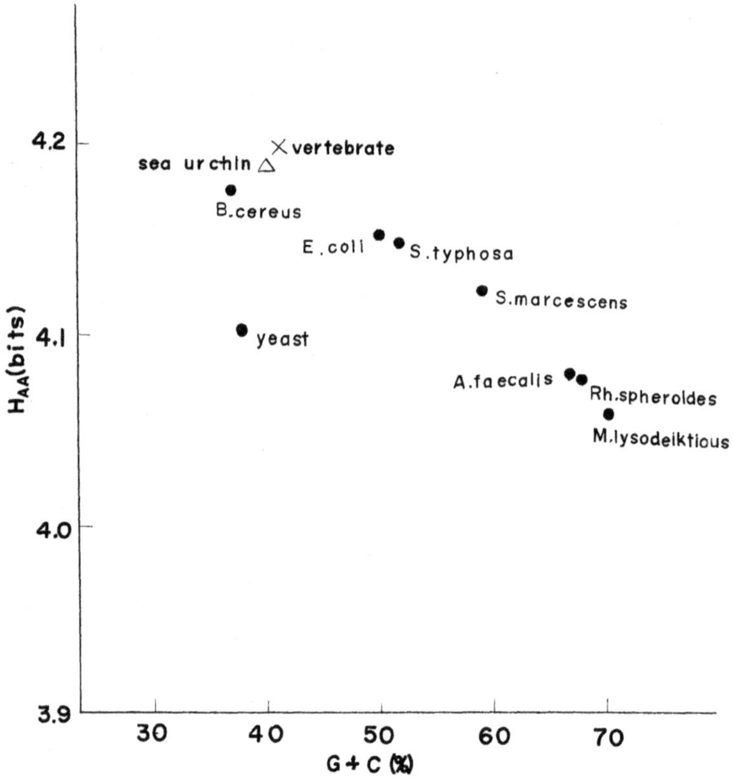


Fig. 4. Entropy $H_{AA}$, which is calculated from the experimental data of the frequencies of amino acids, as a function of C+G content.

this phenomenon may contain very important and prospective problems in the field of molecular evolution.

We wish to acknowledge Prof. Noda for continuous encouragement.

## References

Dayhoff, M. O.: 1972, *Atlas of Protein Sequence and Structure*.

Dayhoff, M. O., Eck, R. V., and Park, C. M.: 1969, in M. O. Dayhoff (ed.), *Atlas of Protein Sequence and Structure*, National Research Foundation, Silver Spring, Md., p. 75.

Gatlin, L. L.: 1972a, *Math. Biosci.* **13**, 213.

Gatlin, L. L.: 1972b, *Information Theory and the Living System*, Colombia University Press, New York.

Hasegawa, M. and Yano, T.: 1972a, *Origin of Life and Evolution*, Kyoritsu Publ. Co., Tokyo (in Japanese), p. 134.

Hasegawa, M. and Yano, T.: 1972b, submitted to *Math. Biosci.*

Josse, J., Kaiser, A. D., and Kornberg, A.: 1961, *J. Biol. Chem.* **236**, 864.

Kimura, M.: 1968, *Nature* **217**, 624.

Kimura, M.: 1969, *Proc. Nat. Acad. Sci. U.S.* **63**, 1181.

King, J. L.: 1972, *Proc. 6th Berkeley Symp. Math. Stat. Prob.* University of California Press, Berkeley.

King, J. L. and Jukes, T. H.: 1969, *Science* **164**, 788.

Miyazawa, S. and Miyata, T.: Personal communication.

Ohta, T. and Kimura, M.: 1971, *Science* **174**, 150.
Smith, T. F.: 1969, *Math. Biosci.* **4**, 179.
Sueoka, N.: 1965, in V. Bryson and H. J. Vogel (eds.), *Evolving Genes and Proteins*. Academic Press, New York, p. 480.
Swartz, M. N., Trautner, T. A., and Kornberg, A.: 1962, *J. Biol. Chem.* **237**, 1961.
Vogel, H. and Zuckerkandl, E.: 1971, in E. Schoffeniels (ed.), *Molecular Evolution II, Biochemical Evolution and the Origin of Life*, North-Holland Publ. Co., Amsterdam, p. 352.
Watson, J. D.: 1970, *Molecular Biology of the Gene*, W. A. Benjamin, Inc., New York, p. 409.
Yano, T. and Hasegawa, M.: 1973, submitted to *J. Mol. Evol.*
Zuckerkandl, E., Derancourt, J., and Vogel, H.: 1971, *J. Mol. Biol.* **59**, 473.

# PHYSICAL FOUNDATIONS OF THE PROBABILITY OF BIOGENESIS

C. A. BOGDANSKI

*Laboratoire d'Evolution des Etres Organisés, Centre National de la Recherche Scientifique, 75006 – Paris VI, France*

**Summary.** For biogenesis – a particular kind of systemogenesis – to occur, certain physical and informative requirements were indispensable, especially:

(1) the selforganization of new kind of negative feed-backs supported by the trans-substantial channels of information. These were certainly at first organized only on the submolecular level, each of them consisting of many charge-transfer connections.

The accomplishment of requirement (1) depends on (2) and (3):

(2) the organization of a sufficiently *complex* structure inside the *agglomerated* system. We should mention here the *desagglomerated* inorganic systems according to the archaic models: 'A' (Atoms) and 'P' (astro-Planetary systems). In these prebiotic models the selfregulation background consists of the old kind of negative feed-backs supported by the extra-substantial channels (lines of fields of opposing forces).

(3) the availability of molecules in which the structure *complexity* is sufficiently high to be able to contribute to the organization of the trans-substantial channels.

Biogenesis consists of spontaneous trans-substantialization of information channels in feed-backs. The trans-substantial channels are spontaneously organized in the biotic model, but they are also present in many technical electronic models of systems constructed by man. Therefore, it is no wonder that biogenesis probably occurred at the $10^{-6}$ m size level (compare the diameter of the microspheres of Fox and the concept of Ponnamperuma who mentions the size of the contemporary Micrococcus). Such a system position – inside the biotic unicellular sub-band (extended actually from $10^{-6}$ to $10^{-1}$ m) – is favourable for the organization of the high *complexity* of the structurogenic processes trajectories. It is at the nearest possible level to this region on the dimensional scale where a maximal plurality of the different joining forces exists.

This region (at $10^{-7}$ m) may also have some implications on the other neighbouring region (on its left side on the dimensional scale) where the molecular band ($10^{-10}$ to $10^{-8}$ m) is placed. The most *complex* natural molecules (for example those of DNA) are in fact placed inside such a sub-band of the molecular band which is nearest to this region.

*Conclusion:* The phenomenon of biogenesis can be considered on the spectral plan in the following way: it was accomplished:

(1) at a sufficiently *high level* for the architectural concept of cellularity to be accomplished, and at sufficiently *low level* for the system structure complexity to be at the highest value possible,

(2) from the molecules situated on this 'M' sub-band which is relatively near to the maximal complexity region.

The question of the origin of moderate frequencies in biological rhythms is discussed here on the basis of the behaviour of the very complex biotic structure giving support to an operational network having *internal inputs*, through which the high frequency signals from the 'A' systems are passed; *external inputs*, through which the low frequency signals from the delocalised structural elements of the $P_{Solar}$ system are passed; and also *high information treatment possibilities* including the transformation of the above-mentioned signals into resulting moderate frequency signals.

For the accomplishment of the biogenesis phenomenon, it appears that at least 3 elementary conditions are indispensable:

(1) the structure of the system in question must represent a certain minimum of complexity in order to be able to maintain the living processes, and for that it is necessary that:

(2) the structure of this system is compact (the system must be aggregated) and

(3) the constituent elements of this system have already sufficient structural

complexity in order to be able to assure the maintenance of internal equilibrium of the compact system, this by means of the negative feed-backs of the channels which – in the case of agglomerated system – must be *trans-substantial* ones.

Although such a simplification of the presentation of the conditions of the beginning of life appears to be a truism, the accomplishment of these conditions seems not to be a simple matter, nor hazardous, but results from the very complicated conjunctures of the events conforming to the fundamental laws of physics.

Let us further comment on these conditions in the following under 2, 3 and 1.

*Condition 2*

There exist several kinds of forces joining matter at a distance: gravitational, electrostatic and deriving from this latter: intermolecular. The field intensity of each of these varies as a function of the distance and, consequently the relative measure of importance of each of these – taken as a whole concerning the uniting of the constituent elements of a studied system – varies as a function of the dimensional level of which the system was organised.

In physical reality, diversification of the importance can be seen on the dimensional scale (see Figure 1). It is to be noted that in certain spots of this scale – where practically only a single type of force is in question – the disaggregated systems are organised, namely the systems conforming to 'A' and 'P' models.

'A' (Atom, disaggregated nucleo-electronic system, the representatives of which are relatively *small*, situated inside the $10^{-10}$ to $5 \times 10^{-10}$ m band) where the nucleus with the electrons is practically linked by the electrostatic force alone.

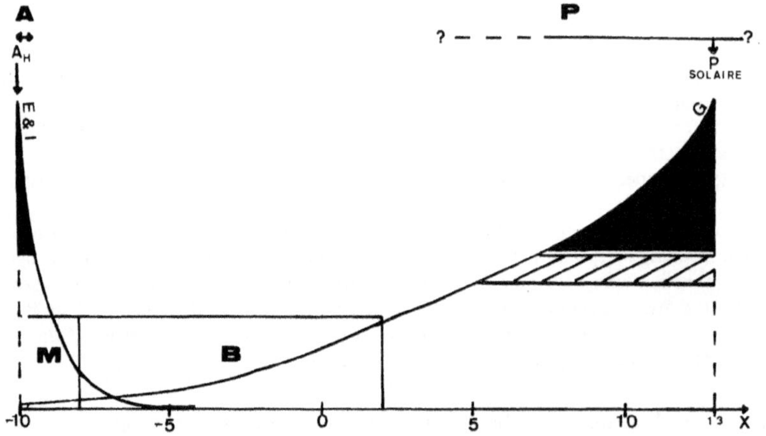


Fig. 1. Apparition probabilities of 'A' and 'P' models (structure disaggregated) as a function of the dimensional level of their organization under the influence of the joining forces favourable conjunction: G – Gravitational; E – Electrostatic; I – Intermolecular. This conjunction can take place only there (see dark peaks) where practically only a single force is represented. X-axis (logarithmic scale): dimensional levels; $10^x$ = size diameter (in meters). Y-axis (logarithmic scale): parametric importance of the dominant force; calibration for G is based on the $P_{Solar}$ sub-model, diameter and for E on $A_{Hydrogen}$ diameter. A – Atomic band; M – Molecular band; B – Biotical band; P – astro-Planetary band.

In the case of this link it is possible not to take the gravitational force into consideration because its role is minimal. For example, in the case of the hydrogen atom, the link existing between its nucleus and its electron is assured by an electrostatic ($F_{el}$) force about 40 orders superior to the gravitational ($F_{gr}$) force. The respective values are the following:

$$F_{el} = 9 \times 10^{-3}\ \text{dyn} \quad \text{and} \quad F_{gr} = 4 \times 10^{-42}\ \text{dyn}.$$

'P' (astro-Planetary, disaggregated system, the representants of which are relatively *large* e.g. the diameter of the $P_{Solar}$ is in the order of $10^{13}$ m) where the planets are practically linked by the gravitational force alone.

In the case of systems conforming to the disaggregated models it should be noted that the internal equilibrium of each of them rests on a relatively high probability of generation [by an adequate cinematic stereotype of movement of the delocalised

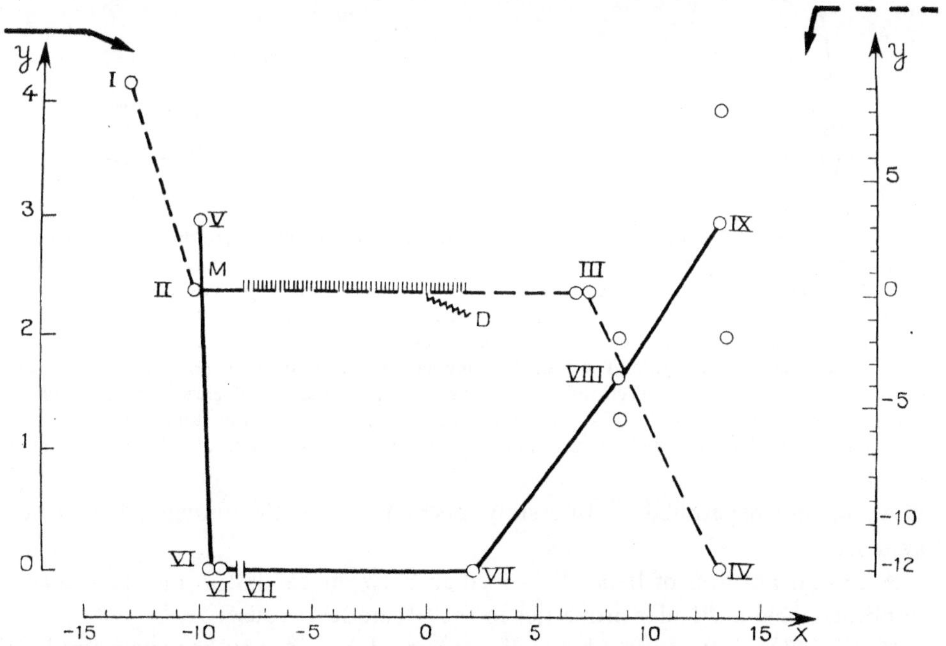


Fig. 2. Density and agglomeration degree. Legend: $10^x$ m = average diameter length. $10^y$ (scale on *left*) = agglomeration degree coefficient: (distance between the $C_{es}$)/(average diameter of the $C_{es}$) represented by the continuous line (———) and where: $C_{es}$ = Constitutive element of the system. $10^y$ (scale on *right*) = density in g $cm^{-3}$ represented by the discontinuous line (- - - -).

Discontinuous line: I – nucleus; II – atoms; II to III valid for: M-molecules – living organisms (with deviation = D for organisms having not very compact structure, for instance: the ramified plants); III – Moon and Earth; IV – Solar system.

Continuous line: V – Atoms; VI–VI – Molecules band; VII–VII – Living systems band; VIII – Earth–Moon system; IX – Sun–Earth system. Remark: In the cases VIII and IX the curve was drafted through the average values. The distance divided by the diameter of the smaller body gives the superior value like that from the division by the larger one. Therefore there are two points: superior and inferior and additionaly an average one. In the case of V the only denominator was the nucleus diameter.

constitutive element of the system] of a repulsive force capable of equilibrating the force of attraction existing between the bodies in question.

It is evident, that the realisation of this principle is always more probable in the case where only a single type of force must be equilibrated (see Figure 1 and consequently Figure 2).

Band '*B*' (Biotical systems; the living systems, always aggregated) is situated – with a range of 10 orders of magnitude – between $10^{-8}$ and $10^{2}$ m (see Figure 3)

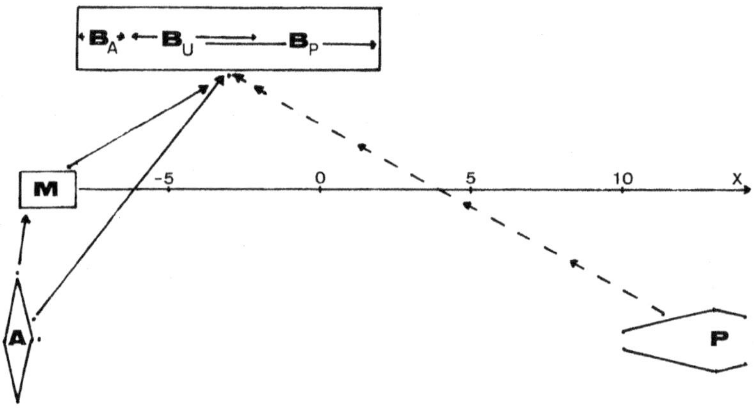


Fig. 3. Positions of the principal models of the self-regulating systems in the dimensional scale. $10^x$ – diameter of the system (in meters). *Below*: The representatives of the models having disaggregated structure: 'A' and 'P'; the latter symbol has been placed where in physical reality is placed $P_{Solar}$ (diameter $10^{13}$ m). *Above*: The representatives of the aggregated systems model 'B', bandlength $10^{-8}$ to $10^{2}$ m, subdivided into three sub-bands corresponding to the three principal sub-models 'B': $B_a$ – Acellular; $B_u$ – Unicellular; $B_p$ – Pluricellular. *On the scale*: The transitory model 'M' (in principle aggregated; however, certain sub-models have delocalised electrons). Note: The most probable genealogical linkages between the models are marked by arrows: ——— in continuous line for the constitutional linkages; – – – – in discontinuous line for spatial interactions (consisting above all of rhythmical radiation signals)

in a zone not penetrated, and possibly inpenetrable, by the disaggregated structure systems.

A certain position of band 'B' – median between bands 'A' and 'P' – seems to implicate some particular properties to the 'B' model because:

(I) the 'A' and 'P' models, being disaggregated, have a very important implicative power (this can be exemplified in the case of rhythmic frequencies*, which are well

* *The low frequencies* implicated by the characteristics of movement of the constitutive element of the $P_{Solar}$ system, and *the high frequencies* by the stereotypes of movement which are immanent to the constitutive elements of representatives of the 'A' model and its derivative 'M' (e.g. vibrations of the functional group of the molecule) of which each representative of the 'B' model is constructed.

The complex structure of the latter becomes an operational network able not only to memorise certain so-acquired rhythms (annual, monthly, daily→circadian), but as well as to transform the messages of *low* and *high* frequencies of which the resultant is constituted of *moderate* frequency rhythms: cardiac, pulmonary, walking.

Since the moderate frequency rhythms are situated inside the human perceptibility band, they have been able to figure amongst the first diagnostic criteria of life.

marked in the systems where the existence of the system depends on stereotypical movement).

(II) a point situated on the dimensional scale between bands 'A' and 'P' manifests the highest dynamic complexity when at maximum distance from both peaks, representing the almost exclusive domination of a single force (Figure 1) and – as this point concerns systems being necessarily agglomerated (Figure 2) – to the plurality of the joining forces at distance must be added the cohesion force.

Consequently at this dimensional level the number of degrees of freedom of movement is also at a maximum, making possible some manifestations of kinetic complexity including those in the structurogenic processes. Consequently here also arises the probability of the organization of a very complex system structure if this is situated inside a band close to the maximal complexity zone ($Z_{mc}$), i.e. about $10^{-7}$ m (see Figure 4).

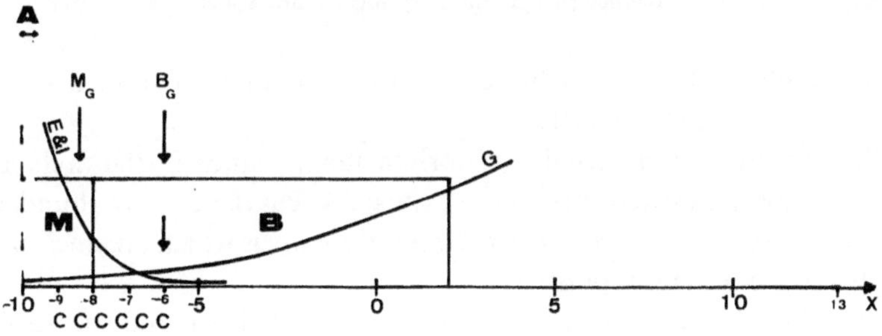


Fig. 4. Position of a high complexity zone due to the non-existence of the absolute domination of a single joining force of matter (see Figure 1). *Legend.* (see also Figure 1): CCCCC – zone of the high complexity where at the point of junction of forces 'G', 'E' and 'I' the domination of any one force is not very pronounced. This zone covers the superior part of the M-band ($M_G$ = large molecules) and the inferior part of the B-band ($B_G$ = position where biogenesis is considered to have taken place) implicating thus cellular level, fundamental for the unicellular organisms and elementary for the pluricellular organisms.

*Condition 3*

As concerns evolution at the molecular level, it should be noted that when molecules become larger, they become more complex not only because the number of atoms is greater, but also because such molecules passing in these sub-bands (of the M= =Molecular band) which approach nearer and nearer the $Z_{mc}$ (see Figure 5).

In reality, the great complexity of the large molecules has already, since long ago, been the object of research in organic chemistry and then in biochemistry. Let us mention especially the molecular submodels responsible for biohomeostasis, without which the biotical system could not live, and also the – very large and complex – molecular sub-model realised in all molecules of DNA.

Speaking about the maintenance of biohomeostasis, we draw attention particularly to the substances capable of carrying out charge-transfers, which are necessary for the formation of information channels at the submolecular level. The feed-backs

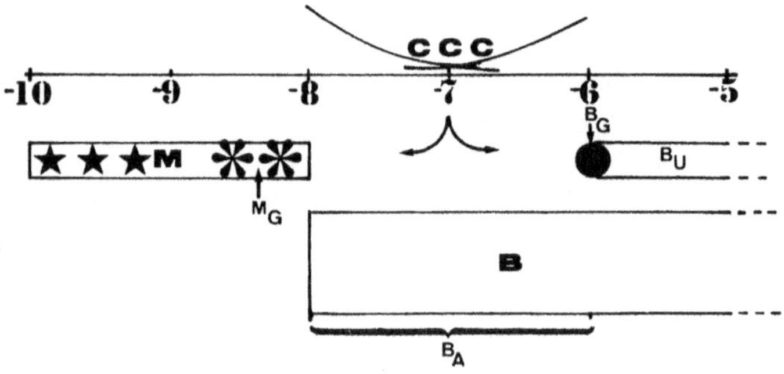


Fig. 5. Relative positions of the 'M' and 'B' bands toward the $Z_{mc}$. $Z_{mc}$ – maximal complexity zone where pluralisation of forces implies the kinematical complexity involving the maximal complexity of of the structorogenic processes (see CCC); M – Molecules band; B – Biotical (living systems) band; $M_G$ – Large molecules sub-band of the M band; $B_a$ – Acellular sub-band of the B band; $B_u$ – Unicellular sub-band of the B band; $B_G$ – probable position of the biogenesis first product; $10^x$ = average system diameter length (meters); logarithmic scale.

between these channels are capable of insuring this kind of homeostasis because they are *trans-substantial* channels.

Note the fact that the internal equilibrium maintenance in the archaic systems 'A' and 'P' – disaggregated structure, dynostatic character – is founded on the negative feedbacks between the *extra-substantial* channels which are lines of opposing force fields: attractive and repulsive.

It follows that the creation of a new type of channel was indispensable for the appearance of the aggregated self-regulating system, the existence of which was dependent on biohomeostasis.

First, let us discuss channels at the submolecular level consisting of lines composed of many charge-transfers occurring between adequate molecules.

These channels have certainly been greatly refined before appearing in the state which we know today. According to the theory of Pullman (1963), concerning evolution at the molecular level*, natural selection of molecules took place. Following this, the organisms which were able to synthesize more useful substances (e.g. which assure the existence of more efficient channels) were more often able to survive. Consequently, there was a high probability that these organisms would be greatly propagated and therefore that in their bodies evolution – at the molecular level – would be achieved.

We mention briefly some trans-substantial channels at the levels superior to submolecular:

at the molecular level (e.g. channels composed by molecules of two hormones, insulin and adrenalin, being secreted, coupled in a negative feed-back which is responsible for the maintenance of the homeostatic level of glucose in the blood.

at the supramolecular level (e.g. couplage of the two opposite muscle actions,

* See also the experimental organic cosmochemistry of Oró (1963).

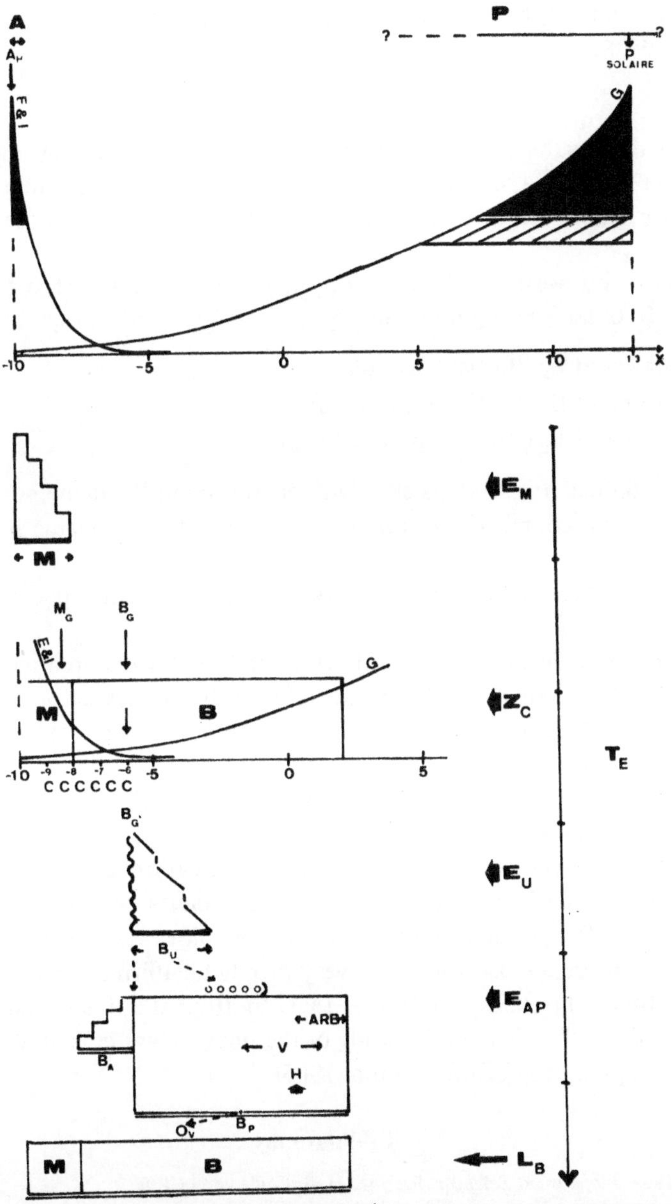


Fig. 6. The evolution of some models of the spontaneous self-regulating systems. *Upper graph*: for the Legend see Figure 1. *Lower graphs series*: Evolution of some models as function of time ($=T_E$; see arrow on the right side with arbitrary time units); $E_M$ – Molecular band evolution; $Z_c$ – Zone of the maximal (dynamic and cinematic) complexity; see Figure 4; $E_U$ – Most probable evolution of the unicellular organisms from $B_G$ (↑); $E_{AP}$ – Most probable evolution of the acellular and pluricellular organisms. ○○○○○○ – eggs, living units being initially unicellular and then transformed into pluricellular units; $B_u$ – Sub-band of the contemporary unicellular organisms; $B_a$ – Sub-band of the contemporary acellular organisms; $B_p$ – Sub-band of the contemporary pluricellular organisms; ARB – trees; V – vertebrates; H – Homo sapiens; $O_v$ – ovulation processes (unicellularisation of the pluricellular organism); $L_B$ – Present position of the 'M' and 'B' bands.

subject of research in biomechanics and more particularly in kinanthropology (Bogdanski, 1972)).

*Condition 1*

According to the results from biogenesis research, the first living organism must have been the size of a modern-day Micrococcus (Ponnamperuma and Gabel, 1968). Therefore it should be situated at approximately $10^{-6}$ m on the dimensional scale.

If we analyse the position of the value $B_G$ in the context or the principal sub-bands of the 'B' band (see Figure 6), i.e.:

sub-band $B_a$ (Biotical-acellular)
sub-band $B_u$ (Biotical-unicellular)
sub-band $B_p$ (Biotical-pluricellular)

we notice that $B_G$ (see Figure 5) is situated* on sub-band $B_u$ at the nearest point to the zone of maximal complexity of joining forces and structurogenic processes.

Thus the phenomenon of biogenesis can be considered on the spectral plan in the following way:

(a) it was accomplished at a sufficiently elevated level for the architectural concept of cellularity to be accomplished, and at sufficiently low level for the number of degrees of system structural complexity (system obviously aggregated) to be at the highest value possible.

(b) from the molecules situated on the 'M' sub-band which is relatively nearest to the maximal complexity zone.

Concluding this – necessarily short** – communication by an affirmation that system dimensional level seems to be an important factor of systemogenesis probability (including 'B'-system organisation, i.e. the biogenesis phenomenon). This affirmation is connected closely with the very prophetic affirmation of the American biologist H. Blum who said – in 1965 – that "putting the living systems into the dimensions used so effectively in the study of the physical systems may give us better perspective of explanation of life" (Blum, 1965).

## Bibliography

Ackerman, E.: 1962, *Biophysical Sciences*, Prentice Hall, Englewood Cliffs.
Blum, H.: 1965, *Nature (London)* **206**, 131–132.
Bogdanski, C.: 1971, *Materials from the First European Biophysics Congress, Vienna* (17–19 September, 1971), VI, pp. 239–244.
Bogdanski, C.: 1972a, *Compt. Rend. Acad. Sci. Paris* **274**, 577–580.
Bogdanski, C.: 1972b, *Compt. Rend. Acad. Sci. Paris* **275**, 199–202.

* Moreover, on this biotical dimensional level, the harm from mechanical vibrations of the ultrasonic band is relatively less pronounced (see tables in Ackerman, 1962).
** For more details see Bogdanski, 1971, 1972a, b, c, d and e.

Bogdanski, C.: 1972c, *Compt. Rend. Acad. Sci. Paris* **274**, 1546–1549.
Bogdanski, C.: 1972d, *Compt. Rend. Acad. Sci. Paris* **275**, 105–108.
Bogdanski, C.: 1972e, *Compt. Rend. Acad. Sci. Paris* **275**, 1395–1398.
Bogdanski, C.: 1972f, *Kinanthropologie (Liège)* **4**, 35–69.
Oró, J.: 1963, *Ann. N. Y. Acad. Sci.* **108**, 464–481.
Ponnamperuma, C. and Gabel, N. W.: 1968, *Space Life Sci.* **1**, 64–96.
Pullman, B.: 1963, *La biochimie électronique*, Presses Universitaires de France, Paris.

# SOME PHYSICAL PARAMETERS CONTROLLING CELL SIZE DURING THE EVOLUTION OF THE PROCARYONS

M. McCABE

*Institute of Medical Chemistry, University of Uppsala, Uppsala, Sweden* *

**Abstract.** Possible factors controlling cell size during the evolution of unicellular organisms have been examined. It has been shown that considerations of osmotic and membrane pressure equilibria will predict minimal cell sizes which are in good agreement with those found in present day microorganisms. It has also been shown that the possibility of random proton 'noise' would not be a limiting factor for even the smallest organisms or structures.

Maximum cell size would be governed by the requirements of diffusion and transport within the cell.

The broadest and most fundamental of all biological generalisations, namely the cell theory, states that all living organisms are composed of cells. Thus from the largest to the smallest organisms there exists an underlying structural similarity which presumably owes more than accident to its stability. What are the forces which control the size of this basic unit of living cells? Of course there do exist variations of cell size over quite wide ranges; for instance an epithelial cell may have a diameter of up to $3 \times 10^{-3}$ cm, bacteria are found down to $10^{-4}$ cm or even less, while finally organisms of the pleuropneumonia groups range in diameter down to $10^{-5}$ cm. Nevertheless the question remains as to why cells appear to exist always within these limits.

## 1. Possible Factors Limiting the Minimum Size of Cells

The question has been raised as to whether or not proton 'noise' within a very small volume might not make accurate pH control of that region impossible (McCabe, 1967). The argument used was a statistical one based on a calculation of the instantaneous concentration of protons within a volume of the order of $10^{-12}$ ml. However it has been shown (Chance, 1967; Mitchell, 1967a) that because the velocity constant of the reaction $H^+ + OH^- \rightleftharpoons H_2O$ is of the order $10^{10}$ to $10^{12}$ l mole$^{-1}$ s$^{-1}$ at 25°, the instantaneous space average of $H^+$ can be replaced by a 'time average' value which can be meaningfully specified in far smaller volumes than the cell or cell compartments. Another way of putting their argument is to say that the cell contains no mechanism capable of responding to the extremely rapid fluctuations of pH and these fluctuations are therefore averaged by the reacting systems which are pH sensitive. A related question has been raised by Baum (1967), who pointed out that if a membrane were to be proton tight then each translocation of a proton must be of a whole proton and not a time average fraction. If the translocation of

* Present address: Dept. of Chemistry and Biochemistry, James Cook University of North Queensland, Townsville, Queensland, Australia.

the proton is into a very small compartment then the time average concentration of protons could still be considerably altered. Mitchell (1967b) has answered this by suggesting that the buffering power of the compartments' constituent proteins could effectively prevent pH fluctuations. Mitchell has calculated that for a volume of $10^{-12}$ ml the presence of a buffer at a concentration of 0.02 M could limit the pH change introduced by the addition of 2 protons, to about 0.1 pH unit. Titration curves show that most proteins have only a small portion of their buffering capacity in the range of intracellular pH, i.e. around 6 to 7. In fact the only amino acid having an ionisable group with a pK in this range is histidine, and most proteins contain relatively few histidine residues, perhaps on average 2% by weight. It can be shown from the calculations of Edsall (1943) that adequate buffering could be provided to a volume of $10^{-12}$ ml by 35 histidine residues, such that the addition of 2 protons would alter the pH of the compartment by less than 0.1 pH unit in the region of pH 6 to 7. For a protein containing 2% by weight of histidine this would only mean a single particle of molecular weight approximately 30000. Assuming a partial specific volume of 0.7, this protein molecule would be accommodated in $7 \times 10^{-21}$ ml, i.e. in a vanishingly small fraction of the cell compartment of even the smallest of the PPLO. This implies that some parameter other than random proton noise must be size limiting.

It seems likely that the earliest of the procaryons would resemble the L forms of bacteria. These forms are devoid of a cell wall and are separated from the environment only by an easily deformable cytoplasmic membrane (Klienenberger-Nobel, 1960). Such an arrangement must have preceded, even if only briefly, the more sophisticated arrangement of a rigid cell wall external to the cell membrane. The subsequent addition of an outer cell wall would provide the advantage of enabling a greater difference in osmotic gradient to exist between the cell and its environment. This would certainly have been an advantage since it would enable the acquisition of a more concentrated and hence potentially more versatile array of cell constituents. It would however limit access somewhat to the cell wall and would therefore diminish the ease of exchange of nutrients and waste products.

Consider a spherical organism separated from the environment only by a deformable cell wall. The surface of the cell cytoplasm is in contact with a separate phase (the lipoprotein layer of the cell wall), and is curved, hence an excess pressure will exist under this surface given by:

$$P = \frac{2(T_1 + T_2)}{r}$$

where $P$ is the excess pressure in dyn cm$^{-2}$; $r$ is the radius of the cell in cm; $T_1$ is the interfacial tension between the inner lipoprotein surface and the cell cytoplasm, and $T_2$ that between the outer lipoprotein surface and the external environment of the cell, both measured in dyn cm$^{-1}$.

Interfacial tensions have been measured for a number of micro-organisms and cells without an outer cell wall, and also for lipoprotein bilayers extracted from cells.

Generally such measurements have revealed that the interfacial tensions of cell membranes are low, the lowest value so far found is 0.08 dyn $cm^{-1}$ (see Table I). It would seem that the low tension is caused by the presence of a monolayer of globulin

TABLE I

Showing results for a variety of measurements of interfacial tensions of cell membranes and cell lipoprotein constituents

| Cell or system | Method | Result | Author |
|---|---|---|---|
| *Amoeba dubia* | Centrifuge microscope | 1–3 dyn $cm^{-1}$ | Harvey and Marsland (1932) |
| *Physarum polycephalum* | Centrifuge microscope plus injected oil | 0.45 dyn $cm^{-1}$ | Vexler (1935) |
| Rabbit macrophages | Centrifuge microscope | 2 dyn $cm^{-1}$ | Shapiro and Harvey (1936) |
| Frog leucocytes | plus injected oil | 1.3 dyn $cm^{-1}$ | |
| *Arbacia punctulata* eggs | Compression method | 0.08 dyn $cm^{-1}$ | Cole and Michaelis (1932) |
| *Arbacia* eggs | Centrifuge microscope | 0.09 dyn $cm^{-1}$ | Harvey (1931) |
| *Arbacia* eggs | Compression method | 0.09 dyn $cm^{-1}$ | Cole (1932) |
| *Arbacia* eggs | Kinetic method | 0.09 dyn $cm^{-1}$ | Sichel and Burton (1936) |
| Lipoprotein bilayers | Langmuir trough | 0.8 dyn $cm^{-1}$ | Danielli and Harvey (1935) |
| Human serum high density lipoproteins | Langmuir trough | 2 dyn $cm^{-1}$ (initially) | Camejo *et al.* (1970) |
| Eggs of *Buscyon conaliculatum* | Sessile drop | 0.5 dyn $cm^{-1}$ | Harvey and Frankhauser (1933) |
| Eggs of *Triturus viridescens* | Sessile drop | 0.1 dyn $cm^{-1}$ | Harvey and Frankhauser (1933) |

type protein. Without this monolayer the interfacial tension is about tenfold larger. If we assume that for the archaic cell the interfacial tension was correspondingly low we can then calculate the excess pressure inside the cell as a consequence of the cell size. This is shown in Figure 1.

Of course if the cell membrane is permeable to water, then the cell will only be stable if the extra hydrostatic pressure generated by the interfacial tension and the radius of curvature is balanced by some other pressure tending to pull the water back into the cell. Thus such living cells must be maintained in equilibrium by a compensating internal osmotic pressure. Such osmotic gradients would be reasonably maintained without the expenditure of large amounts of energy by the presence of macromolecules within the cell. If we suppose that the smallest macromolecule which may not pass the cell wall has a molecular weight of 20000 then the concentration of protein necessary to balance the excess hydrostatic pressure can be calculated. This is shown in the figure. From this it can be seen that the minimum size of present cells requires an osmotic pressure difference between itself and its environment which corresponds quite well with the kinds of soluble protein concentrations we expect to find in cell cytoplasm. It can also be seen that even a relatively small decrease in cell size, to say $2 \times 10^{-6}$ cm radius would require a soluble cytoplasmic protein concentration of about 125 g $l^{-1}$, which would be a concentration difficult for the organism to attain. It should be noted that this decrease in cell size would only be a

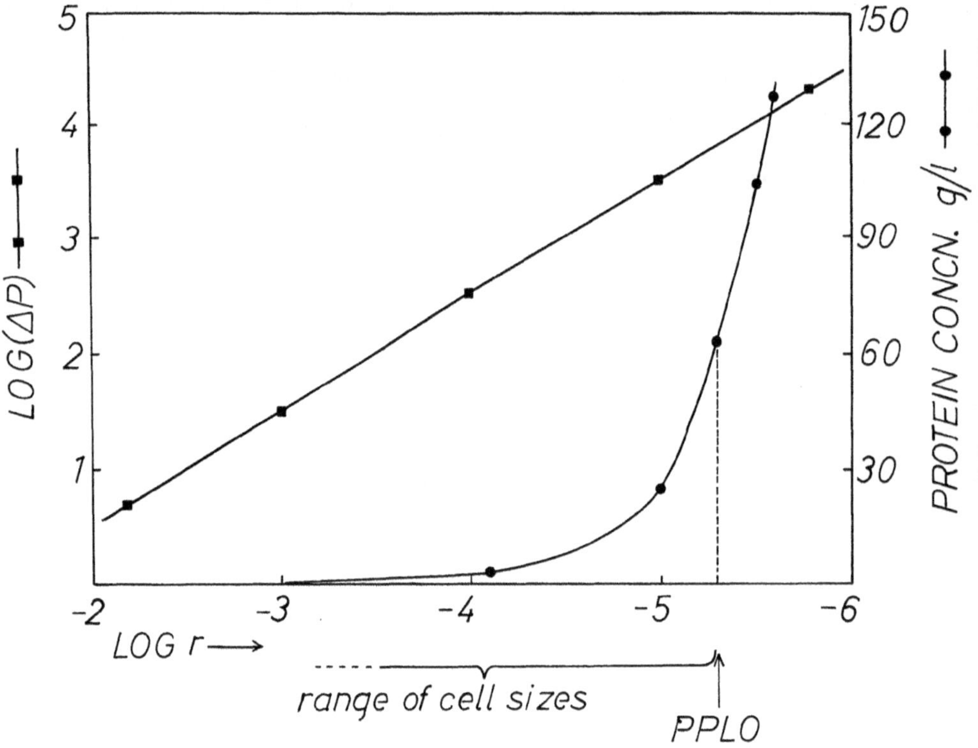


PPLO, pleuropneumonia like organism
$r$, measured in cm
$\Delta P$, measured in dyn cm$^{-1}$
Molecular weight of protein 20000

Fig. 1. Relationship between the radius of curvature of a spherical cell and the excess pressure inside the cell: also between the radius of curvature and the protein concentration which would generate an equivalent osmotic pressure.

further reduction of 40%, thus the PPLO would seem to be living on the borderline of that theoretically possible size which is limited by the solubility of proteins.

## 2. Possible Factors Limiting the Maximum Cell Size

A large cell in the L form, i.e. without a supporting cell wall, would only require a small excess osmotic pressure over its environment, hence the cell membrane would not necessarily be held rigid, thus the larger the cell the more readily it could be deformed by external forces such as gravitational fields and velocity gradients in the supporting medium. Eventually this deformation would lead to rupture and the loss of cell contents. This problem could be partially overcome by the addition of a rigid readily permeable supportive external cell wall. Following this addition the cell could enlarge its volume very considerably, however many of the reactions occurring within the cell are sequential and it can be seen that if the cell were to become very large then the danger of disorganisation of the sequences could become larger since the time taken for a product of one enzyme to diffuse to the next enzyme in the sequence is related exponentially to the diffusion path length. If these distances are made unnecessarily long then the efficiency of the organism at transforming and handling substrate molecules would drop. In other words the consequences of the finite time for diffusion of cell constituents and metabolic intermediates from one enzyme site to the next constantly push the organism towards a minimum size. The cell can partially overcome this problem by raising the equilibrium concentration of the diffusing species since this would increase the rate of impingement of the molecule in question onto the target enzyme. However this presents the cell with the added problem of maintaining a higher concentration gradient of the substance across the cell membrane. Many of these intermediates would be substances of low molecular weight some of which could only be maintained inside the cell by the development of special active transport systems on the cell membrane. Increasing cell size will also present the cell with problems of diffusional exchange of metabolic waste products, which would be further aggravated by the necessity of a rigid external cell wall.

## References

Baum, H.: 1967, *Nature* **214**, 1326.
Camejo, G., Suarez, Z. M., and Munoz, V.: 1970, *Biochim. Biophys. Acta* **218**, 155.
Chance, B.: 1967, *Nature* **214**, 399.
Cole, K. S.: 1932, *J. Cell. Comp. Physiol.* **1**, 1.
Cole, K. S. and Michaelis, E.: 1932, *J. Cell. Comp. Physiol.* **2**, 121.
Danielli, J. F. and Harvey, E. N.: 1935, *J. Cell. Comp. Physiol.* **5**, 483.
Edsall, J. T.: 1943, in *Proteins, Amino Acids and Peptides*, Reinhold, New York, p. 467.
Harvey, E. N.: 1931, *Biol. Bull.* **61**, 273.
Harvey, E. N. and Frankhauser, G.: 1933, *J. Cell. Comp. Physiol.* **3**, 463.
Harvey, E. N. and Marsland, D. A.: 1932, *J. Cell. Comp. Physiol.* **2**, 75.
Kleinenberger-Nobel, E.: 1960, in I. C. Gunsalus and R. Y. Stanier (eds.), *The Bacteria* **1**, Academic Press, New York and London, p. 361.
McCabe, M.: 1967, *Nature* **213**, 280.
Mitchell, P.: 1967a, *Nature* **214**, 400.
Mitchell, P.: 1967b, *Nature* **214**, 1327.
Shapiro, H. and Harvey, E. N.: 1936, *J. Cell. Comp. Physiol.* **8**, 21.
Sichel, F. J. M. and Burton, A. C.: 1936, *Biol. Bull.* **71**, 397.
Vexler, D.: 1935, *Proc. Soc. Exp. Biol. and Med.* **32**, 1539.

# PART V

# EARLY BIOCHEMICAL EVOLUTION

# THE BEGINNING OF PHOTOSYNTHESIS

E. BRODA

*Institute of Physical Chemistry, Vienna University, Vienna, Austria*

**Abstract.** There is no evolutionary continuity between photochemical abiosynthesis and bacterial photosynthesis. Rather, the photosynthetic bacteria are descendants of fermenters that did not use light. Photosynthesis and respiration, both using electron flow coupled with phosphorylation, have a common origin ('conversion hypothesis'), but photosynthesis came first. Anaerobic (nitrate or sulphate) respiration cannot have preceded photosynthesis as neither nitrate nor sulphate existed on the early earth. Sulphate was made first by photosynthetic sulphur bacteria. Nitrate arose even later, namely, in the aerobic biosphere produced by the blue-green algae, the first 'phytotrophs'. Photophosphorylation may have originated through the combination with membrane function of substrate level phosphorylation in reactions of photoproducts. Cyclic photophosphorylation arose while the biosphere was still reducing. It was supplemented later by processes for the light-based production of reducing power (NADH), ATP-powered electron flow, and subsequently light-powered electron flow with ATP production (noncyclic photophosphorylation). These later processes served the assimilation of $CO_2$.

**1.** It is widely thought that the organic compounds in the prebiotic soup were mainly produced by ultraviolet (uv) light. Some authors hold that already the earliest organisms utilized the energy of light (Gaffron, 1965; Granick, 1965). Thus continuity in the action of light from prebiotic to biotic times is suggested.

Fermenters base their energy metabolism, by definition, on dark redox reactions between organic compounds, coupled to phosphorylation on a substrate level. Since a separate origin of the fermenters would imply a polyphyletic origin of life, and therefore appears unlikely, it must be assumed from Gaffron's and Granick's point of view that the fermenters are descended from photosynthesizers.

We shall not accept this. In contrast to the machinery for fermentation, that for photosynthesis always has a complicated solid-phase structure. Photosynthesis in liquid phase is unknown. The components of the photosynthetic electron flow (redox) chain are arranged in spatial order: the loci of electron flow and of photophosphorylation are membranes. Organisms with this degree of structure can hardly have been the first.

More probably, the fermenters preceded the photosynthesizers. Indeed this is the majority view. But what are the relative positions in evolution of the photosynthesizers and the respirers? Like the photosynthesizers, the respirers have ordered electron flow, coupled to phosphorylation. Moreover, both groups – and no other – have cytochromes. Therefore, it has been proposed (Broda, 1970, 1971) that photosynthesis and respiration always evolved one from the other, and never arose independently ('conversion hypothesis'). Which process came first?

**2.** Aerobic respiration clearly was impossible before photosynthesis, as there was almost no free oxygen in the biosphere. The biosphere acquired oxygen (became oxygenic) only through the activity of the most advanced photosynthesizers, the blue-green algae. From this point of view, only anaerobic respiration might have antedated photosynthesis. This would imply a tremendous time interval between the origins of anaerobic and of aerobic respiration. The former would have arisen before, the latter after photosynthesis.

Two groups of anaerobic respirers exist, the dissimilatory nitrate and sulphate reducers. In these organisms, nitrate or sulphate substitutes for oxygen as the terminal electron acceptor. Anaerobic respiration has indeed much in common with aerobic respiration. Electron flow is coupled to phosphorylation, and cytochromes participate in the process.

According to Egami (see Takahashi *et al.*, 1963), nitrate respiration antedated oxygen respiration. These authors think that nitrate respiration arose from a supposed variant of fermentation, nitrate fermentation. However, in absence of free oxygen the synthesis of nitrate would have been highly endergonic:

$$NH_4^+ + 3H_2O = 2H^+ + NO_3^- + 4H_2; \qquad \Delta G'_0 = 144 \text{ kcal mole}^{-1}$$
$$\tfrac{1}{2}N_2 + 3H_2O = H^+ + NO_3^- + 2.5H_2; \qquad \Delta G'_0 = 134 \text{ kcal mole}^{-1}.$$

Thus the anoxygenic biosphere did not contain nitrate. Presumably nitrate was formed later – in the oxygenic atmosphere – by chemolithotrophs, namely, the nitrifiers.

It is a curious feature of Egami's hypothesis that photosynthesis has no role. We have seen that nitrate respiration must have awaited the emergence of free oxygen, and, consequently, of photosynthesis.

**3.** Early sulphate respiration has been assumed by Peck (1966), Klein and Cronquist (1967) and Postgate (1968). But did sulphate exist in the early biosphere? Peck admits that not much sulphate was there before life began. Indeed, little sulphate has been found in the oldest rocks (Cloud, 1968, 1972). This is not surprising, as sulphate is unstable in a reducing biosphere (see Broda, 1971).

According to Peck, sulphate was made in the anoxygenic biosphere by fermenters in reactions unknown in the present world, namely, by oxidation of sulphide with organics. However, such reactions with plausible oxidants are endergonic. For instance, with acetate we have:

$$2CH_3COO^- + HS^- + 2H_2O + H^+ = 2CH_3CH_2OH + SO_4^{2-}; \qquad \Delta G'_0 = 34 \text{ kcal mole}^{-1}$$

and with water (for comparison);

$$4H_2O + HS^- = SO_4^{2-} + 4H_2 + H^+; \qquad \Delta G'_0 = 37 \text{ kcal mole}^{-1}.$$

Realistically, $\Delta G'$ was still more positive than $\Delta G'_0$, as the concentration of $HS^-$ in presence of $Fe^{++}$ is limited by the solubility of FeS. From the solubility product of FeS ($3.7 \times 10^{-19}$) follows that, e.g., in neutral $10^{-3}$ M $Fe^{++}$ the concentration of $HS^-$ must have been less than $3 \times 10^{-8}$ M. Consequently, $\Delta G' - \Delta G'_0 > 10.2$. Of course, the values of $\Delta G'$ could be depressed by assuming lower concentrations of the organic products than of the organic reactants, but this seems unrealistic in respect of the primeval, reducing biosphere.

Thus no sulphate could arise in fermentations. It is proposed (Broda, 1970, 1971) that sulphate was made first by photosynthetic bacteria. Purple bacteria are known which make sulphate as a by-product in the phototransfer of electron from sulphide to ferredoxin. The sulphate produced by early photosynthetic bacteria then served the sulphate respirers as an electron acceptor in 'sulphureta', closed ecosystems. In accordance with the conversion hypothesis the sulphate respirers are assumed to have descended from photosynthetic bacteria.

It is concluded that all respiring bacteria (oxygen, sulphate and nitrate respirers) evolved from photosynthetic bacteria, and that the photosynthetic bacteria evolved from fermenters. Hence chlorophyll type pigments are older than cytochromes.

**4.** How can photosynthesis have evolved from fermentation? Transitional forms are unknown now, and so we are left to mere speculation. What is the most plausible process?

Bacterial photosynthesis, as we know it now, provides both ATP and reducing power (NADH). According to Arnon (1959), the primary function of photosynthesis was the production of ATP. (Of course, ATP may have been preceded by another energy-rich compound, e.g., inorganic pyrophosphate.) Other authors have given precedence to NADH production (Duysens, 1964; Gaffron, 1965). Also the present author (1970) was at one time so inclined.

One argument for the primary role of ATP in photosynthesis is that even now in some conditions only ATP – no NADH – is required by photosynthetic bacteria, e.g., for the photoassimilation of alcohols. This may be an ancient process. A further argument is that NADH could be made, when needed, in a dark reaction by reduction of NAD by $H_2$; the biosphere was still reducing when photosynthesis began.

Through the life processes of the fermenters, the stores of suitable substrates for redox reactions yielding ATP diminished. The first task of light may have been replenishment of suitable redox systems through reversal of the exergonic reactions by light.

Endergonic redox reactions can be forced by light. Thus NAD is reduced by ascorbid acid in pyridine in presence of chlorophyll (Krasnovsky and Voinovskaya, 1952). In early cells, both products of light reactions, reductant and oxidant, were found in the same compartment, and therefore the possibility for useless back reactions was great (see Gaffron, 1965). Systems were only useful where such direct back reactions were slow, and hence back reactions coupled to phosphorylation could compete with them. This idea is not far-fetched. After all, in presence of the necessary catalysts, glycolysis with phosphorylation competes with the direct 'split' of glucose into lactic acid – a reaction slow enough to be purely fictive.

**5.** The yield of ATP in the back reaction could be improved through spatial separation of oxidant and reductant by a barrier, i.e., a membrane. Nowadays all cells have at least a plasma membrane. Quite plausibly the early fermenters, too, were separated by membranes from the aqueous medium, and the membranes were, then as now, the location of the mechanism for transport. Invaginations of the plasma membrane are now common, and the thylakoids of the photosynthetic bacteria surely originated as internal extensions of the plasma membrane.

It seems that a light-powered phosphorylating redox system combined with a necessarily asymmetric transporting membrane, in such a way that the products of the light reaction (oxidant and reductant) appeared on different sides of the membrane. According to Mitchell (1970)

> it is an evolutionarily attractive proposition that the proton-translocating oxidoreduction loop system and the reversible proton-translocating ATPase may have arisen separately... and that the accidental occurrence of both systems in the same cell may then have provided the means of storing the free energy of oxidoreduction in ATP....

Subsequently the membrane could be improved by stepwise insertion of compounds suitably spaced in respect to redox potential, including cytochromes, and by coupling electron flow along this chain to phosphorylation.

Gradually the biosphere ceased being reducing. Further evolution will not be discussed here, but clearly for the mobilisation of the abundant carbon in biospheric $CO_2$ additional reducing power was needed. NADH was probably provided first by ATP-powered noncyclic electron flow ('reversed flow'), and later by light-powered noncyclic electron flow with production of ATP. Increasingly, inorganics (sulphides, ferrous salts) were employed as additional sources of electrons (photolithotrophy). When these sources approached exhaustion, water itself was applied. This was achieved by plants, first by the blue-green algae, through the 2-quanta process. The relationship of bacterial and plant photosynthesis has, of course, been explained in the momentous work of Van Niel (see Van Niel, 1962). The mechanism of photosynthesis in higher plants seem to be essentially identical with that in algae, even blue-green algae. The plant's way of making metabolic energy might merit a separate name: (photo)phytotrophy.

**Note added in proof.** The question of the instability of sulphate in reducing conditions will be taken up again in a forthcoming book (E. Broda, *The Evolution of the Bioenergetic Processes*, Pergamon Press, Oxford). Reference will be made there to the important work of H. D. Holland.

## References

Arnon, D. I.: 1959, *Nature* **184**, 10.

Broda, E.: 1970, *Progr. Biophys. Mol. Biol.* **21**, 146.

Broda, E.: 1971, in R. Buvet and C. Ponnamperuma, (eds.), *Molecular Evolution*, Amsterdam 1971.

Cloud, P. E.: 1968, in Drake, E. T. (ed.), *Evolution and Environment*, New Haven 1968.

Cloud, P. E.: 1972, *Amer. J. Sci.* **272**, 537.

Gaffron, H.: 1965, in S. W. Fox, (ed.), *The Origin of Prebiological Systems and Their Molecular Matrices*, Academic Press, New York, pp. 437–455.

Granick, S.: 1965, in V. Bryson and H. J. Vogel, (eds.), *Evolving Genes and Proteins*, New York 1965.
Klein, R. M. and Cronquist, A.: 1967, *Quart. Rev. Biol.* **42**, 105.
Krasnovsky, A. A. and Voinovskaya, K. K.: 1952, *Dokl. Akad. Nauk S.S.S.R.* **67**, 325.
Mitchell, P.: 1970, in *Membranes of Cells and Organelles*, Symposium XX, Society for General Microbiology, London 1970.
Peck, H. D.: 1966, *Some Evolutionary Aspects of Inorganic Sulfur Metabolism*, Lecture, University of Maryland.
Postgate, J. R.: 1968, *Proc. Roy. Soc. B.* **171**, 67; also private communications.
Takahashi, H., Taniguchi, S., and Egami, F.: 1963, in: M. Florkin and H. Mason, (eds.), *Comparative Biochemistry*, New York 1960 etc., Vol. 3.
Van Niel, C. B.: 1962, *Ann. Rev. Plant Physiol.* **13**, 1.

# EVOLUTION OF PHOTOSYSTEMS OF PHOTO-SYNTHETIC ORGANISMS

N. V. KARAPETYAN

*A. N. Bakh Institute of Biochemistry, U.S.S.R. Academy of Sciences, Moscow, U.S.S.R.*

**Abstract.** It is generally accepted that two photosystems function successively in photosynthetic electron transport chain of plants and algae. The interaction of these photosystems results in the enhancement of photosynthesis. It was suggested that only one photosystem is present in purple bacteria, the most primitive photosynthetic organisms. The functioning of this photosystem is accompanied by absorption changes at 890 nm. Recently new spectral changes were found in *Chromatium* chromatophores under reductive conditions, more favorable for bacterial growth. Some of that spectral changes take place even at liquid nitrogen temperature. It is proposed these absorption changes could be related to other photosystem functioning in low potential region. Such a photosystem is necessary for reduction of NAD in *Chromatium*, for which the reverse electron transport to NAD was not shown. In contrast to photosystems of plants, the bacterial photosystems appear to function independently because the enhancement of bacterial photosynthesis is not found. Apparently the evolution of photosystems involved interaction between independent photosystems, one of them functioning under more oxidative conditions.

The data presented are concerned with light induced absorption changes in the near infrared region of the spectrum of isolated *Chromatium* chromatophores under reducing conditions – the most favourable for bacterial photosynthesis. The properties of these changes as well as their differences from absorption changes ascribed to P890 (bacteriochlorophyll in reaction centre) are considered. The possible mechanisms of NAD reduction in *Chromatium* are discussed. Two independently functioning photosystems in purple bacteria are suggested, one of them capable of reducing NAD. Apparently the evolution of photosystems involved the appearance of interactions between the independent photosystems through the electron transport chain. This would allow one of the photosystems to function in more oxidative conditions using water as the electron donor.

According to modern concepts the photosynthetic electron transport of green plants and algae includes two photoreactions driven by two pigment systems (Duysens, 1964). The functioning of one of them results in water decomposition and oxygen evolution, the other results in NAD photoreduction. Both photosystems act successively and are linked through the electron transport chain. The functioning of either photosystem is advantageous due to various light wave lengths since the photosystems have different bulk pigments. Simultaneous functioning of both photosystems enhances the rate of photosynthesis (Emerson effect). This phenomenon shows the interaction of photosystems at an electron transport chain level.

During light reactions of bacterial photosynthesis the substrate oxidation and NAD reduction take place (Hind and Olson, 1968; Frenkel, 1971). It was assumed that there is only one kind of photosystem in purple bacteria in contrast to plants and algae. Such an assumption is derived from the oxidation by the same reaction centre (P890) of both types of cytochromes, operating in cyclic and noncyclic electron transport chains (Parson and Case, 1970). Besides, the Emerson effect is not found for purple bacteria (Blinks and van Niel, 1963).

It is known that the illumination of purple bacteria results in the photoreduction of NAD (Hinkson, 1965). But isolated *Chromatium* chromatophores, in contrast to *R. rubrum* chromatophores, are incapable of photoreducing exogenous NAD (Vernon, 1963). Because the redox potential of NAD is lower than the potential of the primary electron acceptor from P890 there is a question of the mechanism of NAD reduction in purple bacteria. The NAD molecules could be reduced: (a) by reversed electron transport using ATP formed in cyclic photophosphorylation; (b) by molecular hydrogen evolved by bacteria; (c) in the course of photosynthetic electron transport. However the reversed electron pathway has not been shown for *Chromatium* and reduction of NAD by molecular hydrogen in bacteria is less efficient then the photoreduction of NAD (Buchanan and Bachofen, 1968). Therefore one of the possible pathways of NAD reduction in *Chromatium* chromatophores could be photochemical reduction.

We tried to find for *Chromatium* chromatophores the presence of a photosystem which functions at a lower redox potential region than P890. Because the operation of photochemically active chlorophyll as a rule is accompanied by changes in the absorption spectrum, we studied the light-induced absorption changes in the near infrared region of *Chromatium* chromatophores under reductive conditions – the most favourable for bacterial photosynthesis. The apparatus for measuring difference spectra (light-dark) was described earlier (Karapetyan, 1969).

Cusanovich *et al.* (1968) and Karapetyan (1969) have independently found light-induced absorption increases at 865 and 905 nm for *Chromatium* chromatophores in reducing conditions. These changes in comparison with the changes at 890 nm show some differences. They are observed in a definite region of redox potential, they are saturated at lower light intensities and are more thermolabile. They are also capable of disappearing at low temperatures. Positive absorption changes are more apparent under conditions of phototransformations of low potential cytochrome C-552, functioning in a noncyclic electron transport chain. A similar dependence of light-induced changes at 865 nm and the slowly reversible dark changes at 423 nm, from both a redox potential and light intensity point of view, could indicate a connection between the changes at 865 and 905 nm in noncyclic electron transport (Karapetyan *et al.*, 1973).

There is no doubt that the changes at 865 and 905 nm are due to phototransformations of bacteriochlorophyll molecules. However the nature of these phototransformations is not yet elucidated fully. These changes could be caused: (a) by a spectral shift as a result of conformational changes of the protein carrier of the pigment complex; (b) by a spectral shift of pigment absorption as a result of the action of an electric field originating in the course of electron transport; (c) by redox transformations of the bacteriochlorophyll molecule. In our experiments the insensitivity of the changes at 865 and 905 nm to the action of a denaturating agent (6 M urea) was shown. These changes are not at all due to the conformational changes of the protein carrier (Karapetyan and Krakhmaleva, 1973; Karapetyan *et al.*, 1973). The probability of the electrochromic nature of the spectral shift for *Chromatium* chromatophores would require confirmation.

It was possible to suggest that these changes are due to the phototransformations of bacteriochlorophyll in the reaction centre of a low potential photosystem. However such a suggestion is unacceptable because the changes at 865 and 905 nm disappear at low temperatures (Karapetyan *et al.*, 1973) and furthermore the quantum yield of this shift for *Chromatium* bacteria was calculated to be 10–15 (Vredenberg and Amesz, 1966). The rise time of changes at 865 nm can not be detected because of the intense fluorescence occurring in the same spectral region.

Hence it is difficult at present to interpret with certainty the nature of the processes resulting in the appearance of light-induced changes at 865 and 905 nm. However, the question still remains regarding the presence of two photosystems in *Chromatium*. Attempts to find the low potential photosystem in purple bacteria have been made in other laboratories (Sybesma and Fowler, 1968; Schmidt and Kamen, 1971; Seibert and Devault, 1971). Most interesting are the data obtained by Seibert and Devault (1971) on the light induced absorption changes at 424 nm, which are observed under reducing conditions ($E_{m,7} = -320$ mv) even at $-196\,°C$; the half-rise time of these changes is 75 nsec and the quantum yield is $\approx 1$. Unfortunately the lack of similar absorption changes in the near infrared region do not allow one to ascribe the changes at 424 nm to phototransformations of bacteriochlorophyll. Nevertheless these data are indicative of the possible presence of a low potential photosystem capable of reducing NAD in *Chromatium* chromatophores.

Since we can assume two photosystems for purple bacteria. The question naturally arises as to what possible interaction would take place between them. However the lack of data on action spectra of absorption changes, ascribed to different photosystems, does not allow us to answer this question. The lack of an Emerson effect for bacteria could be explained either by similar bulk pigments in both photosystems or by independent functioning of these photosystems even in the case of various forms of bulk pigments. The preferential operation of either of these photosystems could depend on growth conditions, their age and other factors (Cusanovich *et al.*, 1968; de Klerk *et al.*, 1969).

The possible pathways of evolution of pigment systems and the components of the photosynthetic electron transport chain for photosynthesizing organisms were discussed earlier (Karapetyan, 1971). One of the pathways of the evolution of photosystems could include the appearance of a successive interaction between independently functioning photosystems which are responsible for various photo-processes. It appears that the presence of a successive interaction of two photosystems allowed one of them to operate under higher oxidative conditions eventually using as the electron donor, water which is readily available.

## References

Blinks, L. R. and van Niel, C. B.: 1963, in *Studies on Microalgae and Photosynthetic Bacteria*, Tokyo Press, p. 297.
Buchanan, B. B. and Bachofen, R.: 1968, *Biochim. Biophys. Acta* **162**, 607.
Cusanovich, M. A., Bartsch, R. G., and Kamen, M. D.: 1968, *Biochim. Biophys. Acta* **153**, 397.
Duysens, L. N. M.: 1964, in *Progress in Biophysics*, Pergamon Press, **14**, 1.
Frenkel, A. W.: 1970, *Biol. Rev.* **45**, 569.
Hind, C. and Olson, J. M.: 1964, *Ann. Rev. Plant Physiol.* **19**, 249.
Hinkson, J. W.: 1965, *Arch. Biochem. Biophys.* **112**, 473.
Karapetyan, N. V.: 1969, in *Progress in Photosynthesis Research*, Tubingen, **1**, 778.
Karapetyan, N. V.: 1971, in *Chemical Evolution and the Origin of Life*, North Holland Publishing Company, p. 307.
Karapetyan, N. V. and Krakhmaleva, I. N.: 1973, in *Problems of Photobiochemistry*, Moscow University, p. 210.
Karapetyan, N. V., Krakhmaleva, I. N., and Krasnovsky, A. A.: 1973, *Molec. Biol. (U.S.S.R.)* **7**, 868.
de Klerk, H., Govindjee, Kamen, M. D., Lavorel, J.: 1969, *Proc. Nat . Acad. Sci. US* **62**, 972.
Parson, W. W. and Case, G. D.: 1970, *Biochim. Biophys. Acta* **205**, 232.
Schmidt, G. L. and Kamen, M. D.: 1971, *Biochim. Biophys. Acta* **234**, 70.
Seibert, M. and Devault, D.: 1968, *Biochim. Biophys. Acta* **61**, 1343.
Sybesma, C. and Fowler, C. F.: 1968, *Proc. Nat. Acad. Sci. U.S.* **61**, 1343.
Vernon, L. P.: 1963, in *Bacterial Photosynthesis*, Antioch Press, p. 235.
Vredenberg, W. J. and Amesz, J.: 1966, *Biochim Biophys. Acta* **126**, 244.

# EVOLUTION OF OXYGEN BY PLANTS IN RELATION TO BIOSPHERE EVOLUTION

V. M. KUTYURIN

*V. I. Vernadsky Geochemistry Institute, Academy of Science, Moscow, U.S.S.R.*

**Abstract.** The process of water decomposition by plants is discussed in connection with biosphere evolution. This process consists of two parts: water oxidation and oxygen evolution. The origin of the water oxidation process took place after the synthesis of chlorine-type pigments, the structure of which corresponds to a more oxidized state than the bacteriochlorophyll type. The ability of plants to evolve oxygen is the result of a long evolution process. The capability of decomposing water in the long-wave length spectral region by algae and higher plants, which can be only seen under anaerobic conditions was discovered. This mechanism was suggested to be a reflection of a relict form of plant apparatus having operated under ancient, strictly anaerobic, conditions.

The appearance of blue-green algae on the Earth was one of the most important events of its history. At this time (approximately three billion years ago) the evolution of photosynthetic plants with the capability to decompose water and evolve oxygen took place. The oxygen production by plants caused the development of the ozone screen and of aerobic metabolism. The accumulation of oxygen in the atmosphere greatly changed the character of geochemical processes on the earth's surface.

The data of Table I show the amount of oxygen being evolved by plants per year in comparison with the amount in several natural sources.

TABLE I

The content of oxygen in several natural sources and oxygen production by plants (per year)

| Source | | Oxygen content (tons) |
|---|---|---|
| Atmosphere | | $1 \times 10^{15}$ |
| Hydrosphere | Oxygen dissolved in: | |
| | fresh water | $2 \times 10^{14}$ |
| | sea water | $1 \times 10^{18}$ |
| Sedimentary rocks | | $1.5 \times 10^{15}$ |
| Photosynthesis | | $1 \times 10^{11}$ |

If we assume the same order of oxygen production by plants during the last few million years, we come to the conclusion that all the hydrosphere water undergoes decomposition through photosynthesis during a period of about ten million years.

To connect the origin of water decomposition by plants and its evolution with certain steps of biosphere evolution we must know the mechanism of water decomposition and oxygen evolution. Unfortunately this mechanism belongs to the least developed problem of photosynthesis. Only in 1941 was it shown that photosynthetic oxygen originates from water and not from carbon dioxide (Vinogradov and Teiss, 1941; Ruben *et al.*, 1941). Systematic investigation of the mechanism of water decomposition and oxygen evolution began only ten years ago. A scheme describing this process is shown in Figure 1. It consists of two parts: the first part represents

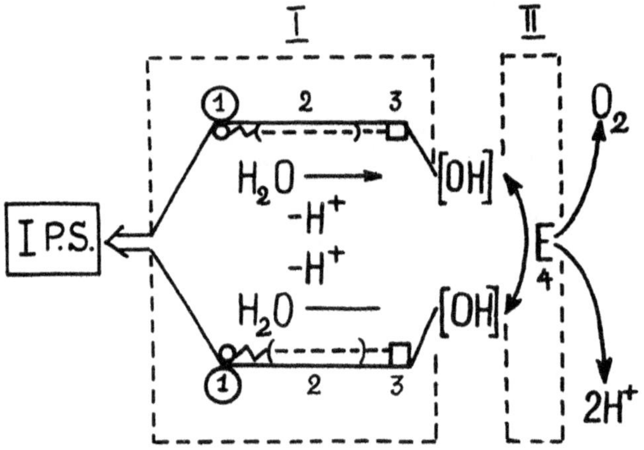


Fig. 1. Scheme for the water decomposition process by plants. I P.S. – the first photochemical system of chloroplasts. I – water oxidation mechanism; II – oxygen formation mechanism; 1 – pool of chlorophyll molecules transferring light energy to photochemical centre, whose mechanism is shown in Figure 2a; 2 – structure-bonding water as shown in Figure 2b; 3 – oxidation-reduction system containing metal ion Me, perhaps manganese, and conjugated to the oxidized form of the chlorophyll as shown in Figure 2c; 4 – unknown enzyme system E evolving oxygen: system 3 may be enclosed in a system 4.

water oxidation, the second one shows oxygen formation. As was shown by Kutyurin *et al.* (1969), water oxidation occurs under special conditions without oxygen evolution. The latter process is more sensitive to the action of detergents than the former, showing its more intimate dependence on chloroplast structure.

Water oxidation may be supposed to be the first stage of the water decomposition process. It only began after the synthesis of chlorine-like pigment such as chlorophylls a and b. Bacterial pigments are not able to oxidize the water molecule, as shown in Table II. Its oxidized forms do not have a sufficiently high potential ($E_0 \geqslant +0.8$ v). (Kutyurin *et al.*, 1968a).

It is interesting to note that this high potential form of oxidized chlorophyll

TABLE II

Oxidation-reduction potentials of chlorophylls and related substances

| Substance | $E_0^{\mathrm{I}\,a}$ | $E_0^{\mathrm{II}}$ | Solvent |
|---|---|---|---|
| Chlorophyll a | 0.77 | 0.90 | Acetone |
| Chlorophyll a | 0.79 | 0.93 | Acetonitrile |
| Chlorophyll b | 0.80 | 0.93 | Acetone |
| Mn-protoporphyrin | 0.85 | – | Phosphate buffer pH 7.5 |
| Chlorobium Chlorophyll | 0.62 | 0.74 | Acetone |
| Chlorobium Chlorophyll | 0.62 | 0.72 | Acetonitrile |
| Bacteriochlorophyll | 0.62 | [b] | Acetone |

[a] $E_0^{\mathrm{I}}$ corresponds to the normal oxidation-reduction potential of the system $chl^+/chl$ and $E_0^{\mathrm{II}}$ to the system $chl^{++}/chl^+$ at pH 7.0.

[b] The transfer of the second electron was not completely reversible under experimental conditions.

forms occurs in lipophilic media or through complex formation with basic amino-acids. This indicates that the formation of a water oxidation mechanism by plants was the result of special binding of chlorophyll with lipoprotein in photosynthetic lamellae. Since the chlorine-type pigments have higher oxidation potentials than the bacterial one, it is clear either that its formation might be a later step of the biosphere evolution than the appearance of photosynthetic bacteria or that there were some regions on Earth with sharply different oxidation-reduction conditions.

It should be emphasized that there are only two possible mechanisms of oxygen formation from the phenomenological point of view:

① $4\,H_2O \xrightarrow{-4e} 4\cdot OH + 4\,H^+$

$4\cdot OH \longrightarrow 2\,H_2O + O_2$

② $2\,H_2O \xrightarrow{-2e} 2\cdot OH + 2\,H^+$

$2\cdot OH \xrightarrow{-2e} 2\cdot O\cdot + 2\,H^+$

$2\cdot O\cdot \longrightarrow O_2$

So the oxygen molecule formation is either the result of dismutation of 4 hydroxyl radicals or the twofold oxidation of an oxygen atom firstly in the form of a water molecule and secondly in the form of the hydroxyl radical (or hydroxyl group connected with a metal ion). The hydroxyl radical is a substance of extremely high reactivity; it exists in a free state for only $10^{-10}$–$10^{-9}$ s. Therefore plants must have a mechanism for collecting and transforming of hydroxyl radicals. We have supposed that this mechanism is connected with the mode of water oxidation which,

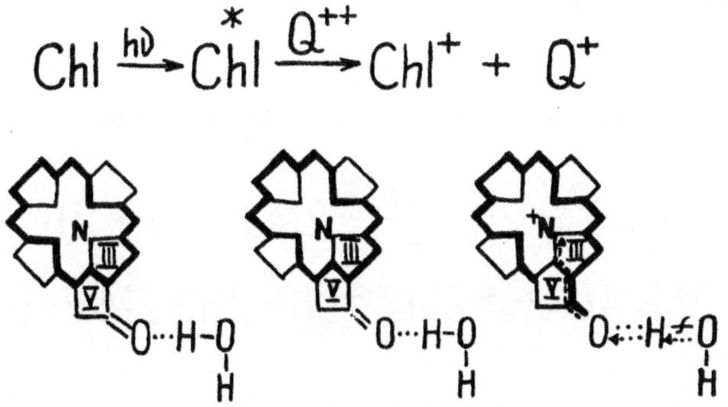


Fig. 2a. A scheme proposed to show the possibility of water oxidation by the oxidized form of chlorophyll. The upper line shows the completion of the conjugated chlorophyll system on the carbocyclic ring when the pigment is oxidized. Chl – chlorophyll, Q – acceptor of electron.

in its turn, is able to separate the reduced form (basic state of chlorophyll) from an oxidized substance (hydroxyl radical) as is shown in Figure 2a. This first step of organization is based on the property of the chlorophyll molecule to attach the water molecule to its carbocyclic ring (Kutyurin, 1972), a hypothesis that was experimentally confirmed in our laboratory.

The second step – a conservation and collection of hydroxyl radicals would be realized by a connection with a metal ion in the manner shown in Figure 2b. This

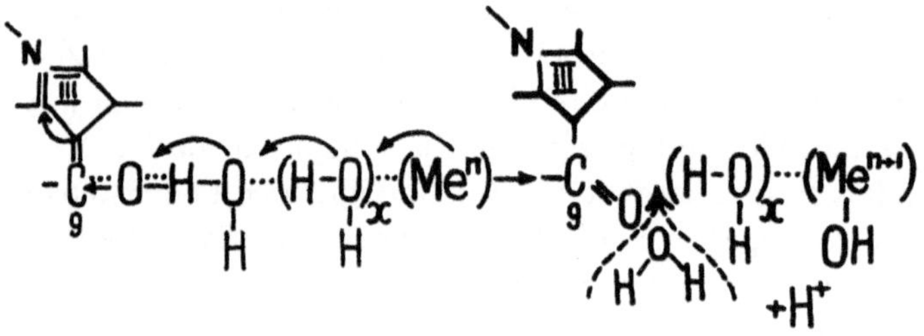

Fig. 2b. Conjugate oxidation of a metal ion linked with the water chain as a result of the primary oxidation of a water molecule by chlorophyll.

mechanism explains the simultaneous spatial partition and cooperation of oxidative action and the result of it. According to experimental results obtained by Kessler (1960) the manganese ion must play an important role in the overall processes leading to oxygen evolution. That is, the manganese ion may be a participant of the mechanism shown above. We did not succeed in finding the binding form of manganese – it occurred neither in porphyrins nor in structural proteins. Manganese may be supposed to be an essential structural factor itself in addition to its participation in the chemical reaction as suggested in Figure 2c (Kutyurin, 1970).

This hypothesis is based on experimental results obtained in Gaffron's laboratory (Homann *et al.*, 1968) which have shown the necessity of two lamellae layers touching, or of a spatial arrangement between them in order to evolve photosynthetic oxygen.

If we assume oxygen evolution to be the result of twofold oxidation of each oxygen atom originating from water we shall expect the participation of two different chlorophyll forms in the mechanism of water decomposition. To investigate this

$$2H_2O \xrightarrow[-e]{Chl_1^+} 2[\cdot OH] \xrightarrow{Me^n} 2Me^{n+1}\text{–}OH + 2H^+$$

$$2Me^{n+1}\text{–}OH \xrightarrow[-e]{Chl_2^+} 2Me^{n+2}\text{–}OH \rightarrow 2H^+ + 2Me^n + O_2$$

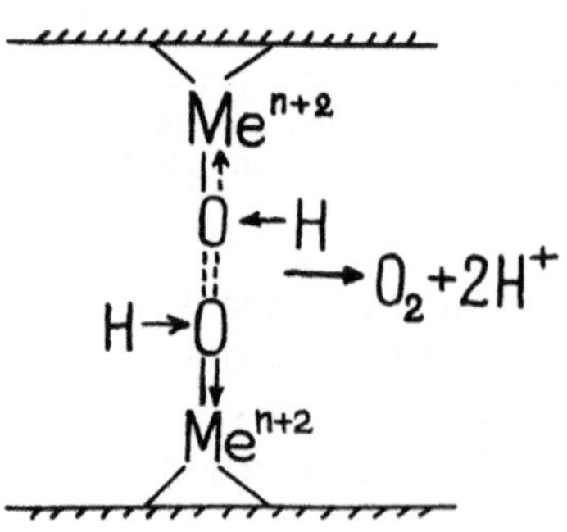


Fig. 2c. Reaction scheme of two-fold oxidation of oxygen atom originating from water and random cooperation of two lipoprotein lamellae with metal ions.

problem we have paid special attention to the oxygen evolution by plants in the long-wave spectral region (λ ⩾ 700 nm). As was shown by Kutyurin *et al.* (1968b) under anaerobic conditions there is a special mechanism of water decomposition in the long-wavelength spectral region in algae which is masked by oxygen uptake under the usual aerobic conditions, as shown in Figures 3a, b. Recently we have confirmed this ability in higher plants (Kutyurin *et al.*, 1973). The sensitivity of this mechanism to a special poison monuron, is greater than that in the short-wavelength spectral region, as is shown in Figure 3c. In searching for the second photochemical

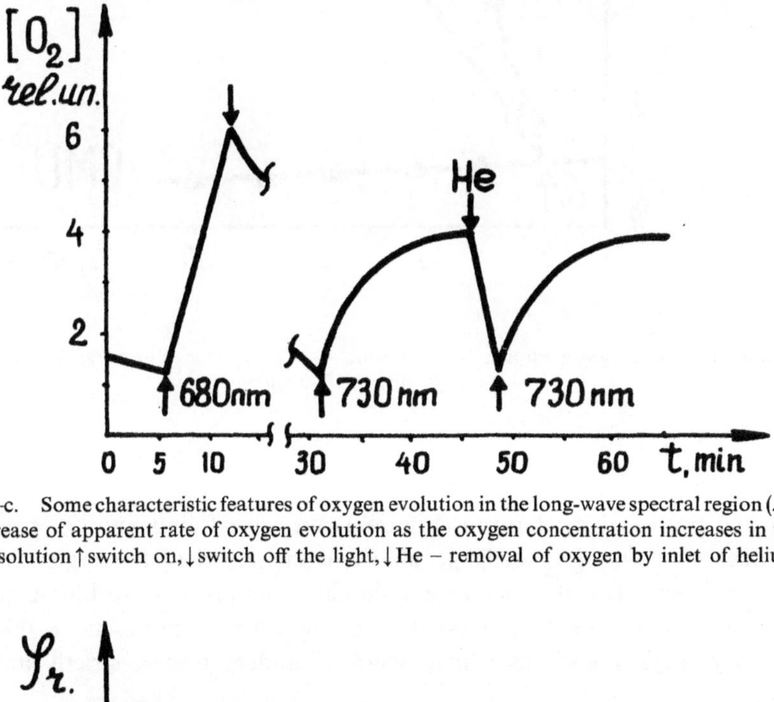


Figs. 3a–c. Some characteristic features of oxygen evolution in the long-wave spectral region (λ ⩾ 700 nm). (a) Decrease of apparent rate of oxygen evolution as the oxygen concentration increases in the nutrient solution ↑ switch on, ↓ switch off the light, ↓ He – removal of oxygen by inlet of helium.

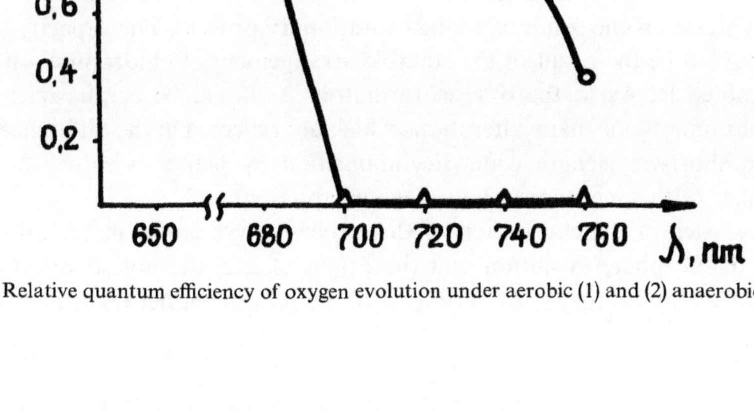


Fig. 3b. Relative quantum efficiency of oxygen evolution under aerobic (1) and (2) anaerobic conditions.

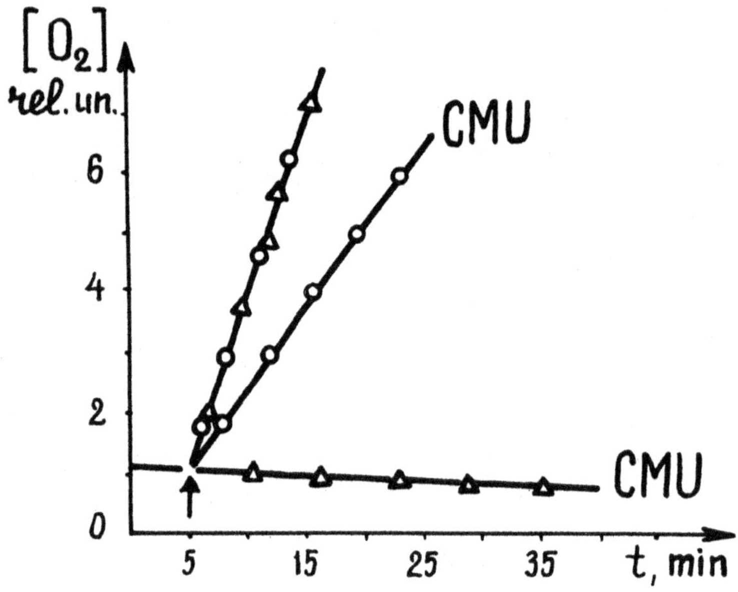


Fig. 3c. Inhibition of oxygen evolution by monuron: -○-○-○- monochromatic light of 680 nm; -△-△-△- light of 730 nm.

centre in the proposed mechanism (Figure 1) we have found another mechanism of water decomposition. Its specific peculiarity is the sensitivity to oxygen concentration in the medium. It is to some extent similar in its property to blue-green algae and this makes it possible to suggest that this mechanism represents a relict form of a photosynthetic apparatus which operated under ancient, strictly anaerobic, conditions.

Space does not permit us to go into detail about the possible correlation between the chloroplast structure of plants belonging to different systematic groups and variability of its oxygen metabolism, as there are few unambiguous facts to be discussed. It may only be said that the mechanism of water decomposition and oxygen evolution is the result of a long evolutionary process. The capacity to oxidize water seems to be the result of the suitable arrangement of chlorophyll on the lipoprotein molecule. As to the oxygen formation it should be emphasized that this process has undergone many alterations which are reflected in the difference between long- and short-wavelength water decomposition by plants as shown above. The first outlines of this mechanism have become apparent.

The next step in the elucidation of this process gives useful information for the problem of biosphere evolution and the origin of life, the investigation of which was greatly advanced thanks to the efforts of the prominent heroes of our Assembly.

## References

Homann, P. H. *et al.*: 1968, *Comp. Biochem. and Biophys. of Photosynthesis*, Tokyo, p. 50.
Kessler, E.: 1960, *Handb. Pflanzenphysiol.* **5**, 923.
Kutyurin, V. M.: 1970, *Izvest. Akad. Nauk SSSR*, ser. biol., **N 4**, 569.
Kutyurin, V. M.: 1972, *Proceed. IInd Internat. Congr. on Photosynth. Research*, **I**, The Hague, p. 93.
Kutyurin, V. M. *et al.*: 1968a, *Dokl. Acad. Nauk SSSR* **171**, 215.
Kutyurin, V. M. *et al.*: 1968b, *Dokl. Acad. Nauk SSSR* **181**, 1270.
Kutyurin, V. M. *et al.*: 1969, *Fiziol. Rast.* **16**, 181.
Kutyurin, V. M. *et al.*: 1973, *Fiziol. Rast.* in press.
Ruben, S. *et al.*: 1941, *J. Amer. Chem. Soc.* **63**, 877.
Vinogradov, A. P. and Teiss, R. V.: 1941, *Dokl. Akad. Nauk SSSR* **33**, 497.

# AMBIGUITY IN THE INTERPRETATION OF ABIOTIC SYNTHESES

A. G. CAIRNS-SMITH

*Chemistry Department, University of Glasgow, Glasgow G12 8QQ, Scotland*

**Abstract.** The main conclusion from abiotic syntheses is that modern organisms prefer molecular units that are easy to make. This could be the result of selection pressures during very early evolution rather than a reflection of the large scale molecular environment within which life first arose. As previously discussed (Cairns-Smith, 1971) there seems to be no good reason to limit our thinking on initial hereditary systems to those with chemical similarity to modern forms.

We should seriously consider the idea that radical design changes took place during the very early evolution of our fundamental biochemical machinery. I do not mean simply that a different set of amino acids might have been used, or that the code was simpler, or that the early systems were a more disorganised version of the present one. I mean that the whole idea of using nucleic acid and protein was probably itself the outcome of a protracted Darwinian process.

Modern genes and enzymes are not only machines, they are advanced machines. They depend for their construction and conditions of operation on a highly sophisticated ongoing technology: they are designed to work in the highly informed environment of the modern cell. It is far from clear that remotely similar structures would have been optimal designs under the most primitive conditions A naive student of our technology might expect to find that the Babylonians had used primitive forms of magnetic tape to record their astronomical data. It is indeed possible that some kind of magnetic tape could have been made: the point is that in the absence of sophisticated ancillary equipment it would have been an inappropriate way of attempting to store information – clay tablets were much more *efficient*. I strongly suspect that early life forms were in a similar position: in the absence of the elaborate matter, energy and information handling subsystems of modern cells quite different structures from nucleic acid and protein would have been the most *efficient* as the stores and executants of primitive hereditary information.

In earlier speculations on the origin of life arguments of this kind were put forward (Pirie, 1951), but since 1953 the opposing 'idea of original biochemical similarity' has almost become a dogma. Is there really overwhelming support for this view? I think not. In particular I suggest that the considerable literature on abiotic syntheses, following Miller's justly famous experiment, does not argue against early fundamental changes in molecular biology.

There are at least three quite plausible explanations for the well founded conclusion that modern organisms show a preference for the kinds of molecules that often turn up under only loosely controlled conditions.

First, there is the now conventional view that the origin of life followed on a large scale pre-biotic stockpiling of molecules and that the substances that happened to be favoured in this process are, as it were, still with us as dominant components of our biochemistry. It is this view that I question.

Second, there is the possibility that early organisms, of some unknown biochemistry, evolved in such a way as to be able to use prevalent abiotic foods.

Third, there is an explanation that does not even require that abiotic syntheses of modern biochemicals took place on the primitive Earth. Easily made molecules are a good idea for a biochemistry anyway. The kinds of molecules that tend to turn up under loosely controlled conditions are, of course, the kinds of molecules that on the whole require minimal control in their syntheses. Such will generally require fewer and simpler catalysts and, for a life system that incorporates them, less genetic information. Such life systems could thus be expected to have a selective advantage. Molecules that are chemically accessible under given conditions will also be, on the whole, genetically cheap under these conditions. If early evolution was flexible enough to allow experiments at the most fundamental biochemical levels this is the kind of set of molecules that you would expect to end up with. What the first Darwinian systems started with would have depended on quite different considerations. Here the problem was to get a hereditary machine off the ground in some way without the pre-existence of the kind of teleonomic organisation that only hereditary machinery can give rise to.

I have proposed that the most likely spontaneously forming replicating apparatus would have been found in the growth processes of inorganic crystals the primitive hereditary information having been written, not as a monomer sequence in a polymer, but as replicable patterns of physical or chemical defects of the kinds that are common in mineral crystals – layer silicates being particularly promising candidates for such primitive genes (Cairns-Smith, 1966, 1971). In this conference Weiss has presented experimental evidence for such printing mechanisms during the synthesis of mica-type layer silicates (Weiss, 1974). We can thus envisage a crystal based life form with no initial requirements for particular organic molecules. Even a $CO_2/CO$ atmosphere might, on this view, have provided a suitable carbon source for silicate based organisms. This is suggested by the results reported in this conference by Hubbard *et al.* (1974) and previously by Hubbard *et al.* (1971). The abiotic experiments of the more traditional kind, in which modern biochemicals are produced under reducing conditions, may be giving us information not about the conditions on earth when the first life forms arose, but about (reducing) conditions generated by the phenotypes of the organisms within which our present life system was eventually perfected.

Apart from the abiotic syntheses, what other reasons are there for the very widely made assumption of original biochemical similarity? There are probably four main ones.

There is *the unity of present day biochemistry*. It is commonly noted that such detailed agreement on biochemical tactics among present day organisms points to a common ancestor. This is indeed so, but it does not say anything about how far from the origin of life the last common ancestor was. Considering its evident sophistication there can be little doubt that it was the product of prolonged Darwinian evolution – that it was far from the origin of life.

Notice particularly that at least as early as this last common ancestor the central biochemical machinery had virtually stopped evolving (otherwise subsequent changes would have destroyed the unity of biochemistry). Evidently for a very long time our system has been unable to tolerate mutations affecting significantly its fundamental machinery: but this can not always have been so: there must have been some period when Darwinian forces moulded the great ingenuities inherent in our system. We have however, no clear view of the situation within which the fundamental biochemistry even could have evolved (Cairns-Smith and Walker, 1974). All that we can say is that the evolutionary processes that are still visible – those concerned with the higher levels of biological organisation – are strongly branching: here there are still great varieties of design available for study. As Darwin noted variety tends to be greatest in the most recently introduced features of organisms. His explanation was not that variety is a new idea but that extinctions have reduced its visible effects on structures that were selected for in the distant past. Now a common argument in favour of a straight line early evolution is that such is the simplest hypothesis – and so is dictated by Ockham's Razor. If Ockham's Razor is appropriate at all to evolutionary questions it should suggest that what we now see to be a complex, richly experimental and devious process was always so. Ockham's Razor is about hypotheses, not processes, and the simplest hypothesis is that the evolutionary process has always been complex.

It often seems to be assumed that the genetic material is too fundamental to change significantly during evolution. We have tried to show that a mechanism for 'genetic metamorphosis' would exist for organisms primitive enough to be still actively evolving their fundamental biochemistry (Cairns-Smith, 1966, 1971) and that this mechanism would actually favour radical changes (Cairns-Smith, 1968; Cairns-Smith and Walker, 1974).

Finally there is the feeling that early radical changes in biochemistry would put the origin of life beyond rational experiments. Indeed the historical question of exactly when, where and how life originated becomes even more remote than it already is: but the possibility of laboratory synthesis of early life forms becomes more immediate. If our view is correct there exists at least one system, much simpler than the present one, that can evolve *in extenso* under natural selection. Such a system would be highly accessible to experiments – the kinds of experiments reported by Weiss (1974) for the first time at this conference.

## References


Cairns-Smith, A. G.: 1966, *J. Theor. Biol.* **10**, 53.

Cairns-Smith, A. G.: 1968, in C. H. Waddington, (ed.), *Towards a Theoretical Biology*, Vol. I, Edinburgh University Press, Edinburgh, p. 57.

Cairns-Smith, A. G.: 1971, *The Life Puzzle*, Oliver and Boyd, Edinburgh.

Cairns-Smith, A. G. and Walker, G. L.: 1974, *Biosystems*, in press.

Hubbard, J. S., Hardy, J. P., and Horowitz, N. H.: 1971, *Proc. Nat. Acad. Sci. U.S.A.* **68**, 574.

Hubbard, J. S., Voecks, G. E., and Golub, E. E.: 1974, Paper presented at the 4th International Conference on the Origin of Life, Barcelona, 1973.

Pirie, N. W.: 1951, in J. D. Bernal (ed.), *The Physical Basis of Life*, Routledge and Kegan Paul, London, p. 71.

Weiss, A.: 1974, *Naturwiss.* in press.

# ON THE QUESTION OF THE ORIGIN AND EVOLUTION OF THE GENETIC SYSTEM

VLADIMÍR NOVÁK and VLASTIMIL LIEBL

*Prebiology Group, Institute of Microbiology, Czechoslovak Academy of Sciences, Praha, Czechoslovaka*

In the present paper we have attempted a Darwinian approach to the question of the origin of heredity and its mechanisms in connection with the problem of the origin of life. Our conclusions are based on Oparin's coacervate theory of evolutionary abiogenesis, on the sociability theory propounded by one of ourselves (Novák, 1967), and on our ideas of the 'coacervate in coacervate' hypothesis, discussed at the first ISSOL meeting in Varna (Novák and Liebl, 1971). We have further tried to utilize data on the structure and behavior of contemporaneous subcellular organisms (viruses, etc.) and analogies with the evolution of organisms at phylogenetically higher levels (from unicellular organisms to human society) and experimental evidence obtained from various model systems of individual components of living matter and similar substances.

According to our view the sequence of the bases in the primordial nucleic acid molecules did not originate as a sudden leap due to chance, but was the outcome of regular and progressive evolution at a given level of chemogenesis that occurred everywhere the necessary chemicals and environmental conditions prevail. Even if the primary abiogenetic proteins originated without a nucleic acid matrix, they presumably had a regular, non-fortuitous base, although they were no doubt far less constant than recent, biogenic proteins. Primary nucleic acid molecules undoubtedly originated where these proteins accumulated in the presence of further necessary materials and other conducive conditions. Both these syntheses, that is to say of proteins and nucleic acids, had to be the most probable chemical reactions under the given conditions, determined by the molecular structure of the relevant substances, although at present we know nothing about their thermodynamic balance.

It is feasible that the stimulus for replication of this abiogenic NA molecule was the only step needed to start the actual evolution of life. It was no doubt enzymatic action of one or perhaps several primary proteins, a 'presynthetase' which probably had enzymatic action far less effective and much slower than the enzymatic activities of present-day proteins. From that moment onward, natural selection came into play by preserving those replicating molecules which were more stable and were more readily reproduced in the given environment. The sequence of their bases thus became a primary genetic code, which afterwards underwent successive changes as a result of individual mutations owing to further selection.

The original abiogenic proteins in which NA-replication first began, preexisted in the form of coacervates. These became the means of accumulation of the low molecular weight substances needed for the synthesis. The replicating nucleic acid molecules spread in these primary coacervates, irrespective of whether they accrued in the form of drops or as continuous layers, and produced their own specific protein as deter-

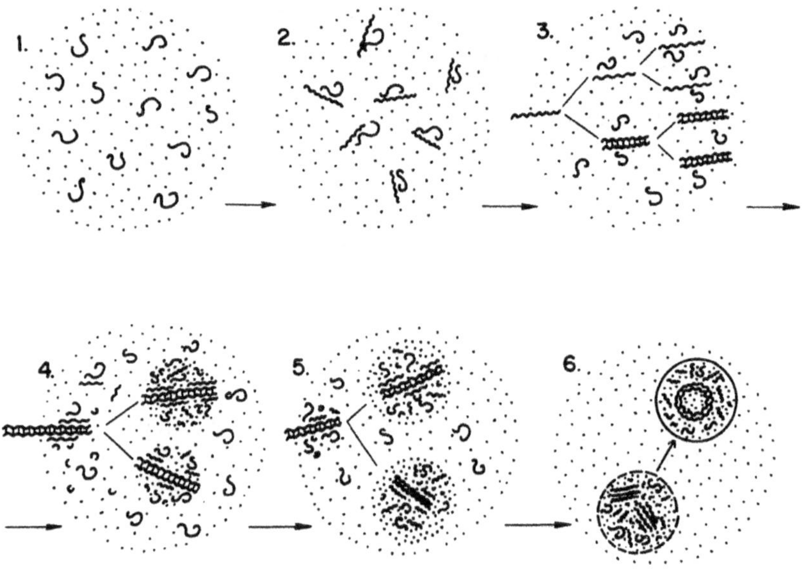


Fig. 1. Six gradual stages of the evolution from the self replicating molecule of nucleic acid to the first primitive cell. (1) – a primary coacervate of protein and low molecular weight substances. (2) – in connection with protein molecules individual single helix nucleic acid molecules originate. (3) – origin of the double helix DNA molecules. (4) – around each double helix DNA molecule a secondary coacervate originates. (5) – the secondary hereditary coacervate divides with each DNA molecule replication being composed of the products of its metabolic activity: transcribed RNA, proteins and low molecular weight substances. (6) – on the surface of the secondary coacervate a surface membrane originates which is later transformed into the cell-membrane of a primitive bacterial cell.

mined by their base sequence. The new protein remained in contact with the reproducing NA molecules on the basis of the same physicochemical forces as the original, primary coacervate, thereby forming new, secondary coacervates in the primary ones, i.e. coacervates in coacervates. As distinct from the primary coacervates, the structure and other features of such secondary coacervates were wholly determined by the sequence of bases of the corresponding NA molecule, i.e. by its genetic code.

Although we do not know the structure of the relevant NA molecule, it may be presumed that it was originally monohelical. It was probably composed of a much shorter chain of bases than the present ones known; it may have resembled recent transfer-RNA as suggested by Ohno (1970). It was likely to have been of a DNA type which developed the characteristic double-helix structure only at later stage, however. Its replication was probably catalyzed by the abiogenic protein of the primary coacervate which had thus the function of a presynthetas. The protein molecules produced (transcribed) might not have been much different from the pre-existing protein or proteins, but they were not, of course, identical with them.

It was these hereditary 'coacervates in coacervate' which underwent further evolution through successive mutations of the genetic code of their DNA molecules. They constitute the necessary intermediate step in evolution from the primary abiogenic coacervates to primary unicellular organisms (with surface membranes) at the level of the phylogenetically most primitive recent bacteria. The structure of various recent viruses and related organisms, from tobacco mosaic virus, via influenza and papatachi virus bodies, bacteriophages, and mycoplasms to L-forms of bacteria and to ricketsiae, shows us the stages through which this evolution probably passed.

As evolution progressed, the structure of the genetic code gradually changed and improved. Single strand-DNA molecule was converted to double-strand, messenger RNA was differentiated and in time transfer RNA and ribosomes developed (see Ohno, 1970; Novák and Liebl, 1971). That was only possible after the formation of a semipermeable membrane on the surface of the secondary coacervate and this was in turn determined genetically by mutations of the DNA molecule. Further differentiation and the more complicated evolution of the future cytoplasm took place within the membrane. Here, various tertiary coacervates, or supercoacervates, originated and formed the starting point for the development of cell organelles, such as ribosomes, chromosomes, the Golgi apparatus, different types of tubules and vesicles, etc.

Short as it is, we believe that this brief note shows that all the pieces in the picture of the gradual origin of living organisms on our planet are, bit by bit, falling into place and that it will not be long before science is able to reproduce any of these processes under laboratory conditions.

## References

Crick, P. H. C.: 1968, *J. Mol. Biol.* **38**, 367.
Novák, V. J. A.: 1967, *Žurn. Obšč. Biol.* **28**, 387.
Novák, V. J. A. and Liebl, V.: 1971, *International Symposium on Origin of Life and Evolutionary Biochemistry*, Abstracts, p. 14, Varna, Bulharsko.
Ohno, S.: 1970, *Evolution by Gene Duplication*, Berlin and New York.
Orgel, L. E.: 1968, *J. Mol. Biol.* **38**, 381.
Woese, C. R.: 1967, *The Genetic Code*, New York.
Woese, C. R.: 1972, *Exobiology*, N.Holland Publ. Co., Amsterdam-London, pp. 301–339.

# A MODEL FOR THE COEVOLUTION OF THE GENETIC CODE AND THE PROCESS OF PROTEIN SYNTHESIS: REVIEW AND ASSESSMENT

JAMES C. LACEY, JR., ARTHUR L. WEBER and WILLIAM E. WHITE, JR.

*Laboratory of Molecular Biology, University of Alabama School of Medicine, University Station, Birmingham, Ala. 35294, U.S.A.*

**Abstract.** The contemporary genetic code and the process of protein biosynthesis most assuredly evolved from a simpler code and process. We believe that there was obligatory coevolution of the two and that the earlier code and process must have involved a more direct linkage between the amino acids and the informational macromolecule. We propose that an early form of translating existed in which amino acids were attached directly to the 'messenger' RNA along the backbone as 2′OH aminoacyl esters. These esters then condensed with each other on the RNA backbone yielding a peptide covalently attached to the RNA, without the use of tRNAs and ribosomes. This presentation is concerned with experimental data which indicate that such a simple translation system is possible and must have involved the following steps: (1) formation of the aminoacyl adenylate anhydride, (2) transfer of the amino acid from the adenylate to imidazole, (3) transfer of the amino acid from imidazole to 2′OH groups along the backbone of RNAs, (4) condensation of the amino acids to yield peptides. Steps (1)–(3) have been confirmed in chemical systems. Our preliminary evidence indicates step (4) is also possible. The aminoacylation of polyribonucleotides and the subsequent formation of peptides is a dynamic and experimentally accessible system for studying genetic coding specifities and our present studies are now concentrated on step (4), looking for such specifities.

## 1. Introduction

Although the coding relationship between nucleic acids and proteins is well known and many details of the process of protein biosynthesis are clear, an explanation of the molecular basis of the genetic code is not yet available. However, our understanding of the details of the contemporary process of protein synthesis allows us to imagine similar but simpler evolutionary precursor processes. Certainly if we extrapolate far enough backwards, we arrive at a time when there was no interdependence between the composition of proteins and nucleic acids. There is ample evidence that the synthesis of protein-like materials could have taken place without the intervention of polynucleotides. For example, thermal condensation of amino acids (Fox and Harada, 1960) easily yields protein-like materials, usually termed thermal proteinoids, which exhibit a wide variety of catalytic (Rohlfing and Fox, 1969) properties and which have the ability to form isolated microsystems. However, thermal polymerization does not easily lend itself to control. Rather, polymer composition in thermal experiments is dictated by which amino acids are available and the conditions employed. Precursor peptide synthesizing systems that might be more readily amenable to regulation would be based on chemically active intermediates such as the aminoacyl adenylate anhydride or the aminoacyl imidazolide. Both of these intermediates are capable of forming peptides under the proper conditions (Krampitz and Fox, 1969; Weber and Lacey, 1974).

If we assume that the raw materials for these active forms, and indeed the active forms themselves, were present in the prebiotic milieu, then how might the presence of a polynucleotide influence the composition of a peptide being actively synthesized? The two fundamental requirements of the polynucleotide are:

(1) It must cause a selective linear organization of amino acids, and

(2) It must facilitate amino acid polymerization.

There are basically two ways in which the polynucleotide might exhibit these effects:

(1) The active aminoacyl monomers approach the polynucleotide, are recognized and positioned in sequences dependent upon the sequence of nucleotides. The organized monomers then polymerize with each other yielding a peptide without ever forming a covalent linkage with the polynucleotide. This type of process might be called direction of synthesis by *association*, and the polynucleotide in this event is a positioning catalyst.

(2) The active aminoacyl monomers approach the polynucleotide, are recognized and positioned in sequences and then react with the polynucleotide forming a covalent linkage. The amino acids covalently linked with the polynucleotide then react with each other to form a peptide. This type of process might be called direction of synthesis by *association and reaction*. The polynucleotide in this event would not only position the monomers but would also cause selected combinations of amino acids and nucleotides to be more reactive.

We can begin to assess the two possibilities by seeking information about how active acyl compounds behave in the presence of polynucleotides. Stuart and Khorana (1964), working with DNA, found that acetic anhydride in aqueous solution principally attacked the terminal 3′OH group but not the nucleotide base rings. Gottikh *et al.* (1970) found, with mononucleotides in aqueous solution, that aminoacyl imidazolides transferred their aminoacyl residues to the 2′ (or 3′) OH groups, but not to the nucleotide $-NH_2$ groups. However, in organic solvents some substitution on the rings was found. These results suggest that in aqueous solution active acyl groups prefer to react with the ribose hydroxyl groups rather than with the base rings. These results are consistent with the fact that oxygen is more electronegative than nitrogen and might thus be a better nucleophile. This is also in accord with the fact that in the contemporary biosynthesis of proteins, the amino acids are esterified to the 2′ (or 3′) —OH groups of the ribose of the terminal adenosine residues in tRNAs.

These results suggest that active acyl monomers positioned along a RNA molecule would form covalent linkages with the 2′OH groups; however, placed along the backbone of a DNA molecule, there would be no covalent linkage with the polynucleotide. In fact, Beljanski *et al.* (1968) worked for a number of years with a bacterial enzyme system which did attach amino acids to the 2′OH groups along the backbones of RNAs. These aminoacylated RNAs were then shown to yield peptides upon incubation. The function of such a peptide synthesizing system was never clear. However, its very existence, coupled with the behavior of active acyl species in reacting with —OH groups, and the fact that protein synthesis is directed by RNA, suggested that we explore the possibility that an early form of protein synthesis involved the second of the two alternatives mentioned above, namely, the polymerization of aminoacyl esters along the backbone of RNAs. Such a system would require neither ribosomes nor many of the other cofactors of contemporary protein synthesis. We have previously discussed this model elsewhere (Lacey and Fox, 1972; White *et al.*, 1973), and the primary purpose of the present paper is to discuss the model in more detail and review the data relative to it.

The idea of considering model protein synthesizing systems in relation to understanding the origin of the genetic code is certainly not new. Before there was any detailed understanding of the molecular basis of protein biosynthesis, a number of suggestions were made regarding possible mechanisms. Some of these bear a relation to the model we are exploring. Dounce (1952) suggested a translation system in which the amino group of each amino acid was covalently linked to the phosphate moieties along the backbone of RNAs. He imagined that these amino acids reacted with each other yielding peptides. Similarly, but without necessarily considering covalent linkage to the nucleic acid, Gamow (1954) considered that protein synthesis might occur directly on DNA. Rich (1962) discussed a possible system in which aminoacyl esters of mononucleotides might simultaneously polymerize to yield both polynucleotides and peptides. Woese (1966) has discussed the importance of considering the evolution of the process of translation in understanding the genetic code. More recently Fox (Krampitz and Fox, 1969; Nakashima *et al.*, 1970; Nakashima and Fox, 1972) has emphasized the importance of studying the origin of the genetic code using chemical systems in which peptides are being actively synthesized. Nakashima and Fox (1972) were even able to establish a system which gave codonic incorporation of amino acids into particles consisting of homopolynucleotides and thermal proteinoids. More specifically, as already mentioned above, the model we are considering is principally suggested by the work of Beljanski (1968).

We emphasize that a principal reason for being interested in the coevolution of the genetic code and translation is that the only known biological expression of the genetic code occurs in the molecular system responsible for the synthesis of proteins.

## 2. Experimental Approaches

The principle questions to be asked of such a system as described above are:

(1) Is it possible chemically to transfer amino acids to 2′OH groups along the backbone of RNAs and not just to the 3′ terminus?

(2) Are these amino acids reactive and will they form peptides?

(3) Is there any specificity between amino acids and polynucleotides?

Beljanski's work (1968) has conceptually answered the first two questions affirmatively, but we would like to understand the mechanism whereby esterification of the 2′OH groups and subsequent peptide formation occur.

Amino acids will not spontaneously esterify the ribose —OH groups in aqueous solution, but must first be activated. In our early experimental approaches, we tried to transfer amino acids from the amino acyl adenylate anhydrides to the polyribonucleotides directly, but were not successful. Other early experiments, however, showed that if the reactive species is the aminoacyl imidazolide, then the amino acid is readily transferred to polyribonucleotide (Weber *et al.*, 1971). In these early experiments, N-acetyl amino acids were used in order to prevent peptide formation. Of the four homopolyribonucleotides, polyuridylic acid (poly U) was by far the most easily aminoacylated with all amino acids tried. These results are shown in Figure 1.

Because the aminoacyl adenylate is used as the universal activated form of amino acids, we felt it necessary to show that the adenylate anhydride could be converted to the imidazolide from which intermediate we knew it could be transferred to the polynucleotides. Jencks (1957) had previously shown that the conversion of acetyl adenylate to acetyl imidazole takes place readily. Still working with an N-blocked amino acid, N-acetyl glycine, we showed that under proper conditions the adenylate could be almost quantitatively converted to the imidazolide. The experiment and results are shown in Figure 2. Subsequently we showed that starting with the adenylate anhydride and adding imidazole and homopolyribonucleotides, the N-blocked amino acid could be transferred to the 2′OH groups along the backbone of the poly-

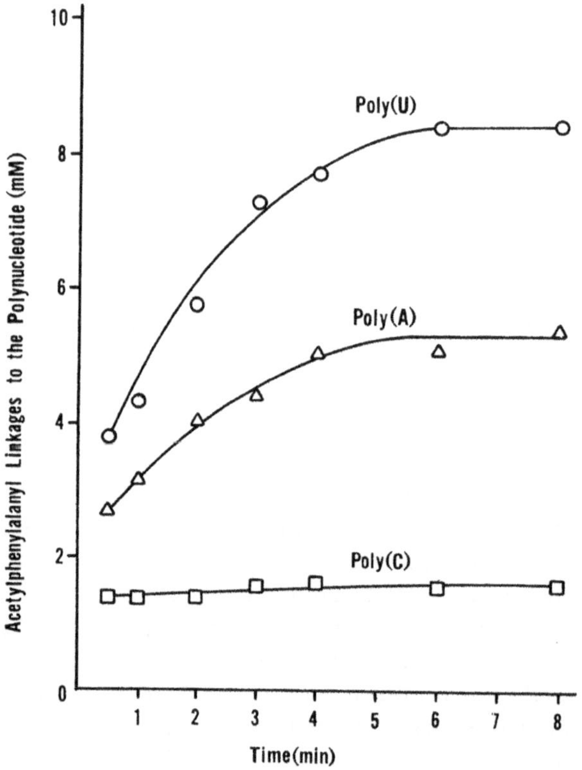


Fig. 1. Rates of acetylphenylalanylation of homopolyribonucleotides by N-(acetyl-phenylalanyl)-imidazole. The reaction was carried out in a 0.4 M imidazole buffer (pH 5.7) with 27 mM polyribonucleotide and 95 mM initial imidazolide. Analysis for active acyl on the polyribonucleotide was carried out by precipitating the polynucleotides with 1 M $BaCl_2$ and assaying the precipitate with hydroxylamine in a manner similar to Lipman and Tuttle (1945) (Weber *et al.*, 1971).

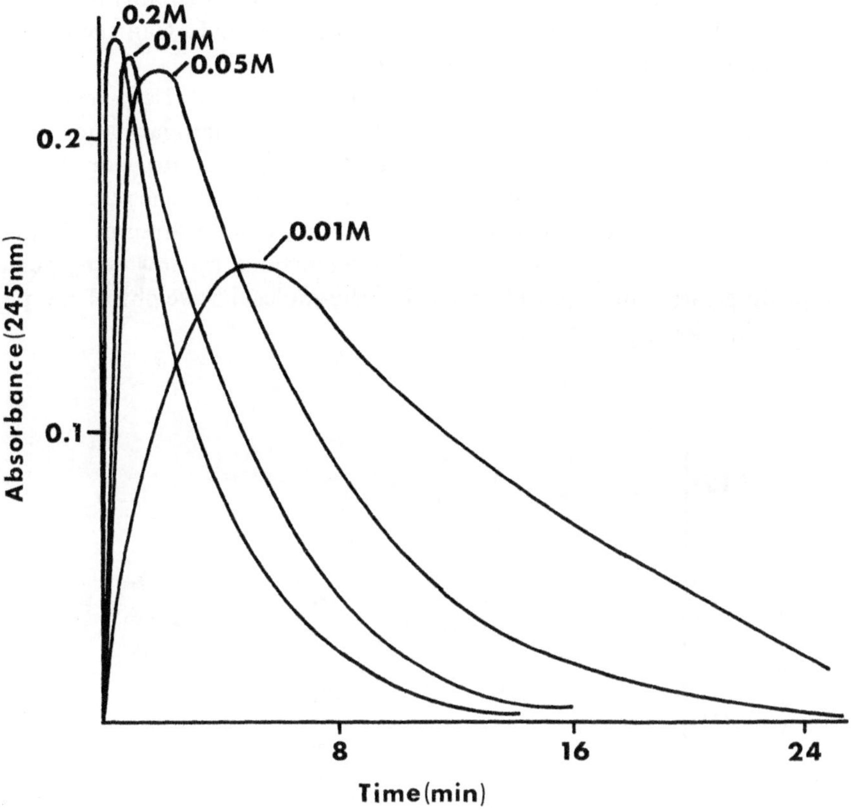


Fig. 2. Formation and disappearance of N-acetylglycyl imidazolide from N-acetylglycyl adenylate (0.0014 M starting concentration) and imidazole (at various concentrations shown), pH 7.0 and room temperature (25 °C). Reaction was followed by monitoring the absorbance changes at 245 nm. At imidazole concentrations of 0.05 M and greater, conversions to the imidazolide were near 100%. A Cary 15 spectrometer was used with 1 mm light path length cells. The blank contained adenylate but no imidazole (Lacey and White, 1972).

nucleotides in one reaction mixture. These data are presented in Figure 3. Again we found that poly U was most easily aminoacylated, and that poly A was next. Poly G and poly C were not aminoacylated under these conditions, nor were poly U and poly A when imidazole was omitted. Subsequently, we proposed the sequence of steps shown in Figure 4 (White *et al.*, 1973).

Further study of the transfer of the amino acid from the imidazolides to polyribonucleotides always showed poly U to be the easiest to aminoacylate (Weber and Fox, 1973), and both N-blocked or free amino acyl imidazoles could be used. Generally, the reaction was easier to control with the N-blocked amino acids because with the free aminoacyl imidazolide, the side reaction of peptide formation can proceed either in solution or on the polynucleotide backbone.

The above experiments readily demonstrate the chemical feasibility of charging the 2′OH groups along the backbone of RNAs with amino acids. Can such chemically prepared aminoacylated RNAs yield peptides?

Using poly U and glycyl imidazolide (preparation described elsewhere), we aminoacylated the poly U with radioactive glycine. The extent of aminoacylation was about one out of three 2′OH groups being involved. The aminoacylated poly U was then incubated under various conditions and then electrophoresed. The results are shown in Table I, and show that at pH 8.5 there is an increase in the amount of tripeptides and higher and a decrease in free glycine. These reactions are now being pursued at higher loading ratios in hopes of increasing the molecular weight of the peptides formed and in showing specificity.

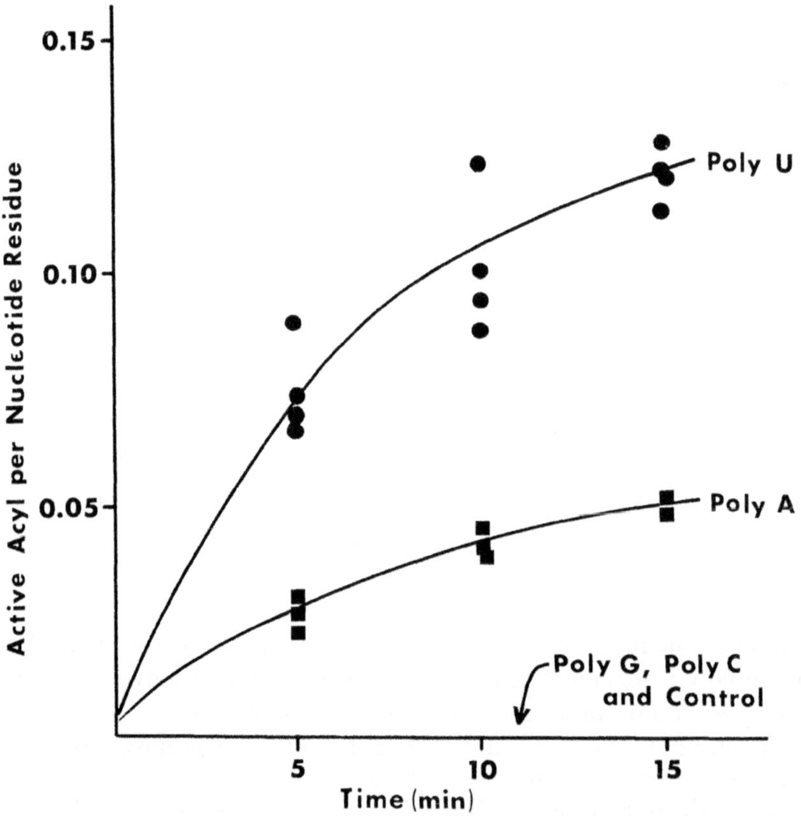


Fig. 3. Transfer of N-acetylglycyl from the adenylate anhydride to the 2′OH groups along the backbone of homopolyribonucleotides using 0.5 M imidazole as a catalyst, at pH 7, 5°C with starting adenylate concentration about 0.1 M. Approximately 20 mg of polynucleotide was present in 1.2 ml reaction volume. Analysis for active acyl on the polynucleotide was carried out by precipitating the polynucleotide with 1 M $BaCl_2$ and assaying the precipitate with hydroxylamine essentially according to Lipman and Tuttle (1945) (White *et al.*, 1973).

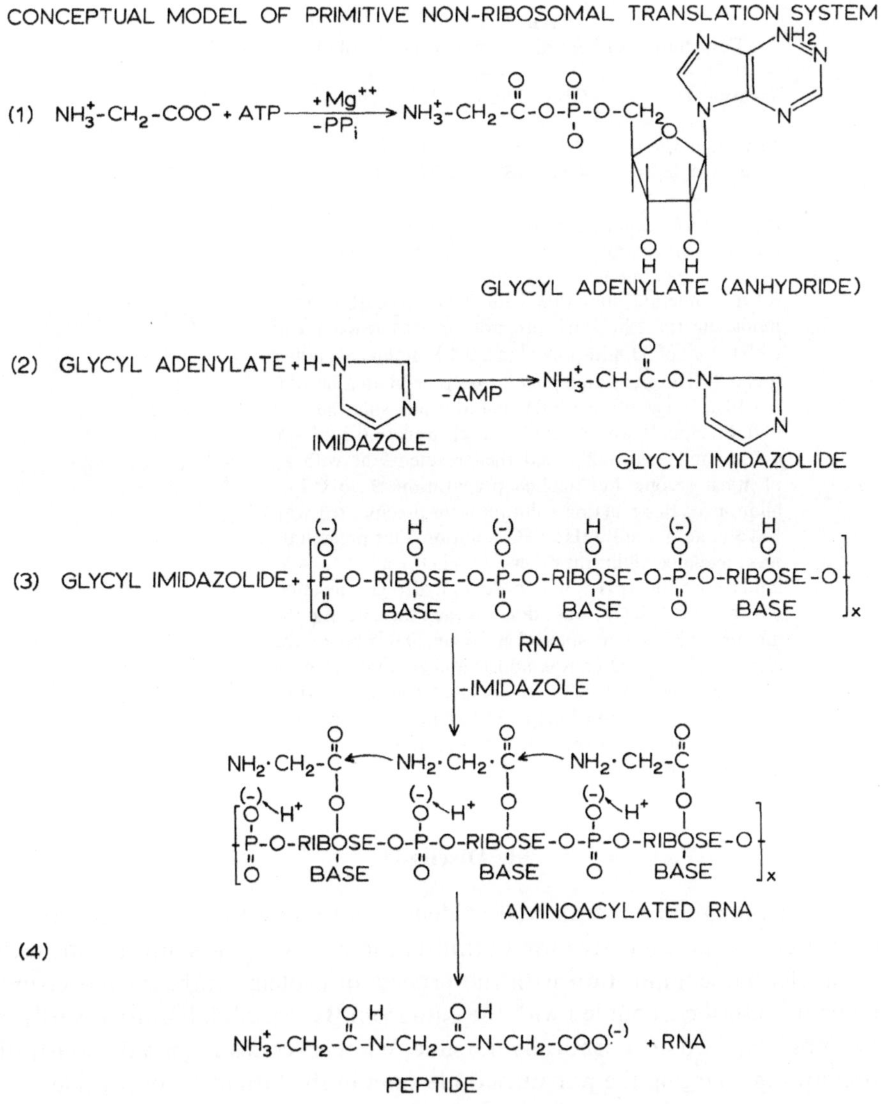


Fig. 4. Model sequence for generation of peptides directly on the backbone of an RNA molecule. Using glycine as an example the steps are:

(1) initial activation of the amino acid by reaction with ATP and formation of the adenylate
(2) conversion of the aminoacyl adenylate to the aminoacyl imidazole
(3) transfer of the aminoacyl groups to the 2′OH groups along the RNA backbone, and
(4) reaction of the aminoacyl groups with each other, perhaps facilitated by the phosphate group abstracting a proton from the $\alpha$-amino group of the amino acid.

Both imidazole and RNA are regenerated in the process and can be considered as catalysts (White *et al.*, 1973).

TABLE I
Distribution of peptides from glycylated poly U

| Treatment | Gly | Digly | Trigly and larger |
|---|---|---|---|
| Incubated 5 min | 58% | 31% | 11% |
| Incubated 48 hr | 45% | 28% | 27% |

Peptide formation via condensation of the glycyl esters of the hydroxyl groups of poly U. The imidazolide was prepared by reacting 4 M $^{14}C$— glycinium trifluoroacetate 0.1 ml dimethyl sulfoxide with 3 M 1,1′-carbonyl diimidazole for 2 hr. This preparation was reacted 2 hr with 0.4 ml of 30 mM poly U in a 0.4 M imidazole buffer (pH 6.0). The glycylated poly U was precipitated by adding 10 volumes of an EtOH: 0.4 M NaCl solution (1:4 v/v). Precipitate was isolated and taken up in 0.3 ml of a 0.4 M imidazole buffer and then reacted 2 hr with an identical second imidazolide preparation. Two 0.2 ml aliquots of this reaction solution were precipitated with 10 volumes of the EtOH: NaCl solution. One precipitate was incubated 48 hr, the other 5 min in 0.1 ml of a 0.4 M imidazole buffer (pH 8.5). In order to hydrolyze the ester, 0.03 ml of 5 N KOH was added to each mixture and the mixtures allowed to stand 3 hr. Then 0.075 ml of the electrophoretic buffer was added and aliquots of each mixture subjected to electrophoresis by using a formic acid: water (1:5 v/v) buffer. The radioactive products were quantitated by radiochromatographic scanning of the electrophoretograms.

## 3. Discussion

We feel that the above experimental evidence shows that the system illustrated in Figure 4 was possibly an early form of translation involving the simplest intersection of polynucleotide composition with the process of protein synthesis. The chemical experiments described, coupled with the data from Beljanski's laboratory, verify that all the steps proposed are chemically possible. What is yet missing is a demonstration of specifities in charging the polynucleotides and in the formation of peptides.

The preferential aminoacylation of poly U by all amino acids tested probably reflects the relative degree of structuring of the homopolynucleotides in solution. Poly U is the least structured of the four homopolymers (Yang and Samejima, 1969) and its 2′—OH groups would be expected to be most available for reaction. Therefore, we intend to use copolymers of polynucleotides having more uniformity of structure in looking for specifities. Gottikh (1970) working with the mononucleotides and aminoacyl imidazolides obtained highest yields with UMP and explained this preference as being due to the basicity of the rings.

How reasonable is the model we propose?

The polymerization of amino acids on a polyribonucleotide backbone offers these obvious advantages. The aminoacyl ester of nucleotides is still an active species. This increased activity might be attributed to the presence of the neighboring phosphate group as found by Shaborova *et al.* (1962). As shown in Figure 3, the 2′OH aminoacyl ester places the amino acid side chain in such close proximity to the nucleotides that 'recognition' is easily imaginable. The nearness of an amino acid on the neighboring 2′OH group allows peptide formation. We suggest that the forced proximity of the plus charged amino group and the negatively charged phosphate might greatly alter the behavior of the aminoacyl esters. The polyribonucleotide can be viewed as a natural proximity catalyst for peptide bond formation.

The demonstrations that imidazole can mediate aminoacyl transfer suggest that this primitive method of peptide synthesis could be facilitated by an enzyme(s) having histidines in their active sties. The enzyme could attach at one end of the RNA and then proceeded down the chain inserting the amino acid dictated by the nucleotides. Beljanski (1968) found a requirement for an intact 3′ terminus —OH groups in his bacterial system which suggests that the enzyme might attach the first amino acid to the 2′(3′) OH groups at the 3′ end of the RNA and progress in the 3′→5′ direction leaving the peptide attached as an ester at the 3′ end. However, this kind of growth would require additions at the amino end of the peptide. This is the opposite direction of growth to that employed in the usual method of protein biosynthesis.

One obvious disadvantage for this proposed method of synthesis of proteins is that, if an affinity exists between the amino acids and the nucleic acids, then the protein would remain associated with the RNA. The proposed model system would not be useful for synthesizing enzymes and other functional proteins that are required to be free of nucleic acids. The only proteins that might be synthesized on RNA backbones in contemporary biological systems would be those whose function was to be complexed with their 'messengers'. The obvious candidates are the ribosomal proteins. Due to the large number of ribosomal proteins, (56 in *E. Coli*), it is not possible that rRNA codes for all of them. However, it is possible that rRNA codes for one or two of the ribosomal proteins responsible for the initial formation of the protein-rRNA complex. But are any of the ribosomal proteins synthesized 'backwards'? Iwata and Kaji (1971) indicate that perhaps some *E. Coli* ribosomal proteins are synthesized in a reverse manner. Then it is possible that this simple method of protein synthesis has been retained for the synthesis of one or two ribosomal proteins and perhaps is the system which Beljanski has described in bacteria.

The early system could have employed 1/1 coding, or if every 2′OH was esterified, the coding could have been overlapping. The 16s rRNA (1500 nucleotides) of *E. Coli* could then potentially code for 1500 amino acids. However, evidence indicates that most of the ribosomal proteins are synthesized in a conventional manner (Flessel, 1970). Zimmerman *et al.* (1969) have reported that there is incorporation of amino acids into protein in nucleoli of HeLa cells, and Mullins (1973) in our laboratory has recently confirmed these results. He finds that this incorporation is not inhibited by cycloheximide, an antibiotic which inhibits cytoplasmic protein synthesis in eukaryotes. Since nucleoli are the site of synthesis of rRNA and partial assembly of the ribosome, we believe that the incorporation we are seeing is due to some odd method of protein synthesis as we are describing. We are presently isolating and studying the proteins that are synthesized by isolated HeLa cell nucleoli (Mullins, 1973).

Although not mentioned in the model (Figure 4), it is possible that instead of activating all amino acids with ATP and transferring them to the RNA 2′OH groups via an imidazole intermediate, different classes of amino acids might be activated by different nucleotides. For example, the highly hydrophobic amino acids (e.g., phe, leu, etc.) might be activated by ATP, the charged amino acids (e.g., lus., asp., glu) by UTP and intermediate polarity (e.g., ser., thr., pro.) by GTP. The variously activated aminoacyl nucleotides could then complex with the proper bases on the 'messenger' and be transferred by imidazole to the messenger 2′OH groups where they would polymerize.

Specific complexes between mononucleotides and polynucleotides have been shown to form (Howard *et al.*, 1964, 1966; Ts'o *et al.*, 1968), and these complexes have been used in origin of life studies. For example, formation of complexes of adenylic acid derivatives has been used as the basis for experiments showing the formation of oligoadenylates on a poly U template (Sulston *et al.*, 1968) and acetyl transfer from acetyl adenylate to adenosine while both are complexed with poly U (Chung *et al.*, 1971).

In this model the mononucleotides carrying the activated amino acids are envisioned as evolutionary precussors of tRNAs. In order to increase the fidelity of reading all 20 amino acids, the evolution of tRNAs was required and the codon was expanded to a 3/1 ratio and became non overlapping.

Beljanski's data support this latter type of reading because all four nucleotide triphosphates are required and there are apparently four enzymes, one for each nucleotide.

Our research efforts are now aimed at getting stronger evidence for the peptide forming step and looking for specificities in the principal model discussed above.

## Acknowledgements

This work was supported in part by a National Institutes of Health General Research Support Grant (No. 5-501-RR-05300-10), a Public Health Service Grant (No. CA-08888) from the National Cancer Institute and a Grant from the National Aeronautics and Space Administration (No. NGR-01-010-001).

## References

Beljanski, M., Fischer-Ferraro, C., and Bourgarel, P.: 1968, *Euro. J. Biochem.* **4**, 184–189.
Chung, N. M., Lohrmann, P., and Orgel, L. E.: 1971, *Biochim. Biophys. Acta* **228**, 536–543.
Dounce, A. L.: 1952, *Enzymologia* **15**, 251–258.
Flèssel, C. P.: 1970, *Biochim. Biophys. Acta* **209**, 587–588.
Fox, S. W. and Harada, K.: 1960, *J. Am. Chem. Soc.* **82**, 3745–3751.
Gamow, G.: 1954, *Nature* **173**, 318.
Gottikh, B. P., Krayevsky, A. A., Tarussova, N. B., Purygin, P. P., and Tsilevich, T. L.: 1970, *Tetrahedron* **26**, 4419–4433.
Howard, F. B., Frazier, J., Lipsett, M. N., and Miles, T.: 1964, *Biochem. Biophys. Res. Comm.* **17**, 93–102.
Howard, F. B., Frazier, J., Singer, M. F., and Miles, H. T.: 1966, *J. Mol. Biol.* **16**, 415–439.

Iwata, S. and Kaji, H.: 1971, *Proc. Natl. Acad. Sci. U.S.* **68**, 690–694.
Jencks, W. P.: 1957, *Biochim. Biophys. Acta* **24**, 227.
Krampitz, G. and Fox, S. W.: 1969, *Proc. Natl. Acad. Sci. U.S.* **62**, 399–406.
Lacey, J. C., Jr. and Fox, S. W.: 1972, in A. Swartz (ed.), *Theory and Experiment in Exobiology*, Vol. 2, Wolters-Nordhoff Pub., Groningen, pp. 35–63.
Lacey, J. C., Jr. and White, W. E., Jr.: 1972, *Biochem. Biophys. Res. Comm.* **47**, 565–573.
Lipman, F. and Tuttle, L. C.: 1945, *J. Biol. Chem.* **159**, 21–28.
Mullins, D. W.: 1973, unpublished results.
Nakashima, T. and Fox, S. W.: 1972, *Proc. Natl. Acad. Sci. U.S.* **69**, 106–109.
Nakashima, T., Lacey, J. C., Jr., Jungck, J. R., and Fox, S. W.: 1970, *Naturwissenschaften* **57**, 67–68.
Rich, A.: 1962, in M. Kasha and B. Pullman (eds.), *Horizons in Biochemistry*, Academic Press, N.Y., p. 115.
Rohlfing, D. L. and Fox, S. W.: 1969, *Adv. Catalysis* **20**, 373–418.
Shaborova, Z. A., Hughes, N. A., and Baddiley, J.: 1962, *Biochem. J.* **83**, 216–219.
Stuart, A. and Khorana, H. G.: 1964, *J. Biol. Chem.* **239**, 3885–3892.
Sulston, J., Lohrmann, R., Orgel, L. E., and Miles, H. T.: 1968, *Proc. Natl. Acad. Sci. U.S.* **59**, 726–733.
Ts'o, P. O. P. and Huang, W. M.: 1968, *Biochem.* **8**, 2954–2962.
Weber, A. L. and Fox, S. W.: 1973, *Biochem. Biophys. Acta* **319**, 174–187.
Weber, A. L. and Lacey, J. C., Jr.: 1974, *Biochem. Biophys. Acta*, in press.
Weber, A. L., Lacey, J. C., Jr., and Fox, S. W.: 1971, *Fed. Proc.* **30**, 1102.
White, W. E., Jr., Lacey, J. C., and Weber, A. L.: 1973, *Biochem. Biophys. Res. Comm.* **51**, 283–291.
Woese, C. R.: 1967, *The Genetic Code*, Harper and Row, Publishers, N.Y., p. 188.
Yang, J. T. and Samejima, T.: 1969, *Nucleic Acid Res.* **9**, 223.
Zimmerman, E. F., Hackney, J., Nelson, P., and Arias, I. M.: 1969, *Biochem.* **8**, 2636–2644.

# PART VI

# MISCELLANEOUS PAPERS

# HOT HYDROGEN ATOMS: INITIATORS OF REACTIONS OF INTEREST IN INTERSTELLAR CHEMISTRY AND EVOLUTION* **

KONG-YI HONG, JANE-HUEY HONG, and RALPH S. BECKER
*Dept. of Chemistry, University of Houston, Houston, Tex. 77004, U.S.A.*

**Abstract.** Photochemically generated hot hydrogen atoms initiate reactions with simple molecular substrates including methane to produce organic alcohols, amines, acids, amino acids, and other compounds. The typical quantum yields for the formation of amino acids are $2 \times 10^{-5}$ to $4 \times 10^{-5}$. Hot hydrogen atoms may be important initiators of reactions in interstellar space and in planetary atmospheres.

Considerable effort has been made to elucidate the pathways by which biologically significant molecules arose prior to the existence of living things [1]. Many of the techniques consisted of conventional electric discharge and ultraviolet irradiation or the heating of mixtures of various simple molecules such as $NH_3$, $CH_4$, and water. In most cases, prime areas of interest included the chemical synthesis of amino acids, purines, and pyrimidines, and ultimately polypeptides and polynucleotides. In a more recent effort, $H_2S$ irradiated in the presence of a mixture of simple gases yielded certain amino acids [2]. In addition, there is considerable interest in the mechanism of formation of molecules in interstellar space. We report here the formation of some amino acids and other organic compounds initiated by 'hot' hydrogen atoms in mixtures of simple gases chosen from among $CH_4$, $C_2H_6$, $H_2O$, and $NH_3$. The hot hydrogen atom donors include both $H_2S$ and $CH_3SH$.

Hot hydrogen atoms are those with kinetic (or translational) energy in excess of that to be expected if they were in thermal equilibrium with the surroundings. In our case, they are generated by photolysis of the donor molecules as

$$CH_3SH \xrightarrow{h\nu} CH_3S\cdot + H\cdot \text{ (hot)} \tag{1}$$

The excess kinetic energy of hydrogen from HR is

$$E_k(\mathrm{H}) = (h\nu - D - I)\frac{M_{\mathrm{R}}}{M_{\mathrm{HR}}} \tag{2}$$

where $h\nu$ is the energy of light, $D$ is the bond dissociation energy, $I$ is the internal energy remaining in the fragment R, and $M$ is the mass. For the donors used, the mass ratio is essentially equal to 1 and the amount of internal energy left in R is approximately 5 to 17% of the maximum kinetic energy of the hydrogen atom over the wavelength band used for the photolysis [3] (220 to 280 nm). Based on the fore-

* A complete discussion regarding molecular formation in interstellar space, hot hydrogen atom reactions, meaning of temperature etc. will appear elsewhere, *J. Molec. Evolution*, in press.
** Reprinted by permission from *Science* (1974) © American Association for the Advancement of Science.

going, the average kinetic energies of the hot hydrogen atoms range from approximately 17 to 32 kcal mole$^{-1}$ (onset of $H_2S$ absorption at 266 nm). These hot hydrogen atoms lose the excess kinetic energy by either non-reactive or reactive collisions with the surrounding molecules [4]. In our case the chemical reaction of interest, based on the nature of the substrates and energies of the hot hydrogen atoms, is

$$RH + H(hot) \rightarrow R\cdot + H_2 \tag{3}$$

All gases used were chemically pure grade (Matheson), and the $H_2O$ was both distilled and deionized. The $C_2H_5OH$ was absolute pure grade (U.S. Industrial Company). In one reaction, $CH_4$ of ultrahigh purity (minimum purity 99.97%, Linde Company) was used to provide assurance that the products were not formed by reactions of contaminant $C_2H_6$. Figure 1 shows the reaction vessel and accompanying circulation system. The total volume was approximately 6 l. The temperature of the heated side (coil A) was kept at 150 °C, and the remainder of the system was maintained at 30° or 50 °C. When water was used, 3 ml was introduced into trap B and carefully degassed. The pressure of each gas reactant was measured with a Bourdon

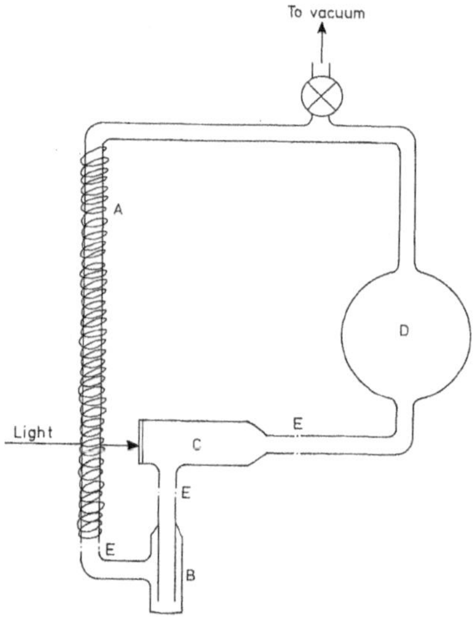


Fig. 1. Experimental apparatus; *A*, heating coil; *B*, cold trap; *C*, reaction vessel; *D*, 5-liter flask; and *E*, O-ring joint.

gauge (mercury gauges can introduce mercury vapor and cause mercury photosensitized reactions). After all gases had been introduced, heating element A was turned on and the gases were allowed to mix for about 12 h. A small portion of gas was removed for analysis prior to reaction. After the completion of a reaction, a small portion of gas was transferred to storage bulbs and trap B was cooled with ice while the gas mixture was circulated for several hours. Nitrogen was then introduced to a total pressure of 1 atm, and the trap was disconnected; one to two drops of product were removed from the trap for analysis. For those reactions for which amino acids were expected, the entire system was washed with water; the wash water was then filtered and concentrated by rotary evaporation at 50 °C. The concentrated solution was passed through a Dowex 50-X4 ion-exchange resin. The eluent was dried and analyzed for amino acids with a Technicon amino acid analyzer. Organic acids were analyzed according to the method of Gerhke and Goerlitz [5] except that the black silver salt was isolated from the solution for esterification instead of the water solution. All reactants and products (except amino acids) were analyzed by means of a combined gas chromatograph-mass spectrometer (LKB model 9000).

The light sources for the photolysis of the $H_2S$ or $CH_3SH$ were 1000-W xenon or 500-W mercury lamps coupled with a filter system [water, Corning filter 9–54, and $Cl_2$ (5-cm cell) at 1 atm]. The band pass was from 220 to 280 nm with a maximum of 54% transmittance at 252 nm. In some cases, a yellow film would form on the window but would disappear if the light source was turned off and the gases were allowed to circulate for some period.

Table I presents a summary of all experiments completed to date with the results. Brief comments on some of the experiments follow.

Experiment II was carried out in the absence of light, but we utilized the same procedures as for all other experiments for which amino acids were analyzed. No amino acids were detected, and this experiment thus provides a blank test.

Experiment IV was done to provide some insight into the mechanism and intermediates of the reaction giving amino acids. The products $C_2H_5OH$ (or $CH_3OH$), $CH_3COOH$, and $C_2H_5COOH$ were detected.

Experiment V was carried out because $C_2H_5OH$ was produced in experiment IV. A larger number of amino acids were produced in experiment V than in experiment III. The real significance of this comparison between experiments III and V regarding the number of amino acids is not clear because of a change in the light source. The quantum yield of amino acids was $\sim 2 \times 10^{-5}$ to $\sim 3 \times 10^{-5}$.

Experiment VI provided an interesting result. The primary difference between this experiment and experiment V was the addition of water plus some liquid $C_2H_5OH$. However, the consequence was quite marked. Roughly equal amounts of the various amino acids were produced in experiment V, whereas in experiment VI approximately 95% was glycine (0.05 mg). However, the number and nature of the amino acids was the same in both experiments. The quantum yield of formation of amino acids was $\sim 3 \times 10^{-5}$ to $\sim 4 \times 10^{-5}$.

Experiment VIII is important in that a substrate containing only one carbon atom, $CH_4$, was sufficient for the production of amino acids of the same nature and number as from $C_2H_6$. Very pure $CH_4$ (99.97% minimum purity) was used so that contaminant $C_2H_6$ could be ruled out as the carbon source.

Experiment VII shows that alkyl amines can be produced in hydrocarbon-$NH_3$ systems induced by hot hydrogen atoms.

Several important conclusions can be drawn from the results of these experiments. (i) Hot hydrogen atoms can initiate reactions among simple molecules to produce

TABLE I

Summary of the results of experiments on hot hydrogen atoms

| Reactants[a] (pressure in cm-Hg) | Lowest temperature | Irradiation time | Light source | Amino acids detected | Other products detected |
|---|---|---|---|---|---|
| | | | *Experiment I* | | |
| $CH_3SH = 42.5$<br>$C_2H_6 = 12.5$<br>$NH_3 = 12.5$<br>$H_2O = 9$ | 30 °C | 14 days | 1-kW Xe lamp | Aspartic acid<br>Glycine<br>Cysteine | $CH_3SSH$, $CH_3SSCH_3$, and amino compounds tha either more acidic or basic than the normal a acids |
| | | | *Experiment II* | | |
| $CH_3SH = 42.4$<br>$C_2H_6 = 12.5$<br>$NH_3 = 12.5$<br>$H_2O = 9$ | 50 °C for 14 days | None | None | None | Other amino compounds acidic than all the p amino acids but no p amino acids[b] |
| | | | *Experiment III* | | |
| $H_2S = 47.5$<br>$C_2H_6 = 12.5$<br>$NH_3 = 12.5$<br>$H_2O = 9$ | 50 °C | 14 days | 1-kW Xe lamp | Aspartic acid<br>Glycine<br>Glutamic acid<br>Alanine<br>Cysteine | Sulfur and maybe one a compound more acidic the protein amino acid possibly some more basi the protein amino acids |
| | | | *Experiment IV* | | |
| $CH_3SH = 37.5$<br>$C_2H_6 = 37.5$<br>$H_2O = 9$ | 50 °C | 14 days | 500-W, high-pressure Hg lamp | None expected | $CH_3COOH$, $C_2H_5COOH$, $C_2H_5OH$,[c] $CH_3SSCH_3$, $CH_3SSH$ |
| | | | *Experiment V* | | |
| $H_2S = 25$<br>$NH_3 = 12.5$<br>$C_2H_5OH = 7$ | 30 °C | 21 h | 500-W, high-pressure Hg lamp | Serine or threonine<br>Glycine<br>Alanine<br>Aspartic acid<br>Valine<br>Glutamic acid<br>Leucine<br>Isoleucine<br>Proline | Sulfur and possibly some a compounds more basic the protein amino acids |
| | | | *Experiment VI* | | |
| $H_2S = 25$<br>$NH_3 = 12.5$<br>$C_2H_5OH = 7$ (2 ml)<br>$H_2O = 3.2$ | 30 °C | 37 h | 500-W, high-pressure Hg lamp | Glycine<br>Serine or threonine<br>Aspartic acid<br>Valine<br>Leucine<br>Isoleucine<br>Glutamic acid<br>Proline (trace amount)<br>Probably alanine | Sulfur and some other a compounds that are more acidic or more basi normal amino acids |

(continued)

| ts[a] (pres- m-Hg) | Lowest temperature | Irradiation time | Light source | Amino acids detected | Other products detected |
|---|---|---|---|---|---|
| | | | *Experiment VII* | | |
| H=25 ₄=25 ₃=25 | Room temperature | 14 days | 1-kW Xe lamp | None expected | $CH_3SSH$, $CH_3SSCH_3$, $CH_3NH_2$[d] |
| | | | *Experiment VIII* | | |
| S=25 ₃=25 ₄=25[e] O= 9 | 50°C | $113\frac{1}{2}$ h | 500-W, high-pressure Hg lamp | Aspartic acid, Serine or threonine, Glutamic acid, Proline, Glycine, Alanine, Valine, Leucine, Isoleucine | Sulfur and possibly some amino compounds either more basic or more acidic than the protein amino acids |

experiments, the first molecule listed ($H_2S$ or $CH_3SH$) is the source of the hot hydrogen atom by photolysis ng to Eq. 1.
arefully determined blanks as follows. In one case, after the thermal reaction the same procedure was followed ll other cases except for passage through the Dowex column. In this case, four peaks were detected prior to the n of aspartic acid and other protein amino acids. In another blank, water plus $NH_3$ used to wash the apparatus e hot hydrogen atom reactions was passed alone through the Dowex column. Three peaks were detected prior etection of aspartic acid and other protein amino acids. Finally, the reference amino acid solution was carefully d and one peak was detected prior to the detection of aspartic acid and other protein amino acids. In all cases ny of these peaks apparently appeared from the hot hydrogen atom reactions, they were five to ten times more than in the blanks but still were not considered as confirmed reaction products.
₄ is used, $CH_3OH$ is produced but analyses for carboxylic acids were not carried out.
H₆ is used, $C_2H_5NH_2$ is produced.
high pure grade, minimum purity 99.97% (Linde Company).

biomolecules of significance. (ii) The one-carbon reactant $CH_4$ can participate in reactions leading to the formation of amino acids containing more than one carbon atom. This is important because large amounts of $CH_4$ existed on the primitive Earth, are now present near some planets, and may occur in interstellar space (clouds). The two-carbon reactant $C_2H_6$ is not a necessary reactant. (iii) In addition to amino acids, other important molecules can be produced by reactions initiated by hot hydrogen atoms such as $CH_3OH$, $C_2H_5OH$, $CH_3COOH$, $C_2H_5COOH$, and organic amines. Furthermore, it appears that some of these could be intermediates in the production of the amino acids. (iv) The first intermediate formed after attack by the hot hydrogen atom is a free radical. These in turn attack other reactants giving other free radicals. In some cases, the radicals combine to give products such as $C_2H_5OH$, $C_2H_5NH_2$, or $CH_3NH_2$, but in other cases, such as for the amino acids, the mechanism is not known. (v) The ambient temperature in interstellar space and clouds is very low so that chemical reactions either would not occur or would be extremely slow. However, the conditions for the formation of molecules in interstellar space are substantially displaced from thermodynamic equilibrium [6]. It is just this possibility of a nonequilibrium condition in which some atoms are hot (0.5 to 50 eV) and not in thermal equilibrium that would permit reactions of relatively high activation energies to occur which ordinarily could not [7]. The importance of such a possibility is not limited to these conditions but is also important in planetary atmospheres where the number of such reactions is still higher because of the greater density of molecules. Hydrogen sulfide is present in interstellar space and was probably present on the primitive Earth [8]. However, any other hydrogen atom (hot) source would be equally effective.

We wish to comment briefly on some of these results relative to other comparable published work [2]. We have demonstrated that amino acids can be produced from a hydrocarbon substrate with only one carbon atom, $CH_4$, whereas Sagan and Khare [2] have stated that the two-carbon substrate $C_2H_6$ is required for the production of amino acids. They have also suggested that the amino acids from such experiments are very likely not primarily made free but result from the hydrolysis of nitriles or polypeptides [2]. Our conditions never resulted in the formation of observable amounts of polymer, nor were acid or basic hydrolysis conditions ever employed. Furthermore, we observed the production of amino acids in two experiments utilizing $C_2H_5OH$ as one of the substrates and, considering the nature of the experiments, it would appear that no nitrile-containing molecules could be generated. We detected no HCN or nitrile-containing molecule at the completion of any experiment either in the gas phase or in solution (water trap), using gas chromatographic–mass spectrometric analysis. In view of these facts, we believe that the major yield of amino acids is produced not from the hydrolysis of a polymer but directly by a unique mechanism or mechanisms not hitherto considered.

## Acknowledgements

This work was supported under National Aeronautics and Space Administration, Grant NGR 44-005-091. We thank Drs G. Holzer and W. Parr for assistance in obtaining the amino acid analysis.

**Note added in proof.** An experiment similar to experiment VI was repeated and the products subjected to esterification and acylation to form *N*-trifluoroacetyl isopropyl esters. Gas chromatography–mass spectrometry analysis positively identified the protein amino acids glycine, alanine, valine, leucine, aspartic acid, and glutamic acid plus some other nonprotein amino acids. This constitutes the first verification of protein amino acids from such reactions.

## References

[1] S. L. Miller, *J. Am. Chem. Soc.* **77**, 2351 (1955); J. Oro and A. P. Kimball, *Arch. Biochem. Biophys.* **94**, 217 (1961); *ibid.* **96**, 293 (1962); H. C. Urey, *The Planets: Their Origin and Development* (Yale

Univ. Press, New Haven, Conn., 1952); P. H. Abelson, *Proc. Natl. Acad. Sci. U.S.A.* **55**, 1365 (1966); S. W. Fox, Ed., *The Origins of Prebiological Systems and Their Molecular Matrices* (Academic Press, New York, 1965); M. Calvin, *Chemical Evolution* (Univ. of Oregon Press, Eugene, 1961); C. Ponnamperuma and N. W. Gabel, *Space Life Sci.* **1**, 64 (1968).

[2] C. Sagan and B. N. Khare, *Science* **173**, 417 (1971).

[3] G. P. Sturm and J. M. White, *J. Chem. Phys.* **50**, 5035 (1969).

[4] C. Vermeil, *Israel J. Chem.* **8**, 147 (1970); J. C. Biordi, Y. Rousseau, and G. J. Mains, *J. Chem. Phys.* **49**, 2742 (1968); R. Wolfgang, *Progr. React. Kinet.* **3**, 97 (1965); F. S. Rowland, in *Molecular Beams and Reaction Kinetics*, C. Schlier, Ed. (Academic Press, New York, 1970), p. 108; K. Y. Hong and G. J. Mains, *J. Phys. Chem.* **76**, 3337 (1972).

[5] C. W. Gerhke and D. F. Goerlitz, *Anal. Chem.* **35**, 76 (1963).

[6] P. M. Solomon, *Phys. Today* **26**, 32 (March 1973).

[7] More details concerning especially the mechanism for molecular formation in interstellar space as well as the experiments and reactions of hot hydrogen atoms will appear elsewhere (R. S. Becker, K. Y. Hong, and J. H. Hong, in preparation).

[8] H. D. Holland, in *Petrologic Studies* (Geological Society of America, New York, 1962), p. 447.

# ROLE OF WEAK BASES ON THE PREBIOTIC FORMATION OF HETEROCYCLIC COMPOUNDS

S. YUASA and J. ORÓ

*Depts. of Biophysical Sciences and Chemistry, University of Houston, Houston, Tex. 77004, U.S.A.*

**Abstract.** The catalytic action of an alkaline earth metal oxide (MgO) on HCN in aqueous solution has been investigated. The reaction mixture was extracted with ether to give diaminomaleonitrile (DAMN). Acid hydrolysis of this mixture resulted in glycine, alanine, aspartic acid and unidentified ninhydrin-positive compounds. Acid hydrolysis of the material remaining after ether extraction (both water-soluble and water-insoluble components) resulted in adenine, aminoimidazole-carboxyamidine, and the amino acids glycine and alanine.

When the weak base magnesite ($MgCO_3$), which is present in carbonaceous chondrites, was substituted for MgO in the reaction DAMN was also produced thus suggesting similar catalytic action on HCN.

## 1. Introduction

Divalent metal oxides such as MgO, CaO and BaO show high catalytic efficiency in the synthesis reactions of monosaccharides (e.g. Oró, 1965). No further study of the possible role of these catalytic actions in prebiotic chemistry has been carried out, even though these metallic compounds are present in significant amounts in the earth and extraterrestrial bodies (Wiik, 1956; Hayes, 1967). The oxides and/or carbonates found in the earth include brucite ($MgO \cdot H_2O$), calcite ($CaCO_3$), magnesite ($MgCO_3$), dolomite ($CaCO_3 \cdot MgCO_3$), and periclase (MgO) (Lange, 1967). It has previously been shown that MgO may be as effective as ammonia in catalyzing the polycondensation of HCN (Yuasa *et al.*, 1970). Accordingly, it seemed important to further investigate the catalytic action of divalent metal oxides and carbonates on the polycondensation of HCN particularly in the formation of purine rings.

In the present paper, we will describe the formation of adenine and related compounds from HCN in the presence of MgO as a catalyst. The catalytic activity of $MgCO_3$ was also evaluated in order to generalize the effects of weak bases on HCN polycondensation.

## 2. Experimental

In a typical experiment 4 g of MgO which had been kept in an oven at 100 °C for 24 hr to eliminate any contamination from ammonia was dispersed in 200 ml of water (500 ml flask). 30 ml of HCN, generated by a conventional method (aqueous sodium cyanide and concentrated sulfuric acid) and trapped in the liquid form, was gently added to the above suspension through connecting glass tubing from the trap. Immediately after mixing, the mixture was refluxed at 50 °C for 15 hr according to the method of Oró (1960). In the present experiments, the maximum formation of diaminomaleonitrile (DAMN) was obtained about 10 hr after starting reflux. However, reflux was sustained for 15 hr in order to maximize other reaction products.

After the reaction, the sample was fractionated according to the procedure shown in Figure 1. Paper chromatography was the method most often used to identify the compounds synthesized or detected in the hydrolyzate. See Table I for paper chromatographic parameters.

## 3. Results and Discussion

### 3.1. DAMN

Crude DAMN was obtained by extracting most of the crude reaction product with two volumes of ether. It was purified by recrystallizing from a water system. The recovered DAMN was analyzed by UV spectroscopy [absorbance ($\lambda_{max}$: 296 nm)], mass spectrometry [$M/e = 108$ (13%), 81 (15%), 69 (7%), 52 (100%), 43 (17%)], and paper chromatography (Table I). The analyses, which were carried out by comparing the sample with an authentic standard sample (PCR Inc. Lot No. 5078), showed that DAMN is predominantly formed under the present experimental conditions (yield: 25%).

When $MgCO_3$ was used instead of MgO under the same conditions, DAMN was also formed, which suggests that both play a similar role in HCN polycondensation in aqueous systems.

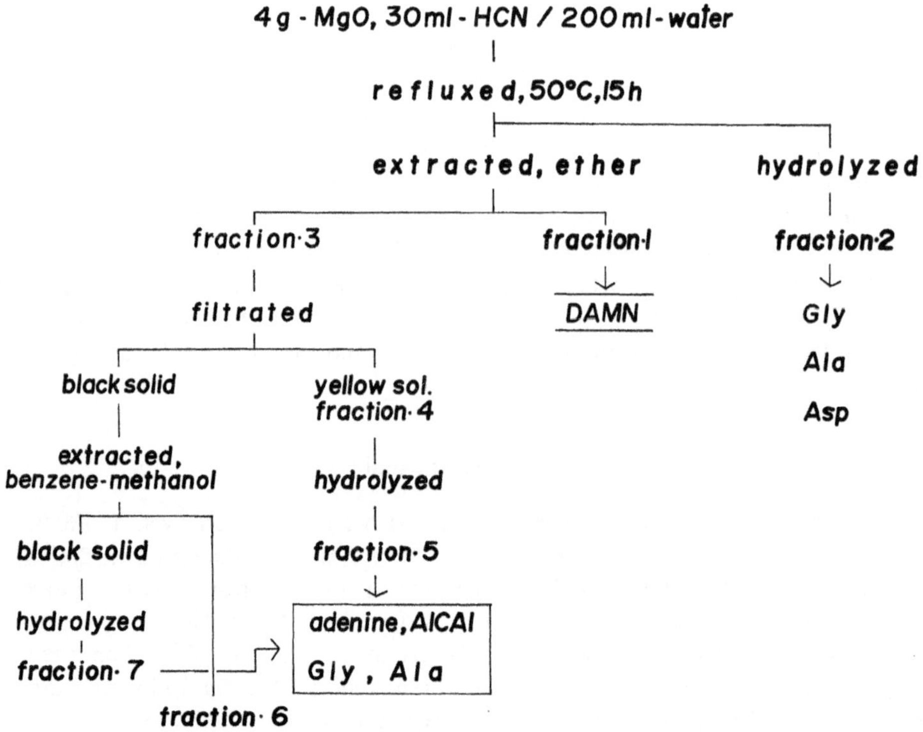


Fig. 1. *Preparative Scheme.* The compounds in parentheses show the confirmed compounds. Detail descriptions were made in the text.

TABLE I

Paper chromatographic analysis of compounds synthesized or detected in hydrolyzates.

Solvents: 1 – *n*-butanol-acetic acid-water (4:1:1, v/v)
2 – water saturated *n*-butanol
3 – *n*-propanol-1N ammonia (3:1, v/v)
4 – *n*-butanol-acetic acid-water (2:1:1, v/v)
5 – ethanol-conc $NH_4OH$-water (90:5:5, v/v)

Paper: Whatman #3MM. For the separation of DAMN, descending method, 18 hr at 20 °C; for amino acids, two dimensional ascending method, 10 hr (1st direction) and for 8 hr (2nd direction); for heterocyclic compounds, descending method, 18–23 hr at 20 °C. Every value in parentheses refers to the authentic standard.

| Fraction | Compounds | Solvents 1 | 2 | 3 | 4 | 5 |
|---|---|---|---|---|---|---|
| 1 | DAMN | 0.75<br>(0.76) | 0.72<br>(0.69) | | | |
| 2 | Gly | | | | 0.38<br>(0.35) | 0.24<br>(0.23) |
| | Ala | | | | 0.40<br>(0.43) | 0.39<br>(0.37) |
| | Asp | | | | 0.38<br>(0.37) | 0.07<br>(0.07) |
| | Unidentified<br>(ninhydrin positive) | | | | 0.40<br>0.36 | 0.43<br>0.43<br>0.12 |
| 5 | Gly | | | | 0.35 | 0.21 |
| | Ala | | | | 0.44<br>(0.42) | 0.36<br>(0.37) |
| | Adenine | 0.60<br>(0.57) | 0.44<br>(0.45) | 0.57<br>(0.59) | | |
| | AICAI | 0.26<br>(0.24) | 0.16<br>(0.19) | 0.48<br>(0.48) | | |
| | Unidentified<br>(UV-absorbing) | 0.50<br>0.47 | 0.35<br>0.22<br>0.16<br>0.11 | 0.50<br>0.42 | | |

### 3.2. Amino acids

A part of the reaction mixture was hydrolyzed with 6N HCl for 22 hr to give Fraction 2 which was analyzed using two dimension paper chromatography (Table I). As shown in the table, Fraction 2 gave several nihydrin positive components among which Gly, Ala and Asp were identified and confirmed by ion-exchange column chromatography. In Fraction 5 and 7, Gly and Ala were identified. The results obtained for Fraction 5 are shown in Table I. The same amino acids have been reported to be formed in ammoniacal HCN system (Oró *et al.*, 1961). The formation of amino acids in the present experiments may suggest a reaction mechanism involving hydrogen cyanide condensation to DAMN similar to those described by different authors (Sanchez *et al.*, 1966, 1967a and b; Matthews *et al.*, 1966; Yuasa *et al.*, 1970).

### 3.3. Heterocyclic compounds

Fraction 5 and 7 gave adenine, 4-aminoimidazole-5-carboxyamidine (AICAI) and other UV absorbing components. Their yields were very low but were considerable in comparison with the HCN—$NH_3$ system (Sanchez *et al.*, 1966).

Fraction 5 and 7 were first analyzed by paper chromatography using three different solvents. Fraction 5 showed quite similar chromatographic behavior to Fraction 7. The results obtained for Fraction 5 are also given in Table I and show that the Rf values of the synthesized samples are comparable with those of the authentic standard samples.

After measuring Rf values, each UV-absorbing spot containing the sample and the authentic standard and a spot from the blank paper chromatogram were cut off in small pieces, and extracted in test tubes with 3 ml of 0.1 N HCl for 24 hr and the UV absorption spectra were measured. Figures 2a and b give the spectra corresponding to two of the chromatographed compounds in comparison with that of the authentic adenine and AICAI.

The above results show that adenine and AICAI were formed in the present

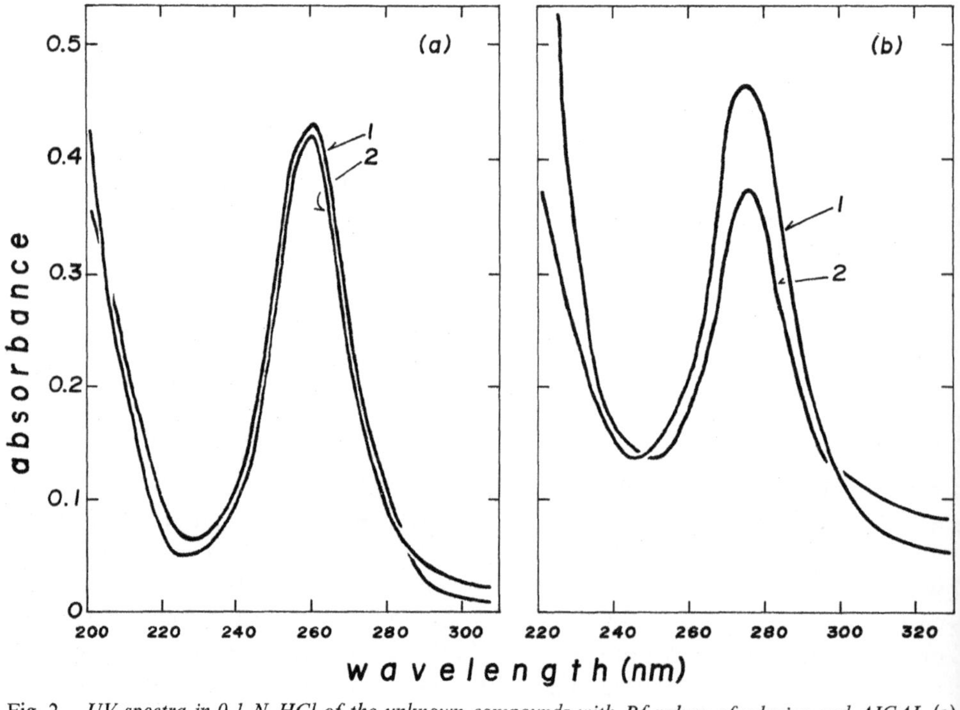


Fig. 2. *UV-spectra in 0.1 N HCl of the unknown compounds with Rf values of adenine and AICAI.* (a) unknown compound with Rf value of adenine (1) and the authentic adenine (2). (b) unknown compound with Rf value of AICAI (1) and the authentic AICAI (2).

experiments. Since DAMN was produced, it is highly conceivable that adenine and AICAI were formed from DAMN, together with formamidine, in our aqueous system, although the reaction mechanisms are still unclear.

Meanwhile, more than three UV-absorbing spots were observed besides adenine and AICAI (Table I), but because of low yields they have not yet been identified. In order to determine if 4-aminoimidazole-5-carboxyamide, and 4-amino-5-cyano-imidazole, significant in a prebiotic synthesis to purines, are formed as expected under the present experimental conditions, more effort should be made to identify them.

It is of interest that protein amino acids and heterocyclic compounds, although few in number, are formed under the simplest conditions without the addition of ammonia or other related bases such as amines (Yuasa and Ishigami, 1972) as catalysts. If primitive gases were similar in composition to volcanic gases, $NH_3$ may not have been present initially, but base catalyzed reactions would produce $NH_3$ from HCN. Other important compounds, e.g., nonproteinic amino acids such as $\beta$-Ala, diaminosuccinic, diaminopropionic, aminobutyric, could also be expected to form from aqueous hydrogen cyanide in the presence of weak bases even though ammonia was not initially present. The above could then be an important prebiotic formation process of organic compounds, since weak bases are abundant in the earth and extraterrestrial bodies.

## Acknowledgements

We would like to thank Drs D. Nooner and D. Flory for their suggestions concerning this manuscript. This research was supported in part by the National Aeronautics and Space Administration grant #NGR-44-005-125.

## References

Hayes, J. M.: 1967, *Geochim. Cosmochim. Acta* **31**, 1395.
Lange, N. A.: 1967, *Lange's Handbook of Chemistry* (10th rev. ed.), p. 165.
Matthews, C. N. and Moser, R. E.: 1966, *Proc. Natl. Acad. Sci. U.S.* **56**, 1087.
Oró, J.: 1960, *Biochem. Biophys. Res. Commun.* **2**, 407.
Oró, J.: 1965, in *The Origin of Prebiological Systems* (ed. by S. W. Fox), Academic Press, N.Y., p. 137.
Oró, J. and Kamat, S.: 1961, *Nature* **190**, 442.
Sanchez, R. A., Ferris, J. P., and Orgel, L. E.: 1966, *Science* **153**, 72.
Sanchez, R. A., Ferris, J. P., and Orgel, L. E.: 1967a, *Science* **154**, 784.
Sanchez, R. A., Ferris, J. P., and Orgel, L. E.: 1967b, *J. Mol. Biol.* **30**, 223.
Wiik, H. B.: 1956, *Geochim. Cosmochim. Acta* **9**, 279.
Yuasa, S., Ishigami, M., Honda, Y., and Imahori, K.: 1970, *Proc. Intl. Symposium Hydrogeochem. Biogeochem.* (Tokyo).
Yuasa, S. and Ishigami, M.: 1972, *Protein-Nucleic Acid-Enzyme*, #200, p. 32.

# THE PREBIOTIC SYNTHESIS OF DEOXYTHYMIDINE OLIGONUCLEOTIDES*

E. STEPHEN-SHERWOOD, D. G. ODOM and J. ORÓ

*Depts. of Biophysical Sciences and Chemistry, University of Houston, Houston, Tex. 77004, U.S.A.*

## 1. Introduction

The only successful syntheses so far reported for the prebiotic formation of oligodeoxyribonucleotides involve the condensation, under warm aqueous conditions of deoxythymidine 5′-phosphate, using either cyanamide (Ibañez *et al.*, 1971a, b) or imidazole (Ibañez *et al.*, 1971b, c) as catalysts. In both condensations, oligomers of up to five units in length were obtained, but the total yields were less than 2%. The majority of syntheses which have led to the formation of oligoribonucleotides have been conducted under drying conditions at relatively high or moderate temperatures, the latter simulating the effects resulting from the drying up of primeval lakes and ocean beds (Moravek *et al.*, 1968a, b, c; Tapiero and Nagyvary, 1971; Osterberg *et al.*, 1973). In view of the success of these reactions it was decided to continue our work on the synthesis of oligodeoxyribonucleotides and to study the effects of drying conditions on the formation of these compounds.

First, the condensation of deoxythymidine 5′-phosphate, in the presence of cyanamide, or cyanamide and 4-amino-5-imidazole carboxamide (AICA), was investigated under drying conditions. In both cases the yields of oligodeoxynucleotides were found to be low and in the case of the reaction with AICA and cyanamide, the major products were cyclic dithymidine diphosphate and $P^1,P^2$-deoxydithymidine 5′-pyrophosphate. However, if in addition to deoxythymidine 5′-phosphate, $^{14}C$ labeled deoxythymidine 5′-triphosphate was also included as a reactant, and the reaction was carried out under drying conditions in the presence of AICA and cyanamide, the major products obtained were found to be oligodeoxyribonucleotides of the formula $(pdT)_n$, with $n=2$–4 and possibly higher. Deoxythymidine 5′-phosphate was chosen as a 'primer' or 5′-P-terminal nucleotide, in this prebiotic condensation, since it was considered essential to block the more reactive primary hydroxyl group in order to prevent appreciable formation of 5′,5′-phosphodiester bonds. Further preliminary experiments showed that deoxythymidine 5′-triphosphate did not condense as readily with itself as with deoxythymidine 5′-phosphate to produce oligodeoxynucleotides under the conditions of the reaction being investigated.

## 2. Experimental

### 2.1. Materials

Cyanamide (Eastman Kodak mp 42 °C), 4-amino-5-imidazole carboxamide hydro-

* Reprinted by permission from the *Journal of Molecular Evolution* (1974) © Springer-Verlag, Heidelberg.

chloride grade A (Calbiochemical), deoxythymidine 5′-phosphate disodium salt (Sigma) and deoxythymidine 5′-triphosphate-2-$^{14}C$ tetralithium salt (Schwartz). Both nucleotides were checked in solvent systems I and II (see below) and shown to be pure. The following enzymes were used in the analytical procedure: Alkaline phosphatase from *E. Coli* (Sigma), snake venom phosphodiesterase (Worthington), and spleen phosphodiesterase (Worthington). The standard oligodeoxynucleotides $(pdT)_2$ and $(pdT)_4$ were purchased from Collaborative Research. Other oligodeoxynucleotide standards, namely $(pdT)_3$ and $(pdT)_5$ were synthesized according to the procedure of Khorana and Connors (1966) and separated on a DEAE cellulose column, bicarbonate form, into the appropriate fractions. The $(pdT)_2$ and $(pdT)_4$ fractions were authenticated by chromatography with the commercially purchased standards, in solvent systems I and II.

### 2.2. Chromatography

Thin layer chromatography was carried out on precoated plastic sheets 20 cm × 20 cm (Polygram cel 300 PEI (polyethyleneimide) Brinkmann, Westby, N.Y.) using the following two systems: I. Consecutive chromatography in two different solvents. The chromatogram was developed first in formic acid (20%), allowed to air dry over night, and then developed again in the same direction in ammonium bicarbonate 2%. II. Chromatography in only one solvent, ammonium bicarbonate (2%). These two chromatographic systems were investigated in detail for this work (Odom, 1973), and Rf values of a number of standard compounds in these systems are given in Table I. Paper chromatography was carried out on Whatmann 1-MM and 3-MM paper by the descending technique in the following solvent (III): isopropanol ammonia, water (70/10/20; v/v) (Weiman and Khorana, 1962a). After chromatography, the presence of radioactive compounds was determined by cutting the polyethyleneimide cellulose plates in 0.5 cm strips and counting them in a Packard scintillation counter Model 3380. Paper chromatograms were cut into 1 cm strips and similarly counted.

TABLE I

Rf values of different compounds on PEI cellulose plates

| | Solvent system I | Solvent system II[a] |
|---|---|---|
| Thymidine | 0.95 | 0.95 |
| pdT | 0.85 | 0.57 |
| $(pdT)_2$ | 0.50 | 0.50 |
| $(pdT)_3$ | 0.44 | 0.46 |
| $(pdT)_4$ | 0.39 | 0.41 |
| $(pdT)_5$ | 0.34 | 0.38 |
| ppdT | 0.37 | 0.49 |
| pppdT | 0.17 | 0.26 |

[a] See text.

### 2.3. Procedure

A typical reaction consisted of an aqueous solution (0.8 ml) of deoxythymidine 5′-phosphate (0.1 m mole), AICA (0.4 m mole), cyanamide (0.5 m mole) and $^{14}C$ deoxythymidine 5′-triphosphate (2$\mu$ Ci, 45.5 m Ci/m mole). The pH of the resultant solution was approximately 4.0. The reaction mixtures were heated at 90 °C in open glass tubes for 40 hr, at which time, water (0.5 ml) was added and the reaction products chromatographed on thin layer plates using solvent systems I and II. Deoxythymidine 5′-phosphate (dTMP), deoxythymidine and deoxydithymidine diphosphate $(pdT)_2$ were used as markers. The plates were subsequently strip counted to determine the position of radioactive compounds. Proper controls of the reaction in the absence of cyanamide and AICA were also carried out.

### 2.4. Results

The crude reaction product obtained with AICA and cyanamide produced three radioactive components which chromatographed below dTMP in solvent system I, and which in subsequent tests were demonstrated to be mainly oligodeoxyribonucleotides (Figure 1).

Since the presence of polyphosphates and pyrophosphates would complicate subsequent analyses, this crude reaction product was subjected to a procedure which degrades pyrophosphate bonds, but leaves other phosphodiester and phosphomonoester bonds of oligodeoxynucleotides intact (Moon and Khorana, 1966). After this degradation, chromatography was performed in solvent systems I, II and III, all of which gave mutually consistent results. The separation of radioactive components in solvent system I of the pyrophosphate-free product is shown in Figure 2a. Co-chromatography of this reaction product with authentic samples of $(pdT)_2$, $(pdT)_3$ and $(pdT)_4$ in solvent system I was also performed, and showed that some of the radioactive components which chromatographed below TMP, contained oligodeoxynucleotides of the type $(pdT)_n$: a = $(pdT)_2$; b = $(pdT)_3$; c = $(pdT)_4$.

Final verification that oligonucleotides of the $(pdT)_n$ series are formed in this reaction was obtained by a series of enzymatic degradation reactions outlined in the scheme of Figure 3. In addition, cold $(pdT)_2$ was added to the reaction prior to incubation with alkaline phosphatase. After incubation and chromatography in system III, one radioactive component co-chromatographed with the dithymidine phosphate (dTpdT) which was produced during the enzymatic degradation by alkaline phosphatase thus verifying the formation of $(pdT)_2$ in the reaction.

After pyrophosphate degradation, snake venom phosphodiesterase was found to completely degrade the reaction products which chromatograph below dTMP in solvent systems I and II, to deoxythymidine 5′-phosphate (Figure 2b). This indicates that these products have a 3′-hydroxyl group and that the oligomers must have either 3′-5′ or 5′-5′ phosphodiester bonds. Conversely spleen phosphodiesterase did not cause a significant degradation of the pyrophosphate-free products (Figure 2c). The failure of spleen phosphodiesterase to degrade these products indicates that

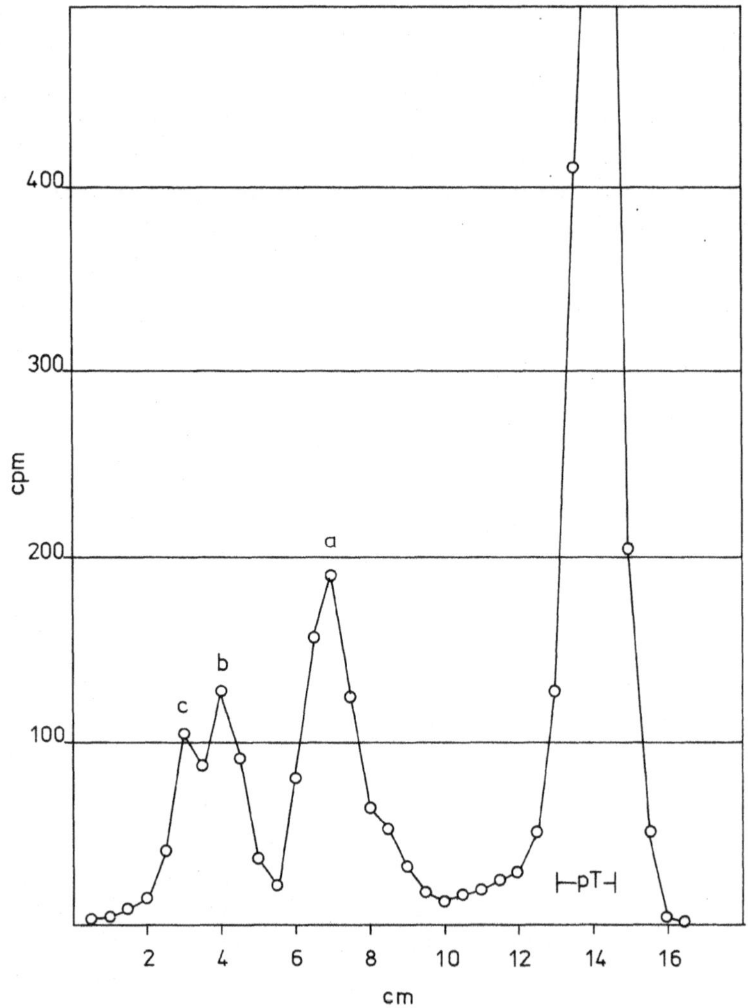


Fig. 1. A strip count of the radioactivity of a thin layer chromatogram on a polyethyleneimide cellulose plate, using solvent system I, of the reaction product before chemical and enzymatic degradation. Peak *a* cochromatographs with a standard sample of $(pdT)_2$. Peaks *b* and *c* represent higher oligomers together with deoxythymidine pyrophosphates. The total counts were corrected for background and normalized to 3500 counts per minute. All the symbols appearing in this figure (e.g. T, pT, etc.) correspond to deoxythymidine or its derivatives.

they do not have free 5′-hydroxyl groups, either due to the presence of 5′-phosphates or the presence of abnormal bonding. Digestion by alkaline phosphatase followed by spleen phosphodiesterase yields mainly thymidine (Figure 2d) indicating that the majority of the oligomers contain only 3′-5′ phosphodiester bonds with a terminal 5′-phosphate group.

The total yield of oligodeoxynucleotide product, determined on the basis of the percent counts which chromatographed below deoxythymidine 5′-phosphate after

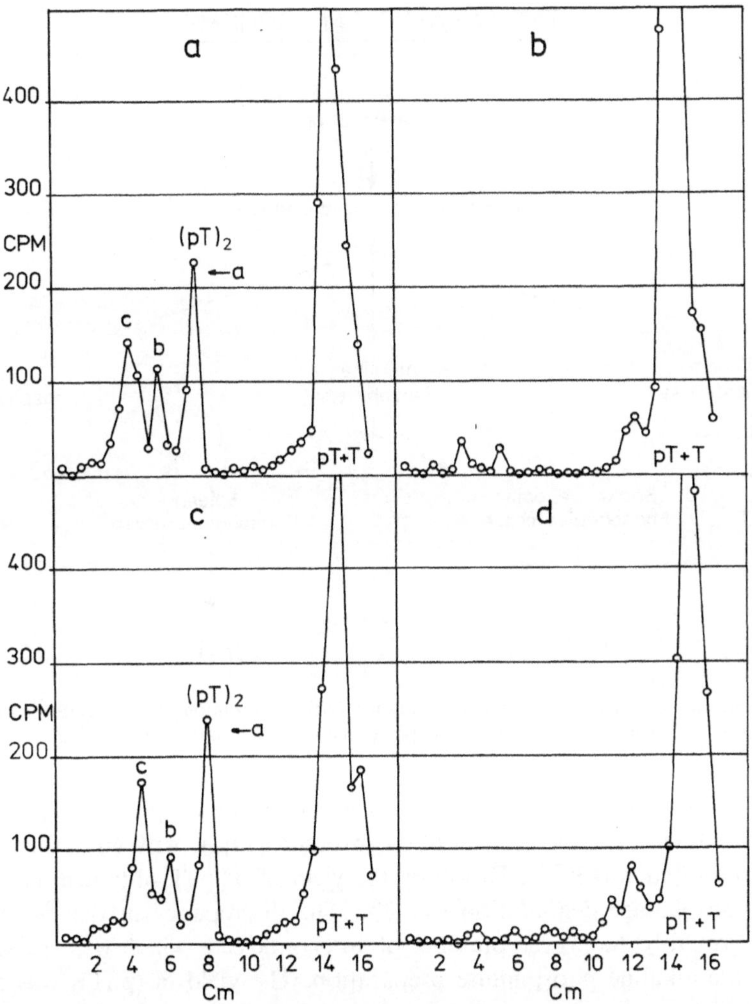


Fig. 2. Strip counts of the radioactivity of a thin layer chromatogram on a polyethyleneimide cellulose plate using solvent system I, of the reaction product; (a) after being subjected to pyrophosphate degradation; (b) after pyrophosphate degradation and treatment with snake venom phosphodiesterase; (c) after pyrophosphate degradation and treatment with spleen phosphodiesterase; (d) after pyrophosphate degradation and treatment with alkaline phosphatase and spleen phosphodiesterase.

The total counts in each case were corrected for background and normalized to 3500 counts per minute. Peak *a* co-chromatographs with a standard sample of a $(pdT)_2$. Peak *b* co-chromatographs with a standard sample of $(pdT)_3$. Peak *c* co-chromatographs with a standard sample of $(pdT)_4$. Rechromatography of this fraction indicated it was a mixture of two components, possibly $(pdT)_4$ and $(pdT)_5$. All the symbols appearing in this figure (e.g. T, pT, etc.) correspond to deoxythymidine or its derivatives.

the degradation of the pyrophosphate bonds, was 32%. On the basis of the results of the enzymatic degradation reactions, 25.6% of the oligodeoxynucleotides (or 80% of the 32%) contained natural 3′-5′ bonding only, while 5.7% contained 5′-5′ bonds, and 0.6% contained 3′-3′ bonds. The total yield of oligodeoxynucleotides of the

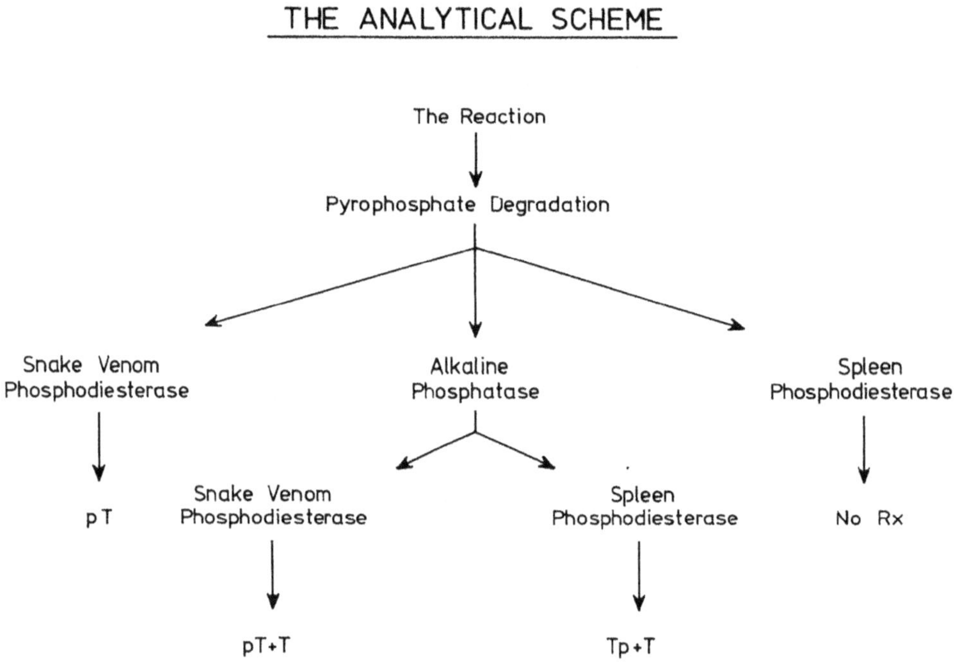


Fig. 3. Separation scheme for the chemical and enzymatic degradation of the reaction crude product. All the symbols appearing in this figure (e.g. T, pT, etc.) correspond to deoxythymidine or its derivatives.

formula $(pdT)_n$ was 21.6%. Another 4% had two terminal phosphate groups. The yield of $(pdT)_2$ estimated as a result of co-chromatography with an authentic sample of this compound was 9.3%. However, the yield of dTpdT obtained as a result of alkaline phosphatase degradation was 3%. The discrepancy in the two yields for $(pdT)_2$ is probably due to the presence of an exonuclease which was shown to contaminate the alkaline phosphatase preparation. The yield of $(pdT)_3$ was estimated at 5.6%.

No significant amounts of radioactive oligomeric products were observed from a control reaction containing [$^{14}$C]deoxythymidine 5′-triphosphate and dTMP under the same conditions of concentration and temperature (89 °C) as described in the general procedure, but in the presence of cyanamide alone. The pH of this control was 8.4. In this case it was found that 98% of the [$^{14}$C]dTTP had degraded to dTMP and deoxythymidine. This demonstrates that during oligonucleotide synthesis, degradation reactions also occur and become extensive at high temperatures. However, at 60 °C, preliminary experiments indicate that oligomeric products are formed when equimolar quantities of [$^{3}$H]deoxythymidine 5′-triphosphate and dTMP are condensed at pH 7.5 in the presence of cyanamide alone. Only traces of oligomeric products are formed under the latter conditions of temperature and concentration and at pH 7.5 in the presence of AICA alone. However, the yield of oligomeric products is considerably improved when both AICA and cyanamide are present in the reaction mixture. A comparison of two identical reactions, one containing [$^{14}$C]dTMP and the other [$^{3}$H]deoxythymidine 5′-triphosphate condensed in the presence of cyanamide and AICA, at either 89 °C or 60 °C indicates that both dTMP and deoxythymidine 5′-triphosphate are incorporated into the oligomeric products. The degree of incorporation of [$^{14}$C]dTMP was considerably lower than that of [$^{3}$H]dTTP. These preliminary experiments [M. More, unpublished] indicate that under appropriately modified conditions, (e.g. lower temperature, pH closer to neutrality, etc.) the synthesis of oligodeoxynucleotides may be further improved.

## 3. Discussion

Though the mechanism of this polymerization is uncertain, it is possible that deoxythymidine 5′-triphosphate, in the presence of cyanamide and AICA, cyclizes to a trimetaphosphate. In an extensive study of the mechanism of internucleotide bond formation in pyridine, using dicyclohexylcarbodiimide (DCC) as a condensing agent, an alkyltrimetaphosphate has been implicated as one active phosphorylating species (Weimann and Khorana, 1962b; Jacob and Khorana, 1964). Further, in the condensation of trialkylammonium nucleoside 5′-phosphates with inorganic phosphate in the presence of DCC, though tetra- and higher polyphosphates were initially produced, ultimately, nucleoside 5′-triphosphates were formed at the expense of these higher polyphosphates. The increased yield of nucleoside 5′-triphosphates over higher polyphosphates was postulated as being due to stabilization of the triphosphate molecule by trimetaphosphate formation (Smith and Khorana, 1958; Weimann and Khorana, 1962b). The above interpretations are consistent with the results obtained in this work. However, a more extensive study is required to firmly establish the mechanism of this reaction.

The role of AICA in this reaction, is probably that of a proton donor which is consistent with the mechanism of catalysis of a number of imidazole derivatives. $\beta$-Imidazole-4(5) propanoic acid or triethylamine hydrochloride catalyze the polymerization of deoxythymidine 5′-phosphate to oligonucleotides in boiling N,N,-dimethylformamide (Pongs and T'so, 1969, 1971). The catalytic function of these compounds has also been proposed as that of proton donor. AICA was probably readily available on the primitive earth since it is formed during the polymerization of ammonium cyanide and is an intermediate in the abiotic formation of purines (Oró and Kimball, 1961). Cyanamide is an accepted prebiotic condensing agent (Oró, 1963; Kenyon and Steinman, 1969), and due to its low melting point, 42 °C, may also aid in forming a melt in which the polymerization reaction can proceed. In addition, AICA and cyanamide have points of structural similarity with histidine and arginine, respectively, two amino acids which have been shown to be present in the active center of DNA polymerase (Salvo, 1973). Although this analogy is somewhat tenuous, it could well be that the guanidine and imidazole groups present in these two amino acids play an important role both in the biological and prebiotic synthesis of oligodeoxyribonucleotides.

This synthesis of oligodeoxyribonucleotides offers a possible method under which these important polymers may have been produced on the primitive earth. The reaction also takes into account the fact that although nucleoside triphosphates were probably available on the primitive earth, their concentrations relative to nucleoside monophosphates were probably low (Neuman *et al.*, 1970).

## Acknowledgements

The results presented here are taken in part from the thesis of D. G. Odom (1973). This work was supported by a research grant NGR-44-005-002 from the National Aeronautics and Space Administration, by an NDEA fellowship to one of us (D.G.O.) and was presented at the 4th International Conference on the Origin of Life, June 25–28, 1973, Universidad Autonoma de Barcelona, Barcelona, Spain. We wish to thank Miss M. A. Moré for her technical assistance in the preparation of this manuscript.

## References

Ibañez, J. D., Kimball, A. P., and Oró, J.: 1971a, *Science* **173**, 444.

Ibañez, J. D., Kimball, A. P., and Oró, J.: 1971b, *Chemical Evolution and the Origin of Life* (ed. by R. Buvet and C. Ponnamperuma), North-Holland, Amsterdam, p. 171.

Ibañez, J. D., Kimball, A. P., and Oró, J.: 1971c, *J. Mol. Evolution* **1**, 112.

Jacob, T. M. and Khorana, H. G.: 1964, *J. Amer. Chem. Soc.* **86**, 1630.

Kenyon, D. H. and Steinman, G.: 1969, *Biochemical Predestination*, McGraw-Hill Book Company, New York.

Khorana, H. G. and Connors, W. J.: 1966, *Biochem. Prep.* **11**, 113.

Moon, N. W. and Khorana, H. G.: 1966, *J. Amer. Chem. Soc.* **88**, 1798.

Moravek, J., Kopecky, J., and Skoda, J.: 1968a, *Collection Czech. Chem. Commun.* **33**, 960.

Moravek, J., Kopecky, J., and Skoda, J.: 1968b, *Collection Czech. Chem. Commun.* **33**, 4407.

Moravek, J., Kopecky, J., and Skoda, J.: 1968c, *Collection Czech. Chem. Commun.* **33**, 4120.

Neuman, M. W., Neuman, W. F., and Lane, K.: 1970, *Curr. Mod. Biol.* **3**, 277.

Odom, D. G.: 1973, 'Thin Layer Chromatographic Analysis of Oligonucleotide Mixtures and Prebiotic Condensation of Mononucleotides', Ph.D. Thesis, University of Houston, Houston, Texas, August, 1973.

Oró, J.: 1963, *Ann. N.Y. Acad. Sci.* **108**, 464.

Oró, J. and Kimball, A. P.: 1961, *Arch. Biochem. Biophys.* **94**, 217.

Osterberg, R., Orgel, L. E., and Lohrmann, R.: 1973, *J. Mol. Evol.* **2**, 231.

Pongs, O. and T'so, P. O. P.: 1969, *Biochem. Biophys. Res. Commun.* **36**, 475.

Pongs, O. and T'so, P. O. P.: 1971, *J. Amer. Chem. Soc.* **93**, 5241.

Salvo, R. A.: 1972, 'The Identification of Catalytically Reactive Amino Acid Residues in the Active Center of DNA-dependent DNA Polymerase I', Ph.D. Thesis, University of Houston, Houston, Texas, December, 1972.

Smith, M. and Khorana, H. G.: 1958, *J. Amer. Chem. Soc.* **80**, 1141.

Tapiero, C. M. and Nagyvary, J.: 1971, *Nature* **231**, 42.

Weimann, G. and Khorana, H. G.: 1962a, *J. Amer. Chem. Soc.* **84**, 419.

Weimann, G. and Khorana, H. G.: 1962b, *J. Amer. Chem. Soc.* **84**, 4329.

# A POSSIBLE STEP IN THE ORIGIN OF THE GENETIC CODE*

L. E. ORGEL

*The Salk Institute for Biological Studies, San Diego, Calif., U.S.A.*

**Abstract.** It is suggested that the earliest coding polynucleotides contained a high proportion of alternating sequences of purines and pyrimidines, and that these sequences coded for polypeptides in which hydrophobic and hydrophylic amino acids alternated. Structural properties of such alternating polypeptides are discussed.

## 1. Introduction

In trying to understand the evolution of the protein-synthetic apparatus, we are faced with a problem of a kind that is familiar to students of other aspects of evolution. Protein synthesis is a complicated process made possible by the cooperation of more than a hundred macromolecules, including both proteins and nucleic acids. It is quite clear that a system of this complexity could not have formed by the random aggregation of prebiological polynucleotides and polypeptides. However, it is equally clear that any system that could have come together by chance would be useless in the context of contemporary protein synthesis.

Dilemmas of this type often arise when attempts are made to explain the evolution of complex biological structures. A rudimentary wing would be of little use to a snake, so how did birds evolve? The strategy that has usually been adopted in tackling such problems is to postulate a series of minimal steps that, occurring in sequence, could have led to the evolution of the complex structure in question. It is then necessary to explain why an additional selective advantage accrued with each step in the proposed sequence. Here, I shall apply this technique to devise a theory of the first step in the evolution of the genetic code.

It has been suggested that contemporary protein synthesis evolved from a simpler process in which classes of amino acid rather than individual amino acids were specified by a primitive genetic code (Woese, 1967). It has also been pointed out that the hydrophobic or hydrophilic character of the side-chain helps to determine the position of an amino acid in the genetic code (Woese, 1967). I shall use these ideas in the development of my theory.

I propose that the function of the most primitive coding mechanism was to direct the synthesis of peptides in which hydrophobic and hydrophilic side-chains tended to alternate. I shall show, first, that such alternating peptides might have facilitated the replication of families of associated polynucleotides. Then, I shall discuss very briefly the abiotic synthesis of polynucleotides, and show how a system of polynucleotides capable of directing the synthesis of alternating polypeptides could have evolved.

In developing my theory I shall make a number of assumptions that have been discussed in detail elsewhere (Crick, 1968; Orgel, 1968). In particular I suppose:

* The material covered in this paper was presented previously in the *Israel Journal of Chemistry*, **10** (1972), pp. 287–292. Dedicated to the Memory of Gerhard M. J. Schmidt.

(1) Some form of polynucleotide replication was taking place at the time the code evolved.

(2) Replication, although inaccurate, obeyed the Watson-Crick base-pairing rules.

(3) A variety of amino acids was present in the environment of the replicating polynucleotides.

(4) The activation of amino acids occurred with sufficient frequency to permit the accumulation of peptides.

(5) The evolution of the code was initiated when an interaction between amino acids and polynucleotides that depended on the structure of the amino acids and the sequences of the polynucleotides permitted some sequences to replicate faster than others.

## 2. Partially Ordered Peptides and the Code

The assignment of amino acids to trinucleotide codons that is specified by the genetic code is not random. The most obvious regularities relate to the abstract structure of the code; quartets of codons XYU, XYC, XYA, XYG are assigned to a single amino acid or distributed between two amino acids according to definite rules. I shall discuss these regularities in detail elsewhere. Here I shall concentrate on the structural representation used in the code, that is, the way in which the chemical charzcteristics of an amino acid are, so to speak, spelt out by the codon.

A glance at the code reveals that all amino acids specified by a codon with U in the central position (U-group for short) have hydrophobic side-chains (Phe, Leu, Ileu, Met, Val). It seems natural, therefore, to seek structural features shared by amino acids of the A-, C- and G-groups. The nature of these features has been recognised by several authors (Woese, 1967) and exhibited in a striking fashion by Dickerson (1971).

Amino acids of the A-group have hydrophilic side-chains (Asp, Glu, Asp N, Glu N, Tyr, Lys, Hist). It is hard to find any other characteristic shared by these amino acids, since the first two are negatively charged, the next three are neutral and the last two positively charged. Amino acids of the C- and G-groups are intermediate between those of the U- and A-groups; the C-group is perhaps more hydrophobic than the G-group. These rules are not perfect, and it is here that Dickerson's generalization is so suggestive. He states that amino acids of the U-group almost always occur inside globular proteins, while those of the A-group almost always occur outside. Amino acids of the C- and G-groups occur frequently in both types of position. Any theory of the origin of the code must explain this important generalization.

Elsewhere, in discussing the evolution of the code, we have been guided by the principle that no postulated step should depend on the fortuitous aggregation of a large number of macromolecules, each with a remarkable chemical specificity (Crick, 1968; Orgel, 1968). Now I shall suppose that the first step in the evolution of the code imposed only a very simple order on peptide sequences, and that even partial ordering conferred some selective advantage. These assumptions, together with a knowledge of the representation used in the code, lead to some surprisingly strong conclusions.

It seems likely that the most primitive code discriminated between two amino acids or groups of amino acids, since all other patterns of discrimination are more difficult. The structure of the code suggests strongly that the discrimination was in fact between groups of amino acids and was based on the hydrophobic or hydrophilic character of the side-chains.

It follows next that the ordering imposed by the code was a local ordering. Any other ordering is more complex. A local ordering, as we shall see, can be achieved with a minimum of 'apparatus', but any other ordering could be realized only by the stepwise progression of a translation apparatus along a long message with a precisely defined sequence.

The degree of local ordering in a linear sequence containing two types of element is specified by the nearest-neighbour frequencies. Partially ordered sequences may deviate from randomness either in the direction of more or less alternation. I assume the code was designed to favour the synthesis of alternating sequences.

What structural feature would differentiate alternating sequences of hydrophobic and hydrophilic residues from random sequences? X-ray studies have shown that the $\beta$-sheets are very common structural elements of globular proteins. Furthermore, there is a tendency for alternating sequences of hydrophobic and hydrophilic residues to form parts of $\beta$-sheets (A. Mc. Glochlen, private communication). Dickerson's classification suggests, therefore, that the initial purpose of the code was to produce peptides that tended to form asymmetric $\beta$-sheets with one hydrophobic and one hydrophilic surface.

I can give only general reasons for believing that the formation of asymmetric $\beta$-sheets would have conferred selective advantage on a system of polynucleotides and polypeptides. In aqueous solution, regions of asymmetric $\beta$-sheets would come together with their hydrophobic surfaces in apposition. Alone, or in association with lipids, they would have formed colloidal aggregates with a hydrophobic interior and a hydrophilic exterior. Of course, some random polymers would also have formed colloidal particles, but less efficiently.

Many authors have emphasized the importance of colloidal aggregates (coacervates and membranes) in the evolution of life (Kenyon and Steinman, 1969). The first function of the code may have been to direct the synthesis of peptides that aggregated more readily than random polypeptides to form water-soluble colloidal particles. These colloidal particles would have played an important role in biochemical evolution by holding together families of cooperating polynucleotides.

A novel feature of this model is the form of the relationship that it suggests between the accuracy of coding and its effectiveness. Since even random peptides could have facilitated polynucleotide replication to some extent, the selective advantage conferred by ordering would have been a continuous, monotonically increasing function of the degree of ordering. Since no threshold existed that had to be overcome before selection could begin, no improbable accidents would have been required to initiate the evolution of the code.

This type of code also has the advantage that the consequences of errors of all types are local and cumulative. Provided selection was strong enough, a code of this kind would have evolved continuously to reduce the error frequency to the lowest value compatible with the specificity of polynucleotide-amino acid recognition.

The proposed mechanism would have been effective only if all amino acids had the same optical configuration. The insertion of a single hydrophobic D-amino acid in an alternating $\beta$-sheet composed of L-amino acids, for example, would place the side-chain on the hydrophilic side of the surface. This provides an explanation, although not a unique explanation, of another feature of protein synthesis.

## 3. Alternating Polynucleotides

We have argued in detail elsewhere that molecules with structures similar to those of tRNA are likely to be formed during the abiotic synthesis of polynucleotide (Crick, 1968; Orgel, 1968) and that alternating polynucleotides Py Pu Py Pu ..., where Py and Pu are pyrimidine and purine respectively, are likely to replicate faster than other polynucleotides and hence to have dominated the prebiotic soup (Sulston and Orgel, 1969). If these ideas are correct, they throw some light on the origin of alternating peptides.

Let us first make a highly restrictive assumption, namely that the prebiotic soup contained perfectly alternating polynucleotides and no others. Then all messengers would have consisted of an alternating sequence of Pu Py Pu and Py Pu Py codons; all tRNA-like molecules would have had Pu Py Pu or Py Pu Py anticodons. If hydrophobic amino acids were assigned to tRNA's with Py Pu Py anticodons and hydrophilic amino acids to tRNA's with Pu Py Pu anticodons, the synthesis of alternating polypeptides would have been guaranteed.

A code of this type would have considerable stability against mutational error if, as seems likely, the most frequent errors of abiotic nucleotide replication are transitions caused by G-U (wobble) mispairing. By assigning the two sets of eight anticodons (Py Pu Py and Pu Py Pu) to two sets of amino acids (hydrophobic or hydrophilic), the effects of Py-Py or Pu-Pu replacements are neutralized.

It should be noted that in a model of this type, long alternating polypeptides could be formed from short messengers. In the absence of a ribosome, it seems likely that tRNA would occasionally slip between equivalent positions on a messenger, since base-pairing between a trinucleotide anticodon and its codon is not adequate to immobilize a tRNA. However, a given loaded tRNA could move only to a position of the appropriate class (Pu Py Pu or Py Pu Py) so that the mobility of tRNA's would not interfere with the production of alternating polypeptides. In principle, alternating polypeptides could be produced on an oligonucleotide only seven units long.

In this oversimplified model, the evolution of a translation mechanism would require only the appearance of two tRNA-like molecules, one associating with hydrophobic and the other with hydrophilic amino acids. Mutations in the anticodon, brought about by wobble pairing, would guarantee that each tRNA soon 'spread' over the eight members of the appropriate set.

The discrimination between hydrophobic and hydrophilic amino acids is among the least demanding that we can imagine. Furthermore, even a very imperfect discrimination would immediately confer selective advantage, for example if the distribution factors were at first only 55:45 in favour of the correct assignments. Thus, insofar as selective interactions between amino acids and oligonucleotides do occur at all (Ruska and Mandel, 1971), they should particularly favour the evolution of the code of the type envisaged.

Next we must examine the effect of irregularities in polynucleotide messages caused by pairs of purines or pyrimidines occurring together. In the absence of appropriate tRNA's with non-alternating anticodons, Py Py Pu or Pu Pu Py codons would be translated according to their central base, since codon-anticodon interaction is dominated by base pairing at the central position. Later, appropriate tRNA's would have evolved which contained non-alternating anti-codons. This would account for the dominant role of the central base in determining the general character of the amino acid specified by a codon.

The effect of irregularities in the messenger sequence would thus be the same as that of poor discrimination, namely to reduce the degree of alternation; provided some polynucleotide alternation remained, newly synthesized polypeptide would tend to alternate. Selection could begin, therefore, even with messages that were only partially alternating.

Finally, it is worth noting that if alternation is a useful property of polypeptides, selection of alternating polynucleotides could occur, even if there was no tendency for them to form spontaneously. However, it does not seem worth developing this argument in detail, since there is almost certainly a spontaneous tendency for alternating sequences to accumulate.

## 4. Further Development of the Code

The theory that I have outlined is so speculative that I do not feel justified in dis cussing the further development of the code in any detail. I shall only point out that the assignment of amino acids to their codons may have been determined as much by the need to build specific structures such as folded sheets as by any requirement for enzyme-like activity.

## 5. Experimental Programme

This theory makes one very specific prediction, namely that partially ordered polypeptides in which hydrophobic and hydrophilic residues alternate will prove to form colloidal aggregates that are significantly different from those formed by random polypeptides. I hope to test this prediction in the near future, by synthesizing such polypeptides using a modification of the Merrifield technique.

**Acknowledgement**

I am indebted to Dr F. H. C. Crick for his helpful comments.

**References**

Crick, F. H. C.: 1968, *J. Mol. Biol.* **38**, 367.
Dickerson, R. E.: 1971, *J. Mol. Biol.* **57**, 1.
Kenyon, D. H. and Steinman, G.: 1969, *Biochemical Predestination*, McGraw Hill, New York.
A. McGlochlen, Private Communication.
Orgel, L. E.: 1968, *J. Mol. Biol.* **38**, 381.
Ruska, M. and Mandel, M.: 1971, *Proc. Nat. Acad. Sci. U.S.* **68**, 1190 and references therein.
Sulston, J. and Orgel, L. E.: 1970, *Second Conference of the Institute de la Vie*, Paris; *Proceedings of Conference on Origins of Life*, Houston, Texas.
Woese, C.: 1967, *The Genetic Code*, Harper and Row, New York.

# THE EMERGENCE OF CARCINOGENS DURING THE EVOLUTION OF LIVING SYSTEMS

A. GINER-SOROLLA and A. BENDICH
*Memorial Sloan-Kettering Cancer Center, New York, N.Y. 10021, U.S.A.*

**Abstract.** A correlation is explored between the appearance of chemical carcinogens throughout the evolution of living systems and a plausible chronology for the emergence of cancer cells. A model of malignancy in the form of a prebiotic neoplastic cell may have existed before the occurrence of normal cells. It is more likely, however, that malignant cells arose by the interaction of chemical agents with normal cells. In the early stages of Biochemical Evolution, the presence of certain molecules, ranging from inorganic substances to polycyclic hydrocarbons, may have constituted the first abiotic carcinogens (protocarcinogens). A variety of carcinogenic alkylating agents which exist in certain plants (actinomycetales, fungi) as secondary products of their metabolism, may have emerged during the early period of Biological Evolution. Favorable conditions for the first interaction between these substances (paleocarcinogens) and primitive metazoans probably occurred in the early Paleozoic. This probability of interaction reached a maximum in the Mesozoic where the abundance of plants containing toxic and potent carcinogenic substances (ferns, cycads) might have caused poisoning and malignant tumor formation of pandemic proportions in animals. In contrast, there exists only an insignificant proportion of present day angiosperms containing carcinogens of low potency. A parallel is drawn between the possible effect of plant carcinogens in the geological past, and the threat posed by substances of technological origin which abound and proliferate in today's environment.

The difficulties encountered in elucidating the origin of life parallel those found in the search for the cause of the abnormal form of life, cancer. Malignant tumors can be induced by diverse endogenous or exogenous agents which act by mechanisms which are still unclear. It has been suggested that a model of malignancy, in the form of a prebiotic neoplastic cell, may have existed before the occurrence of normal cells and could have represented the first appearance of 'pre-life' (Dawe and Harshbarger, 1969). Whether or not prebiotic sequences which led to the emergence of life were also operative in the genesis of cancer in its present manifestations, is a matter of controversy.

Many plausible physical and chemical pathways for the abiotic synthesis of the necessary components exist which may have led to the formation of living organisms. Similarly, many physical, chemical and, in addition, biological agents can induce malignant cell formation.

Chemical carcinogens, according to Finstad and Good (1969), can induce malignant tumors in all vertebrates, but none has yet been produced in hagfishes or invertebrates. Because of their potency and widespread occurrence, chemical carcinogens with all probability may have constituted the principal cause of cancer in the remote past.

We shall explore a possible correlation between the presence of chemical carcinogens throughout biochemical and biological Evolution and a plausible chronology of the initial process and further expression and development of malignancy. This correlation will be based upon the premise that the process of carcinogenesis has been operative since its inception in the same way as in the present; this conforms with the principle of evolutionary continuity.

## 1. The Occurrence of Carcinogens Through the Earth's History

Environmental conditions for the appearance of cancer cells (ultraviolet, ionizing radiations and carcinogenic chemicals), were at least as propitious in the primitive Earth as they are today. In the remote past, radiation (cosmic, from the Earth's crust radioactivity, and that from a younger sun) was more intense than at the present (Sagan, 1965). In addition, the protective ozone layer which now shields the planet from the solar ultraviolet radiation did not exist (Berkner and Marshall, 1965). One of the postulated components of the primitive Earth atmosphere was hydrazine (formed by short-wavelength solar ultraviolet and ionizing radiation of ammonia [Urey, 1952]). This substance, which has been detected in comets (Donn, 1972), exhibits carcinogenic properties (Severi and Biancifiori, 1968) and is structurally related to other potent natural and synthetic carcinogens.

Evidence for the occurrence of carcinogenic organic substances in sediments or in fossils, which could yield a chronological clue to the emergence of the first cancer cells, would be very difficult to obtain as most of these substances are highly reactive and consequently unstable. The search for bio-organic products has thus far revealed the occurrence of amino acids, alkanes, porphyrins, carotenoids, and fatty acids in fossils and sediments dating back to the Precambrian (Abelson, 1956; Oró *et al.*, 1965; Barghoorn *et al.*, 1965; McCarthy and Calvin, 1967; Knoche and Ourisson, 1967). Compounds of this type are usually not listed among those recognized as carcinogens.

Polycyclic hydrocarbons have been detected in very small amounts in meteorites (Oró and Nooner, 1967; Pering and Ponnamperuma, 1971), in combustion products (Masuda *et al.*, 1967) and in a variety of foodstuffs, (Grimmer and Hildebrandt, 1965), many plants (Graef, 1965) and virgin soil (Stocks and Davies, 1960). The ubiquitous presence of carcinogenic polycyclic hydrocarbons in very minute amounts could represent 'background carcinogenicity' analogous to that of the background radiation. The biochemical and evolutionary significance of their presence (other than a universal pollution by combustion products) remains to be determined (Graef, 1965). The assertion that the domestic use of fire introduced the first carcinogens is erroneous (Burnet, 1970a): carcinogenic agents of physical or chemical origin were present in Nature long before the arrival of man, as was fire. It has been suggested by Oró that the occurrence of aromatic polycyclic hydrocarbons in meteorites is indicative of cosmic occurrence and remote existence of carcinogens (Oró, 1973).

Nucleic acid bases (purines, pyrimidines) have been synthesized by plausible prebiotic routes (Oró and Kimball, 1962; Sanchez *et al.*, 1967). Carcinogenic substances can be obtained by simple modifications of some of these nucleic acid bases by direct (Brown, 1968; Giner-Sorolla *et al.*, 1973) or indirect (Giner-Sorolla and

Bendich, 1958; Giner-Sorolla *et al.*, 1972) oxidation. A prebiotic precursor of these bases, 5-aminoimidazole-4-carboxamide, and its nitrosation derivative, 2-azahypoxanthine have been reported to induce tumors in animals (Beal *et al.*, 1974). A misdirected prebiotic synthesis of nucleic acid components could occur by the interaction of physical or chemical agents leading to carcinogenic analogs. The primitive Earth environment may well have contained compounds of these types.

Whether the first cancer cell on the Earth originated before or after the appearance of the vertebrates is a matter of controversy as reflected in the contemporary debate concerning whether or not true malignant tumors occur in invertebrates (Good and Finstad, 1969; Sparks, 1972) and even in plants (Braun, 1972). The interpretation by Burnet of malignancy as an inevitable manifestation of the same mutability of the genome from which both vertebrates and the system of cell-mediated immunity evolved can be mentioned in this regard. This epochal development may have taken place in jawless fishes during the Paleozoic (Burnet, 1970b). The development of the system of immune surveillance, according to Thomas (1959), constituted the evolutionary challenge to the appearance of cancer cells. In another view, Todaro and Huebner (1972) have proposed that the potential for malignancy was present in the genetic determinants of all vertebrate cells since the beginning of evolution, and that this could be turned on or activated by carcinogenic agents (radiation, chemicals).

The osteomas and multiple myelomas found in fossil herbivorous reptiles from the Upper Cretaceous may constitute the earliest paleontological record of neoplasms, (Hatcher, 1907; Moodie, 1923; Lull, 1933; Swinton, 1971a). Tumors arising in earlier geologic eras would obviously be very difficult to detect.

## 2. Plant Carcinogens and the Emergence of Malignancy

In any comprehensive explanation of the origin of malignancy during biological Evolution (paleocarcinogenesis), the tumor-inducing agents which should be included and perhaps figure prominently are those carcinogens occurring as secondary substances in diverse families of the plant kingdom. It is remarkable that a great proportion of the present-day carcinogen-producing plants consists of species which emerged and were ubiquitous in the remote geological past: fungi, ferns and cycads. These groups of plants are proportionally not as widespread today as in the past; they constituted the most abundant plants on Earth until the Jurassic when the angiosperms started to appear (Meeuse, 1963; Andrews, 1965; Scagel *et al.*, 1966). Paleobotanists call the Mesozoic 'the Age of Cycads' because they were prevalent at that time (Andrews, 1965; Scagel *et al.*, 1966). In contrast to the high proportion of species containing carcinogens in the geological past, the angiosperms, which appeared in the Jurassic and are the prevalent plants today, comprise an insignificant number of species possessing carcinogenic substances, and these are of lower potency than those from gymnosperms and cycads.

As a working hypothesis it is assumed that secondary substances which are carcinogenic and occur in today's plants were also present in the same botanical groups in the past and that the process of carcinogenesis was operative in the same manner as today. This assumption is supported by the existence of the molecular fossils. These are biochemical substances which have persisted in the composition of living organisms throughout Evolution as inferred by the comparison of analytical data of fossil remains and ancient sediments with those of surviving analogous species (Abelson, 1956; Oró *et al.*, 1965; Barghoorn *et al.*, 1965; McCarthy and Calvin, 1967; Knoche and Ourisson, 1967). According to this correlation, the advent of the first paleocarcinogens of plant origin could be traced to the upsurge of the actinomycetales; their modern counterparts contain secondary substances which are potent carcinogens. This would place the appearance of plants capable of producing carcinogens in the Precambrian, well before the advent of sea (Cambrian) and land (Devonian) vertebrates. A period in Evolution must have existed (between the Precambrian and early Paleozoic) in which there was no malignancy, despite the postulated presence of chemical carcinogens of biological origin. Conditions favorable for the first interaction between plant carcinogens and primitive animals leading to the first malignant cell may have occurred in the early Paleozoic. The probability of such interaction reached a maximum in the Mesozoic with the advent of a greater variety and abundance of carcinogen-producing plants.

## 2.1. Gymnosperms

A survey of the sources and chemical structures of carcinogens from contemporary plants reveals that lower gymnosperms such as actinomycetous mycelial bacteria

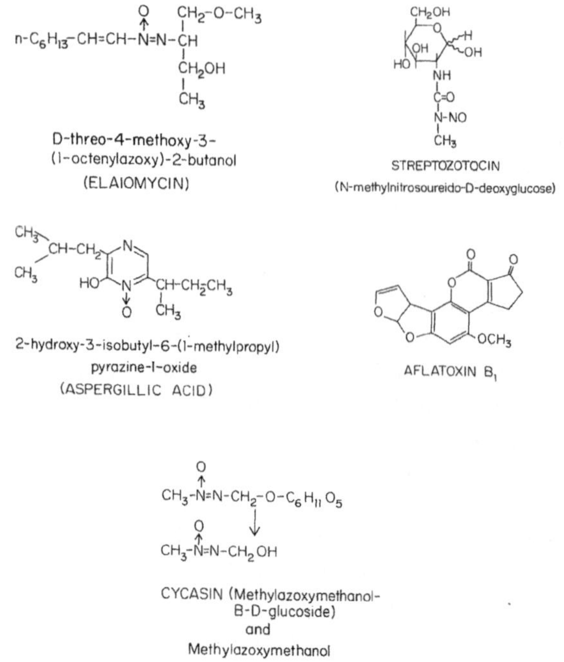


Fig. 1. Carcinogens from Gymnosperms and Cycads.

of the family *Streptomyces*, produce carcinogenic metabolites. Among them, elaiomycin (D-threo-4-methoxy-3 (1-octenylazoxy)-2-butanol) (Herr *et al.*, 1967), from *S. hepaticus* and streptozotocin (N-methyl-nitrosoureido-D-deoxyglucose) (Schoental, 1969) which causes unusual renal tumors (Arison and Feudale, 1967) was isolated from another *Streptomyces* (Figure 1). Several *Penicillium* species produce antibiotics such as patulin, griseofulvin and penicillin which are carcinogenic (Dickens and Jones, 1963; Dickens, 1964). Carcinogens of yet unknown structure from a great variety of fungi pathogenic to man have been described (Blank *et al.*, 1968). Several pyrazine N-oxide antibiotics may be carcinogenic by analogy with other N-oxides (Brown, 1968): emimycin (Terao *et al.*, 1960) isolated from S. *griseochromogenes*, aspergillic acid (Micetich and McDonald, 1964) and pulcherriminic acid (Ohta, 1964) from several *Aspergillus*. The aflatoxins are polycyclic lactones which occur in several species of *Aspergillus* and are among the most potent carcinogens (Schoental, 1967; Wogan, 1973). Bracken (*Pteris aquilinum*), the most widespread fern on the globe today, contains unidentified hepatotoxic and carcinogenic substances which cause tumors in cattle (Pamucku and Price, 1969).

### 2.2. Cycads

The plant division Cycadophyta consists of two orders, the Bennettitales (also called Cycadeoideales) and the Cycadales; both are similar in habitat and leaf organization (Scagel *et al.*, 1966). The Bennettitales appeared in the late Paleozoic, reached a maximum of development in the Jurassic and declined to become extinct in the Upper Cretaceous. The related Cycadales were widespread in the Mesozoic, but only 9 genera with about 100 species exist today thinly dispersed around temperate or tropical zones of the globe.

All surviving Cycads thus far studied contain the toxic and potent carcinogenic substance cycasin (methylazoxymethanol-$\beta$-D-glucoside) (Mickelson *et al.*, 1963); other related methylazoxymethanol derivatives have been reported (Kobayashi and Matsumoto, 1965). Enzymatic hydrolysis by the intestinal $\beta$-D-glucosidase yields the active aglycone methylazoxymethanol (MAM) which is responsible for the potent toxic and carcinogenic properties (Laqueur and Spatz, 1968; Zedeck *et al.*, 1970). A single administration of cycasin or MAM is sufficient to produce a high proportion of neoplasias in rodents; tumors have also been induced in fishes (Stanton, 1966). Cycasin, MAM and related compounds have been used as models for human colon carcinogenesis (Weisburger, 1971). Transplacental tumor induction has also been demonstrated with these substances with equally high incidence of neoplasias in the offspring (Spatz *et al.*, 1968). The teratogenic effects of MAM have been described (Spatz *et al.*, 1967). It is quite plausible that the Cycadales in the past also contained this (or related) toxic and carcinogenic substance. The potency and abundant occurrence of cycasin makes this substance a conspicuous paleocarcinogen.

It is not known why the Bennettitales, one of the most widespread plants in the past became extinct (Meeuse, 1963; Scagel *et al.*, 1966). No evidence exists for an evolutionary interdependence of these plants with the life cycle of large reptiles.

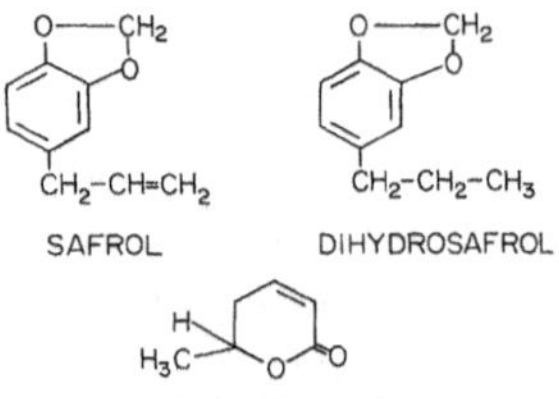


Fig. 2. Carcinogens from Angiosperms.

Their decline and fall may have made the herbivore reptiles highly dependent upon the recently developed Cycadales (Meeuse, 1963), these, as we have seen, contained toxic and carcinogenic substances.

### 2.3. Angiosperms

Although the number of living angiosperm species is about half a million, those which contain carcinogenic substances comprise only a few species. In addition, the carcinogens occurring in these angiosperms are generally weaker than those found in gymnosperms. Among these angiosperms, some species of the family *Sassafras* contain safrole and dihydrosafrole which induce hepatomas in animals (Long *et al.*, 1963) (Figure 2). The mountain ash, *Sorbus aucuparia*, possesses in its berries parasorbic acid, which is a carcinogen (Dickens and Jones, 1963). Pyrrolizidines, a group of plant alkaloids found in about one hundred species of several unrelated families are mild mutagens and have been claimed to be carcinogenic (Schoental, 1968), although this has been disputed by others (Bull *et al.*, 1968; Culvenor *et al.*, 1969).

## 3. Discussion

More complex structures distinguish the type of carcinogens which occur in recently evolved plants from those in plants which were prevalent in remote geological times. Relatively simple structures (aliphatic and/or glucoside derivatives), containing the N-oxide ($R_1R_2N \rightarrow O$), a substituted N-nitroso [$(R_1R_2)$—N—NO] or the azoxy [$R_1$—N=N—(→O)—$R_2$] as active functional groups characterize the known carcinogens from gymnosperms and cycads. The utter simplicity of methylazoxy--methanol, the active moiety of cycasin, is matched only by potent synthetic carcinogens dimethylnitrosamine, diazomethane or methylhydrazines and hydrazine.

The carcinogenic substances from angiosperms are structurally unrelated to those found in gymnosperms. They consist of complex polyalkyl phenolic or lactonic derivatives with homocyclic or heterocyclic nuclei. As a common feature, the tumor inducing activity of all plant carcinogens may be attributable to an alkylating reaction (Miller and Miller, 1965).

The growth cycle, decline and extinction of living species (present and past) depend on several factors (Colbert, 1965; Swinton, 1971b). Although there is evidence that climatic changes may have played a role during the Mesozoic extinctions

(Axelrod and Bailey, 1968; Worsley, 1971; Urey, 1973), the occurrence of plant secondary substances which are toxic and carcinogenic should no longer be overlooked as a contributory factor in the evolutionary process. It has been a widely held view that massive poisoning or infection leading to the extinction of a given animal population in the geologic past is not possible. This hypothesis was tenable only before the initiation of research, carried out during the last decade, which revealed that actinomycetous bacteria, many fungi, a widespread fern, and more important, all surviving cycads, contain toxic and potent carcinogenic substances. Prior to these investigations there was no indication that staple plants in the geological past may have been poisonous to animal species in pandemic proportions. This notion is consistent with the historical record of even the most deadly infectious diseases and their effects on man and animals. However, the continued ingestion of carcinogenic substances widespread in food plants could have resulted in poisoning and malignant tumor formation of pandemic proportions in animal species of the geological past. Rapid and devastating cancer epidemics have been observed in recent times in animals fed on diets contaminated with carcinogens, thus indicating the plausibility of a similar occurrence in the past. Epidemic hepatomas have occurred in large turkey flocks which had been fed ground nut meal contaminated with the fungus *Aspergillus flavus*, which produce the potent carcinogenic aflatoxins. Diets contaminated with aflatoxins as well as other types of carcinogens have also resulted in a high incidence of liver cancers in large trout populations (Kraybill and Shimkin, 1964; Goldblatt, 1969). In both instances the epidemics were brought under control after the causal agent was determined and remedial action taken.

It may be argued that cancer is a disease which strikes mainly animals past their reproductive period and, therefore, could not by itself provoke the demise of a given animal species. However, in addition to the postulated effects of dietary carcinogens on adult animals in the remote past, the transplacental and perinatal carcinogenic effects of cycasin, MAM and related substances may have elicited a high tumor incidence in young animals. It is also likely that the teratogenic effects of these substances (Spatz *et al.*, 1967), as that of many alkylating agents, would lead to severe malformations and high mortality in the progeny (DiPaolo, 1969). These factors add to the plausibility that toxic and carcinogenic substances such as cycasin may have played a contributory role in the decline and extinction of certain animal species in the past.

## 4. Conclusions

The first chemical carcinogens (protocarcinogens) present in the primitive Earth before the process of biopoesis may have been hydrazine, some metallic ions (radioactive or not) and extraterrestrial or abiotic polycyclic hydrocarbons. In a more advanced period in biochemical Evolution, certain biogenic molecules or their precursors could also have undergone transformation into carcinogenic compounds. The chemical carcinogens (paleocarcinogens) which arose after the appearance of living cells, and which most likely played a primordial role in the emergence of malignancy were substances derived from actinomycetales in the Precambrian. Paleocarcinogens with possible evolutionary significance undoubtedly occurred as secondary substances in ferns and cycads which were staple food for reptiles in the Mesozoic.

The realization of the possible effects of plant carcinogens in the process of Biological Evolution, should lead to reflection upon a similar situation that might arise in the present industrialized world from the ever-increasing number and amount of artificial carcinogens. The assertion by Boyland (1969) that 90% of human tumors are of endogenous or exogenous chemical origin is sobering indeed. The evolutionary effects of the injurious environment on man and other species are not difficult to predict.

## Acknowledgements

The investigations by the authors referred to in this article were supported in part by the National Cancer Institute Grant CA-08748 and the Atomic Energy Commission Contract AT-[11-1]-3521. We thank Professor E. E. Swinton, Dr G. B. Brown and Dr G. Stöhrer for their valuable discussions. The authors are specially indebted to Professor J. Oró for stimulating discussions and the suggestions indicated in this paper. They gratefully acknowledge the editorial assistance of Mrs V. Rosso Taracido in the preparation of this article.

## References

Abelson, P. A.: 1956, *Sci. Amer.* **195**, 83.
Andrews, H. N., Jr.: 1965, *Ancient Plants*, Comstock Cornell, Ithaca, N.Y., p. 137.
Arison, R. N. and Feudale, E. L.: 1967, *Nature* **214**, 1254.
Axelrod, D. and Bailey, N.: 1968, *Evolution* **22**, 595.
Barghoorn, E. S., Meinschein, W. G., and Schopf, J. W.: 1965, *Science* **148**, 461.
Beal, D. D., Skiba, J. L., and Croft, W. A., Jr.: 1974, *Proc. Amer. Assoc. Cancer Res.*, Abstr. 185.
Berkner, L. C. and Marshall, L. V.: 1965, *Proc. Natl. Acad. Sci.* **53**, 1215.
Blank, F., Chin, O., Just, G., Meranze, D. R., Shimkin, M. B., and Wieder, R.: 1968, *Cancer Res.* **28**, 2276.
Boyland, E.: 1969, *Progr. Exp. Tumor Res.* **11**, 222.
Braun, A. C.: 1972, *Progr. Exp. Tumor Res.* **15**, 69.
Brown, G. B.: 1968, *Progr. Nucl. Acid Res. Mol. Biol.* **8**, 209.
Bull, L. B., Culvenor, C. C. J., and Dick, A. T.: 1968, *The Pyrrolizidine Alkaloids*, North Holland, Amsterdam, p. 221.
Burnet, F. M.: 1970a, *Immunological Surveillance*, Pergamon, Oxford, p. 134.
Burnet, F. M.: 1970b, *Progr. Exp. Tumor Res.* **13**, 1.
Colbert, E. H.: 1965, *The Age of Reptiles*, Norton, New York, p. 249.
Culvenor, C. C. J., Downing, D. T., Edgar, J. A., and Jagu, M. V.: 1969, *Ann. N.Y. Acad. Sci.* **163**, 837.
Dawe, C. J. and Harshbarger, J. C. (eds.): 1969, *A Symposium on Neoplasms and Related Disorders of Invertebrates and Lower Vertebrate Animals*, Natl. Cancer Inst., Monograph No. 31, p. 31.
Dickens, F.: 1964, *Brit. Med. Bull.* **20**, 96.
Dickens, F. and Jones, H. E. H.: 1963, *Brit. J. Cancer* **17**, 100.
DiPaolo, J. A.: 1969, *Ann. N.Y. Acad. Sci.* **163**, 801.
Donn, B.: 1972, in C. Ponnamperuma (ed.), *Exobiology*, North Holland, New York, p. 439.
Finstad, J. and Good, R. A.: 1969, in Dawe and Harshbarger (eds.), *op. cit.*, p. 51.
Giner-Sorolla, A. and Bendich, A.: 1958, *J. Amer. Chem. Soc.* **80**, 3932.
Giner-Sorolla, A., Burchenal, J. H., and Bendich, A.: 1972, in E. D. Bergman and B. Pullman (eds.), *The*

*Purines, Theory and Experiment*, 4th Symposium on Quantum Chemistry and Biochemistry, The Israel Academy, Jerusalem, p. 478.
Giner-Sorolla, A., Longley-Cook, J., McCravey, M., Brown, G. B., and Burchenal, J. H.: 1973, *J. Med. Chem.* **16**, 365.
Goldblatt, L. A. (ed.): 1969, *Aflatoxins, Scientific Background, Control and Implications*, Academic Press, New York.
Good, R. A. and Finstad, J.: 1969, in Dawe and Harshbarger (eds.), *op. cit.*, p. 41.
Graef, W.: 1965, *Med. Klinik* **60**, 561.
Grimmer, G. and Hildebrandt, A.: 1965, *Z. Krebsforschung* **67**, 272.
Hatcher, J. B.: 1907, *The Ceratopsia*, Monog. U.S. Geol. Serv. **49**, p. 149.
Herr, R. R., Jahnke, H. K., and Argoudelis, A. D.: 1967, *J. Amer. Chem. Soc.* **89**, 4809.
Knoche, H. and Ourisson, G.: 1967, *Angew. Chem. Intl. Ed.* **6**, 1085.
Kobayashi, A. and Matsumoto, H.: 1965, *Arch. Biochem. Biophys.* **110**, 373.
Kraybill, H. F. and Shimkin, M. B.: 1964, *Adv. Cancer Res.* **8**, 191.
Laqueur, G. L. and Spatz, M.: 1968, *Cancer Res.* **28**, 2262.
Long, E., Nelson, A. A., Fitzhugh, O. G., and Hansen, W. H.: 1963, *Arch. Pathol.* **75**, 595.
Lull, E.: 1933, *Mem. Peabody Mus. Nat. Hist.* **III**, Pt. 3, p. 131.
Masuda, Y., Mori, K., and Kuratsune, M.: 1967, *Gann.* **58**, 69.
McCarthy, E. D. and Calvin, M.: 1967, *Nature* **216**, 642.
Meeuse, A. D. J.: 1963, *Acta Biother.* **16** (III), 9.
Micetich, R. G. and McDonald, J. C.: 1964, *J. Chem. Soc.*, 1507.
Mickelson, O., Whiting, M. G., and Kurkland, L. T.: 1963, *J. Nat. Cancer Inst.* **31**, 919.
Miller, J. A. and Miller, E. C.: 1965, *Cancer Res.* **25**, 1292.
Moodie, R. L.: 1923, *Palaeopathology*, Univ. Illinois Press, p. 162.
Ohta, A.: 1964, *Chem. Pharm. Bull.* **12**, 125.
Oró, J.: 1973, personal communication.
Oró, J. and Kimball, A. P.: 1962, *Arch. Biochem. Biophys.* **96**, 293.
Oró, J. and Nooner, W.: 1967, *Nature* **213**, 1082.
Oró, J., Nooner, D. W., Zlatkis, A., Wilkström, S. A., and Barghoorn, E. S.: 1965, *Science* **148**, 77.
Pamucku, A. M. and Price, J. M.: 1969, *J. Nat. Cancer Inst.* **43**, 275.
Pering, K. L. and Ponnamperuma, C.: 1971, *Science* **173**, 237.
Sagan, C.: 1965, in S. W. Fox (ed.), *The Origins of Prebiological Systems*, New York, p. 274.
Sanchez, R. A., Ferris, J. P., and Orgel, L. E.: 1967, *J. Mol. Biol.* **30**, 223.
Scagel, R. F., Baudon, R. J., Rouse, G. E., Schofield, W. B., Stein, J. R., and Taylor, T. M. C.: 1966, *An Evolutionary Survey of the Plant Kingdom*, Wadsworth, Belmont, Cal., p. 587.
Schoental, R.: 1967, *Ann. Rev. Pharmacol.* **7**, 343.
Schoental, R.: 1968, *Cancer Res.* **28**, 2237.
Schoental, R.: 1969, *Nature* **221**, 765.
Severi, L. and Biancifiori, C.: 1968, *J. Nat. Cancer Inst.* **41**, 331.
Sparks, A. K.: 1972, *Invertebrate Pathology: Non-Communicable Diseases*, Academic Press, New York, p. 271.
Spatz, M. and Laqueur, G. L.: 1968, *Proc. Soc. Exptl. Biol. Med.* **127**, 281.
Spatz, M., Dougherty, W. J., and Smith, D. W. E.: 1961, *Proc. Soc. Exptl. Biol. Med.* **124**, 476.
Stanton, M. F.: 1966, *Fed. Proc.* **25**, 661.
Stocks, P. and Davies, R. I.: 1960, *Brit. J. Cancer* **14**, 8.
Swinton, E. W.: 1971a, *The Dinosaurs*, Allen and Unwin, London, p. 268.
Swinton, E. W.: 1971b, *op. cit.*, p. 273.
Terao, M., Karasawa, K., Tanaka, N., Yonehara, H., and Umezawa, H.: 1960, *J. Antibiotics* **13A**, 401.
Thomas, L.: 1959, in H. S. Lawrence (ed.), *Cellular and Humoral Aspects of the Hypersensitive State*, Hoeber, New York, p. 529.
Todaro, G. J. and Huebner, R. H.: 1972, *Proc. Natl. Acad. Sci. U.S.A.* **69**, 1009.
Urey, H. C.: 1952, *The Planets*, Yale Univ. Press, p. 152.
Urey, H. C.: 1973, *Nature* **232**, 32.
Weisburger, J. H.: 1971, *Cancer* **28**, 60.
Wogan, G. N.: 1973, *Methods in Cancer Res.* **7**, 309.
Worsley, F. R.: 1971, *Evolution* **230**, 318.
Zedeck, M. S., Sternberg, S. S., Poynter, R. W., and McGowan, J.: 1970, *Cancer Res.* **30**, 801.

# INDEX OF SUBJECTS

GPSR Compliance

*The European Union's (EU) General Product Safety Regulation (GPSR) is a set of rules that requires consumer products to be safe and our obligations to ensure this.*

*If you have any concerns about our products, you can contact us on ProductSafety@springernature.com*

In case Publisher is established outside the EU, the EU authorized representative is:

Springer Nature Customer Service Center GmbH
Europaplatz 3
69115 Heidelberg, Germany

**Batch number: 10761599**

Printed by Printforce, the Netherlands